# 5G AND BEYOND

5G communications technologies will transform entire industries around the world and are already a core element of the mobile communications and automotive ecosystems. 5G *and Beyond* brings together some of the world's leading thinkers in law, economics, and competition policy, drawn from academia, government, and industry, to lay the intellectual foundation for sound innovation and competition policy in wireless-enabled environments. Contributors include former heads of the United States Patent and Trademark Office, commissioners of the US Federal Trade Commission and International Trade Commission, distinguished academics, and industry leaders. Chapters provide economically grounded and empirically informed analyses of the innovation policy issues involved in the development and adoption of 5G-enabled computing and communications technologies in the Internet of Things. This title is also available as open access on Cambridge Core.

Jonathan M. Barnett specializes in antitrust, intellectual property, and corporate law. He has published widely in scholarly and policy journals, with a focus on monetization strategies and transactional structures in innovation markets. He is the author of *Innovators, Firms, and Markets: The Organizational Logic of Intellectual Property* (2021).

Seán M. O'Connor's research and law practice focus on intellectual property and business law, especially the role of general counsel for start-ups and commercializing innovation in technology and the arts. He is Faculty Advisor for the Center for Intellectual Property x Innovation Policy (C-IP$^2$) at Antonin Scalia Law School, George Mason University.

# 5G and Beyond

## INTELLECTUAL PROPERTY AND COMPETITION POLICY IN THE INTERNET OF THINGS

Edited by

JONATHAN M. BARNETT
University of Southern California

SEÁN M. O'CONNOR
George Mason University

CAMBRIDGE
UNIVERSITY PRESS

CAMBRIDGE
UNIVERSITY PRESS

Shaftesbury Road, Cambridge CB2 8EA, United Kingdom

One Liberty Plaza, 20th Floor, New York, NY 10006, USA

477 Williamstown Road, Port Melbourne, VIC 3207, Australia

314–321, 3rd Floor, Plot 3, Splendor Forum, Jasola District Centre, New Delhi – 110025, India

103 Penang Road, #05-06/07, Visioncrest Commercial, Singapore 238467

Cambridge University Press is part of Cambridge University Press & Assessment, a department of the University of Cambridge.

We share the University's mission to contribute to society through the pursuit of education, learning and research at the highest international levels of excellence.

www.cambridge.org
Information on this title: www.cambridge.org/9781009274272

DOI: 10.1017/9781009274289



First published 2024

*A catalogue record for this publication is available from the British Library.*

*Library of Congress Cataloging-in-Publication Data*
NAMES: Barnett, Jonathan M., editor. | O'Connor, Sean, editor.
TITLE: 5G and beyond : intellectual property and competition policy in the internet of things / edited by Jonathan Barnett, University of Southern California, Sean O'Connor, George Mason University Antonin Scalia Law School.
OTHER TITLES: 5G and beyond (Cambridge University Press)
DESCRIPTION: New York, NY : Cambridge University Press, 2023. | Includes index.
IDENTIFIERS: LCCN 2023025407 (print) | LCCN 2023025408 (ebook) | ISBN 9781009274272 (hardback) | ISBN 9781009274302 (paperback) | ISBN 9781009274289 (epub)
SUBJECTS: LCSH: 5G mobile communication systems–Industrial applications. | Business communication. | Industrial property. | Intellectual property. | Competition. | Internet of things.
CLASSIFICATION: LCC TK5103.25 .A1145 2023 (print) | LCC TK5103.25 (ebook) | DDC 384.5/3–dc23/eng/20230821
LC record available at https://lccn.loc.gov/2023025407
LC ebook record available at https://lccn.loc.gov/2023025408

ISBN 978-1-009-27427-2 Hardback

*This book is dedicated to the memory of Alexander Galetovic, who passed away in July 2022. Alex was a dear friend and pioneering scholar who challenged conventional wisdom on patent and competition policy through the meticulous pursuit of knowledge, truth, and evidence.*

# Contents

# Contributors

**Jonathan M. Barnett** is Torrey H. Webb Professor of Law at the Gould School of Law, University of Southern California.

**Mark A. Cohen** is Distinguished Senior Fellow and Director of the Asian IP Project at the Berkeley Center for Law and Technology.

**Alexander Galetovic** was Senior Fellow and Professor at the Universidad Adolfo Ibáñez, Santiago, Chile, Research Fellow at the Hoover Institution, Stanford University, and Research Associate at the CRIEP.

**Thomas D. Grant** is a fellow at the Lauterpacht Centre for International Law at the University of Cambridge, a former designee to the Permanent Court of Arbitration (US National Group), and a former senior advisor for Strategic Planning in the Bureau of International Security and Nonproliferation at the US Department of State.

**Stephen Haber** is A.A. and Jeanne Welch Milligan Professor in the School of Humanities and Sciences, Professor of Political Science, Professor of History, and Professor (by courtesy) of Economics, Stanford University.

**Bowman Heiden** is Executive Director of the Tusher Initiative for the Management of Intellectual Capital Management at the Haas School of Business, University of California, Berkeley, and Codirector for the Center for Intellectual Property at the University of Gothenburg.

**Fabian Hoffmann** is Judge at the Federal Court of Justice (Bundesgerichtshof) in Karlsruhe, Germany.

**Andrei Iancu** is a partner at Sullivan & Cromwell LLP and former Undersecretary of Commerce for Intellectual Property and Director of the United States Patent and Trademark Office.

**David J. Kappos** is a partner at Cravath, Swaine & Moore LLP and former Undersecretary of Commerce for Intellectual Property and Director of the United States Patent and Trademark Office.

**F. Scott Kieff** is Stevenson Bernard Professor at the George Washington University Law School, and former Commissioner of the US International Trade Commission.

**Igor Nikolic** is Research Fellow at the European University Institute.

**Maureen K. Ohlhausen** is a partner and Global Section Chair of the Antitrust & Competition Law Practice Group of Baker Botts LLP and former Acting Chair and Commissioner of the US Federal Trade Commission.

**Kristen Osenga** is Austin E. Owen Research Scholar and Professor of Law at the University of Richmond School of Law.

**Jorge Padilla** is Senior Managing Director at Compass Lexecon.

**Ruud Peters** is CEO at Peters IP Consultancy B.V., and former Chief IP Officer and Executive Vice President at Koninklijke Philips NV.

**Jana I. Seidl** is Senior Associate in the Washington, DC office of Baker Botts LLP.

**J. Gregory Sidak** is Chairman of Criterion Economics, Washington, DC.

**David J. Teece** is Thomas W. Tusher Professor in Global Business, Faculty Director of the Tusher Initiative for the Management of Intellectual Capital, Institute for Business Innovation, Haas School of Business, University of California, Berkeley, and Founder and Executive Chairman of the Berkeley Research Group.

**Nikolaus Thumm** is Senior Scientific Advisor at ETH Board, Switzerland, and former Chief Economist of the European Patent Office.

**Andrew Tuffin** is Senior Policy Advisor at Compass Lexecon.

**Lew Zaretzki** is Managing Director at Hamilton IPV.

# Foreword

## *Why Patents Are Critical for Standards-Based Technologies*

*Andrei Iancu*

On the morning of February 7, 1904, not far from Washington, DC, a dry goods store in downtown Baltimore, Maryland, burst into flames. The fire alarm sounded, and firefighters from several units throughout the city rushed to answer the call. As they smashed through the burning building, explosions shot embers through the broken windows and onto neighboring structures. Before long, the fire – believed to have been ignited by a discarded cigarette in the building's basement, near a pile of wood shavings – fueled a blaze that would go on to destroy much of central Baltimore.

Firefighters from other cities, including Washington, DC and Philadelphia, were sent to help battle the inferno, but they quickly encountered a serious problem. Because there were no national standards for firefighting equipment in those days, firefighters from one city could not effectively use the equipment from another city. Poorly matched hoses emitted weak streams of water. And so, Baltimore burned. All in all, this was the most destructive conflagration in the United States since the Great Chicago Fire of 1871. A real tragedy, aggravated by the lack of standards.

But as often happens with crises, powerful lessons were learned. When the fire first started that fateful February morning, the US patent system had no shortage of firefighting innovations. Indeed, as of 1904, nearly 1,000 patents relating to firefighting, including those for fire hydrants, hoses, and connectors, were in force. But there were no standards. That was about to change.

Within two months, legislative bills and conferences were held to standardize fire hoses, and many cities began replacing their fire hose couplings. The National Fire Protection Association (NFPA) and National Board of Fire Underwriters (NBFU) established certain standards, such as thread size, in an effort to prevent further incidents like the one faced by the out-of-state fire units during the Baltimore Fire. And though it took time for these standards to truly catch on, today we have the National Standard Thread, along with standardized hydrants, as well as hose adapters that firefighters carry to avoid another disaster like the Great Baltimore Fire.

Government mandates are one way to ensure standardization of technology. But in a free-market economy, depending entirely on the government taking action is neither feasible nor desirable. The United States greatly benefits from private industry investing resources and developing technology on its own, without government mandates. This is particularly true for technology that eventually becomes standardized.

Our country's founders realized early on the value of patents as drivers of innovation. This is why intellectual property rights are enshrined in the Constitution itself, giving Congress the right "[t]o promote the progress of science and useful arts, by securing for limited times to authors and inventors the exclusive right to their respective writings and discoveries." Backed by the patent system envisioned by the Constitution, American technology has progressed in the last two and a half centuries at rates that are unparalleled at any other time in human history or in any other place on Earth. As Thomas Jefferson observed, "patents have given a spring to invention beyond any comprehension." And Abraham Lincoln explained why: "the patent system adds the fuel of interest to the fire of genius."

Developing technology that might get adopted into a standard, however, is risky and expensive, and without the proper incentives and protections, industry might not choose to invest in it. First of all, and like all new technologies, it might not work. By definition, disruptive new technologies are untried and have no proven track record of success. These innovations can fail for myriad reasons, such as technical failures or market rejection.

On top of all this, for standards-based innovation there is risk even if the technology does work. In the standards world, it is often the case that multiple solutions are proposed by different companies for a particular problem to be solved by the standard. Only one of those solutions will typically be adopted into the standard. If your technology is not adopted, even if it works, your investment and development is largely for naught. Furthermore, even if an innovative technology has merit and is adopted into a standard, that standard may never gain traction in the market.

And if your technology is adopted into a widely implemented standard, multiple implementers will certainly use it – that is the whole point of having a standard in the first place. In the standards context, therefore, unlike many other inventions, the inventor is not assured of exclusive use. To the contrary, the hope is that the standard will be successful and the inventions incorporated into the standard will be broadly used.

This is why patents are more important for technology to be used in standards than in almost any other field. In order to overcome these risks, inventors and investors need to be assured that if their technology is in fact adopted into the standard and broadly used, they will be appropriately compensated and their investment will be protected. Patents can and should serve that role.

Patents historically provide a quid pro quo arrangement between the inventor and the public. The public gets the benefit of the invention that is described in the patent document and brought to market for use by and for the benefit of the public. In exchange, the inventor gets a period of market exclusivity to commercialize her invention. Among other things, the inventor can license her technology to companies who want to implement the technology in their products. For technology used in standards, this is the best tool to ensure that successful inventors can be rewarded for the risks they took, the investments they made, and the technical contributions they brought.

But this is true only if such patents are reliable and meaningfully enforceable. In other words, the patents issued by the Patent Office need to be robust enough to withstand challenge down the line – that is, the original examination should be thorough enough so that courts and other tribunals that review patents years after their issue will find them to have been correctly issued in the first place. Plus, the system must enable patent owners to enforce their patents if others infringe them. A patent serves little purpose if others can ignore it and the owner cannot practically stop them or secure timely and adequate compensation.

The American patent system has grown increasingly complex over the past few decades, adding hoop after hoop that patent owners need to jump through to enforce their patents. For example, there are now multiple tribunals where the validity of a patent can be challenged, leaving patent owners to defend their patents again and again, drastically increasing expenses and time to resolution. In addition, many courts take a very long time to bring a case to trial, often delaying resolution for years. And even if the patent owner prevails, it is now very difficult to obtain an injunction to enforce the promised market exclusivity a patent is meant to convey. These and many other hoops make it practically difficult to enforce any patent, thereby diminishing its value.

Standard-essential patents (SEPs) add another complexity to an already complex field. When a standard is adopted, patent owners usually agree to offer to license their patents to potential implementers on fair, reasonable, and nondiscriminatory (FRAND) terms. Some have interpreted this promise to mean that patent owners can never get an injunction for SEPs. The practical inability to exclude makes voluntary license agreements even more difficult to secure, thereby increasing the likelihood of litigation. After all, one accused of infringement has little to lose if they refuse to license a patent when the worst that happens after years of litigation is to pay the same royalty the patent owner offered during initial license negotiations.

On the other hand, it is difficult to argue that patent owners who have agreed to submit their technology to a standard and made a commitment to license their patents on FRAND terms should be able to exclude those who actually want to implement the technology under license and pay FRAND royalties. This is especially true if the implementer negotiates in good faith for a FRAND license, yet the patent owner refuses. After convincing a standards setting organization to adopt its patented technology, a patent owner should work in good faith to ensure reasonable access to that technology by those who want to implement it.

Patent policy, therefore, must be carefully balanced to protect the interests of both patent owners who contribute their innovations to a standard, and those who seek to implement those technologies into actual products that are brought to the market. On the one hand, our IP systems should be robust enough to incentivize, protect, and remunerate fairly the developers of standards-based technology. On the other hand, our IP systems should not unduly burden users of such technology with overvalued royalties or threats of unjustified injunctions.

Our IP systems should also be balanced to incentivize good-faith negotiations between innovators and implementers. In the end, a well-functioning IP-backed standards system encourages voluntary transfer of the technology adopted into the standard. That can only happen if innovators and implementers negotiate in good faith toward a voluntary license. Licensors should not unreasonably deny a license ("hold up"), and licensees should not unreasonably refuse to take a license ("hold out"). Government policy and the law should make clear that the presence or absence of good faith during negotiations on both sides is important and will be taken into account.

Our policymakers need to ensure that the United States remains the best place in the world for innovation that will be adopted into worldwide standards, and the best place in the world for implementing that innovation. Maintaining a proper balance of incentives will benefit the United States and humanity in general.

# Preface

The global technology ecosystem stands at a historical point of transition between the 3G- and 4G-enabled wireless technologies that launched transformative changes in audio–video communications and 5G-enabled technologies that will enable transformative changes across the entire Internet of Things (IoT). Thus, beyond general communications, broad industries such as transportation, health care, and industrial production are likely to be transformed. The vast increases in bandwidth – and decreases in latency – enable novel business models that cultivate dense networks of person-to-person, person-to-machine, and machine-to-machine pathways. In 2021, the Federal Communications Commission's (FCC) "C-Band Auction 107" sold 280 megahertz of airwaves for $81 billion, reflecting the immense value the market places on the adoption and deployment of 5G technologies.

This book contains the dedicated efforts of a distinguished group of scholars, former government officials, and industry practitioners to set forth a theoretical and empirical basis for sound policymaking in the vital 5G and IoT sector. The chapters are drawn from a conference held in December 2021 by the Center for Intellectual Property x Innovation Policy (C-IP²) at Antonin Scalia Law School, George Mason University. The contributions are informed by not only economic and legal concepts but also a practically informed perspective on the challenges of securing returns on innovation – an asset that is inherently exposed to expropriation – and the realities of enforcing and licensing IP rights in real-world technology markets. This point is of considerable importance since, in our view, scholarly and regulatory discussions in this area often rely on theoretical models that make little inquiry into "on the ground" conditions in real-world technology markets. Collectively, the contributors to this book bring decades of policymaking experience (at agencies such as the United States Patent and Trademark Office, the US Federal Trade Commission, and the US International Trade Commission), industry experience, and scholarly analysis concerning the legal, economic, and technological issues involved in formulating and implementing efficient "rules of the game" in global 5G-enabled markets.

While the contributors deploy different approaches to, and reach different conclusions on, these complex issues, the project rests on two common foundational principles. First, responsible policymaking must be grounded in economic theory and empirical evidence, rather than rhetoric, narrative, or ideology. Second, meaningful enforcement of property rights and contracts is a critical predicate for enabling the formation of efficient markets in technological innovations. Just as it is widely agreed that these two key legal inputs have supported unprecedented rates of wealth creation in physical-goods industries in market-oriented economies, so too these same legal inputs – with appropriate modification for the intangible-goods environment – are necessary to support wealth creation in the wireless computing and communications markets that will drive IoT. While this perspective may seem elementary to much of the business community, it has been surprisingly overlooked by much of the regulatory and scholarly community.

The book is divided into five parts that complement each other but can be read separately based on a reader's interests.

Part I, Intellectual Property and Competition Policy in Global Wireless Markets, addresses "big picture" issues underlying past and future development of IP and competition policies relating to mobile communications technologies. In "Restoring and Revitalizing Technology Markets for Mobile Wireless: Geopolitical Dimensions of Patented Technology Embedded in Standards" (Chapter 1), David Teece argues that the transition to IoT will demand a renewed appreciation by policymakers of the critical function played by a robust IP infrastructure in facilitating the research and development, standard-setting, and licensing activities of lead innovators in the global wireless industry. This includes standard-essential patents (SEPs). In "Antitrust Convergence on Substantive Norms for SEP Licensing Negotiations: Should and Could It Be?" (Chapter 2), Maureen Ohlhausen and Jana Seidl similarly underscore the importance of robust patent rights (and caution in using antitrust law to limit those rights) in supporting wireless innovation in general, and US technological leadership in particular, as markets make the investments necessary to develop and adopt IoT technologies. The authors describe incremental steps taken by US policymakers that suggest a growing acceptance of legal innovations in Europe that have promoted a more even playing field in licensing negotiations between innovators and implementers of SEP-protected technologies.

Part II, Patent Holdup, Royalty Stacking, and the FRAND Standard, addresses critical empirical questions that must be addressed to provide a reliable basis for policymaking and adjudication concerning SEP licensing and enforcement. In "Cellular SEP Royalties and 5G: What Should Competition Policy Be?" (Chapter 3), Alexander Galetovic, Stephen Haber, and Lew Zaretzki review a transformative body of empirical research (in which the authors have played a central role) that has reassessed the factual basis for widely adopted patent holdup and royalty stacking theories in SEP licensing markets. The authors show that empirical studies have repeatedly failed to find evidence for these theoretical assertions of market failure. Rather, the evidence favors the view that SEP licensing represents a case of exceptional market *success*, as indicated by declining quality-adjusted prices, expanding output, and continuous innovation in SEP-dependent technology markets. In "The Fair Division of Surplus from a FRAND License Negotiated in Good Faith" (Chapter 4), Gregory Sidak takes on a difficult theoretical challenge with practical implications for SEP licensing and litigation: Is it possible to reconcile the standard of "fair, reasonable, and nondiscriminatory" (FRAND) licensing with the efficiency objective that underlies contract law? In a novel analysis, Sidak shows that, under certain behavioral assumptions, the "fairness" principle embodied by the FRAND standard can promote efficiency by truncating the range of "reasonable" royalty terms, which in turn can promote mutually beneficial transactions between innovators and implementers.

Part III, Patent Holdout and the Rise of "Efficient" Infringement, addresses the consequences of the stringent limitations that regulators and some courts have imposed on SEP owners' ability to secure injunctions against infringing users. In "Efficient Infringement in the SEP Space" (Chapter 5), Kristen Osenga documents how theoretical concerns over patent holdup have supported limitations on infringement remedies that encourage infringers, especially the most well-resourced infringers, to engage in patent "holdout" and compel SEP owners to undertake costly and lengthy litigation around the world. The unfortunate result: Successful innovators are increasingly unable to secure positive returns on investments in research and development. In "Restoring Deterrence: The Case for Enhanced Damages in a No-Injunction Patent System" (Chapter 6), Jonathan Barnett and David Kappos propose a policy innovation to deter patent holdout even in a legal environment in which injunctive relief is largely unavailable. Specifically, the authors propose requiring the award of enhanced damages against adjudicated infringers to mimic the deterrence effect of the "missing" injunction, adjusted to reflect potential underenforcement and overenforcement effects. The predicted fortunate result: The market will shift away from value-depleting litigation and toward value-enhancing dealmaking.

Part IV, Transactional Solutions: Redesigning SEP Licensing Markets, leverages theoretical analysis and industry experience to present practical proposals to mitigate the litigation-related and other transaction costs that can encumber SEP licensing negotiations between innovators and implementers. In "Designing SEP Licensing Negotiation Groups to Reduce Patent Holdout in 5G/IoT Markets" (Chapter 7), Bowman Heiden, Igor Nikolic, and Ruud Peters assess recent proposals to enable licensees to negotiate collectively with SEP owners through licensing negotiation groups (LNGs). Whereas LNGs have been proposed to mitigate the risk of patent holdup, the authors argue that LNGs may be a useful tool to mitigate the risk of patent holdout, especially in light of the fact that the licensee population for 5G technologies, which extend across various industries, is expected to be more numerous and heterogenous than has been the case in 3G and 4G wireless technologies (which have mostly been applied in mobile communications). In "How to Create a Smoother SEP Licensing Ecosystem for IoT" (Chapter 8), Ruud Peters, Fabian Hoffmann, and Nikolaus Thumm propose modifications to SEP licensing practices to address the expected increase in transaction costs in the 5G/IoT ecosystem. These modifications seek to mitigate the risk of negotiation failure and ensuing litigation through a suite of mechanisms designed to increase transparency in SEP licensing, to increase assurance that a licensed SEP is valid and essential, to enhance implementers' incentives to negotiate a license (rather than "use and then litigate"), and to increase the likelihood that a licensee ultimately bears a "reasonable" aggregate royalty for use of the total SEP stack.

Finally, Part V, Patent Enforcement, Wireless Markets, and Global Competitiveness, addresses the geopolitical issues that are being increasingly raised by IP and competition policy in wireless communications markets. In "The Geopolitical Implications of Patent Holdout and the Ensuing Race to the Home Court" (Chapter 9), Jorge Padilla and Andrew Tuffin discuss the danger posed to standardization in wireless technology markets by strategic efforts to initiate SEP-related litigation in courts that are perceived to favor the interests of innovators or implementers. These global forum-shopping strategies have been promoted by certain courts' willingness to determine FRAND royalty rates on a global basis and to issue "anti-suit" and "anti-anti-suit" injunctions to interfere with litigants' ability to seek injunctions, or initiate related SEP litigation, in foreign jurisdictions. In "China's Practice of Anti-suit Injunctions in SEP Litigation: Transplant or False Friend?" (Chapter 10), Mark Cohen provides a comprehensive account, using primary Chinese sources, of the proliferating use of anti-suit injunctions by Chinese courts, usually for the purpose of barring SEP owners (typically, foreign companies) from pursuing infringement actions against implementers (typically, Chinese device makers) in courts outside China. Showing how these legal developments are part of a larger and long-standing effort by Chinese policymakers to secure technological independence and leadership in critical industries, this contribution delivers important and novel insights as SEP policy discussions increasingly integrate geopolitical considerations into the conventional focus on competition and innovation policy concerns. Finally, in "Patents and Competition: Commercializing Innovation in the Global Ecosystem for 5G and IoT" (Chapter 11), Scott Kieff and Thomas Grant close out our book with a return to the "big picture" issues with which it starts. In particular, the authors emphasize the enabling function played by a secure IP infrastructure in facilitating surplus-enhancing cooperative activities between the holders of innovation and non-innovation assets in technology markets. This "win-win" enabling effect stands in contrast to the conventional emphasis on the "win-lose" exclusionary effect of IP rights in the litigation context. The constructive transactional role played by patents and other IP rights, in conjunction with contract, is demonstrated by the standardization and licensing structure that supports 3G and 4G wireless markets and is expected to continue and intensify as wireless technologies are applied across a broader variety of markets as the digital economy migrates to the IoT.

We conclude on a sad note. During the editing of this book, our dear colleague, Alexander Galetovic, passed away. Alex's untimely passing has left a hard-to-fill void in the economic and empirical analysis of IP and competition policy issues in global wireless markets. Alex's unparalleled dedication to meticulous empirical scholarship yielded breakthrough results that challenged settled assumptions – widely accepted but never rigorously tested – in this economically and socially critical industry. This achievement has promoted a more balanced discussion of the complex IP and antitrust policy issues raised by wireless communications markets, leading to incremental policy changes by US and European regulators. We hope that this book (including Alex's coauthored contribution) will similarly provide an economically and factually informed foundation on which policymakers and scholars can build when proposing and taking action in this vital sector of the global digital economy.

# Acknowledgments

A book such as this is the work of many dedicated individuals beyond the contributors and editors. We extend our deepest appreciation to the following people who helped us bring the book to the finish line.

We would like to thank our staff at the Center for Intellectual Property x Innovation Policy (C-IP²): Joshua Kresh, managing director; Kristina Pietro, director of operations and events; and Mary Clare Durel, program and communications coordinator. Without their efforts, we would not have been able to complete this book. We would also like to thank the following Antonin Scalia Law School, George Mason University, students and C-IP² research assistants: Cala Coffman, Quyen V. Dang, Sarah Kratt, Jaehee (Amy) Jung, Kyung Min "Rachel" Park, John "Jack" Ring, Tejo Sravani Tunuguntla, and Valerie Yu.

We would like to further thank Matt Gallaway from Cambridge University Press for guidance through the publication process, especially with regards to Cambridge's open access option.

We also would like to thank C-IP² Advisory Board members – especially Claudia Tapia, John Kolakowsi, Kirti Gupta, and Matteo Sabattini – who provided industry insights into topics in the book.

Additionally, we would like to thank many of our contributors who spoke at the conference, Intellectual Property and Innovation Policy for 5G and the Internet of Things, on December 2 and 3, 2021. This conference covered many of the topics in the book, and discussions at the conference helped guide the book's completion. Speakers included editors Jonathan M. Barnett and Seán M. O'Connor, as well as the following book contributors: Mark A. Cohen, Alexander Galetovic, Bowman Heiden, Andrei Iancu, David J. Kappos, F. Scott Kieff, Igor Nikolic, Maureen K. Ohlhausen, Kristen Osenga, Jorge Padilla, J. Gregory Sidak, and David J. Teece. We would also like to thank other speakers who contributed to discussions at the conference: Fabian Gonell, Kirti Gupta, Brian Hinman, Anne

Layne-Farrar, Keith Mallinson, Luke McLeroy, Kathleen M. O'Malley, Jon Putnam, Diane Rinaldo, Jeffrey M. Wilder, and John M. Yun.

Finally, we thank our donors, particularly Qualcomm Incorporated, Ericsson, and Nokia, who generously support C-IP² through unrestricted gifts. Their willingness to support independent scholarly research, even if it does not always align with their views, is vital to our work. In particular, these generous gifts have allowed us to publish this book through the Cambridge University Press Open Access program. The Center for Intellectual Property x Innovation Policy does not receive any financial support from our host university or law school and thus could not operate, let alone complete any significant work, without our donors' support.

PART I

# Intellectual Property and Competition Policy in Global Wireless Markets

1

# Restoring and Revitalizing Technology Markets for Mobile Wireless

## Geopolitical Dimensions of Patented Technology Embedded in Standards

*David J. Teece*

### I. INTRODUCTION

Standards are important to the further development and rollout of technologies, including 5G and the Internet of Things (IoT), self-driving cars, and artificial intelligence (AI). The United States' technological leadership is no longer assured in many frontier technologies. Indeed, America's strategic rivals, and in particular China, are focused on and committed to taking away American, European, and Japanese technological and marketplace leadership in emerging technologies. Standards will play a role in such outcomes. This chapter focuses mainly on the licensing of standard-essential patents (SEPs) in foundational or "enabling" 5G wireless technology.

Many policy issues are at hand, have tremendous geopolitical consequences, and cannot be looked at in isolation. For instance, the US Department of Justice (DOJ) is reviewing its antitrust policies toward SEPs. These issues are complicated.

The best way forward with technology development is to incent business enterprises operating in the United States and in allied nations to invest heavily in research and development (R&D), as several of them have done in the past. Success in this regard will promote competition and is the best chance the liberal democracies have to maintain technological leadership – and, along with it, achieve long-term economic growth and national security while advancing long-term consumer welfare.

Given US industrial weakness in manufacturing, it is especially important that the research-intensive sector of the US economy remains viable and robust. The development of the foundational technologies stage of the value chain has anchored US competitive advantage in recent decades, and this advantage needs to be sustained and enhanced, if possible. Doing so will require maintaining the viability of the open innovation model in technology development under guidance from the

European Telecommunications Standards Institute (ETSI).[1] Competition authorities tend to applaud open innovation models,[2] because they favor new entrants (as compared to the vertically integrated model, which is in effect a closed model because research by the implementer is done in-house).

This chapter endeavors to scope the true nature of SEP issues. It will focus almost entirely on ETSI, whose intellectual property (IP) policy, and its functioning as a standards development organization (SDO), is important to the world economy. In particular, the development and future evolutions of 5G (and 6G that may follow it[3]) have considerable economic and geopolitical implications for the United States, Europe, and Japan. The chapter also explores the implications for US global technology leadership, competitiveness, and national security of taking a step backward by reinjecting specious antitrust concepts into the analysis.

## II. INTELLECTUAL PROPERTY RIGHTS MATTER

In policy circles, there is an all-too-common failure to understand that the weakening of IP results in less innovation and undermines open innovation approaches, thereby favoring vertical integration. At a time when many policymakers and analysts are concerned about (integrated) Big Tech,[4] it is paradoxical that many of the same individuals favor the weakening of IP. Yet it is the SEP licensing model that enables open innovation and new entry into existing ecosystems. This policy contradiction indicates a lack of clear understanding that standards development, at least for mobile wireless, is an expensive undertaking that requires spending billions of R&D dollars to create new technologies, which get folded into technological ensembles that become next-generation standards technologies available for licensing to industry. The interoperability and interconnections aspect of SEP licensing is just the wrapper. In particular, five issues are often misunderstood and are addressed in this chapter. These issues arose from:

1 David J. Teece & Edward F. Sherry, The IEEE's New IPR Policy: Did the IEEE Shoot Itself in the Foot and Harm Innovation? (Tusher Center for the Management of Intellectual Capital, Working Paper Series No. 13, Aug. 2016), https://businessinnovation.berkeley.edu/wp-content/uploads/2014/07/Tusher-Center-Working-Paper-No.-13.pdf.

2 Open innovation is defined by Henry Chesbrough as the use of purposeful inflows and outflows of knowledge to accelerate internal innovation and expand the market for the external use of innovations. Open innovation eschews the model of closed or vertically integrated R&D where a company relies overwhelmingly on in-house R&D to build its technological capabilities. Henry Chesbrough, Open Innovation: The New Imperative for Creating and Profiting from Technology (Harvard Business School Press 2003).

3 For a proposal for cooperative research efforts on 6G, see Bruce Guile & Albert Pisano, International 6G R&D and Innovation Consortium (BRG Institute, Project Working Paper 2, Oct. 2020), https://static1.squarespace.com/static/5d5f0079ed0caf00014c2fe2/t/5ff4cb44af41aa3cf58d2c19/1609879244797/6G-Case-Statement_10-12-20_v2.pdf.

4 Vertically integrated, not only with respect to the research function but also with respect to design and manufacturing too.

(i) An implicit belief that new standards technology arrives more or less like "manna from heaven." The reality is that royalties from SEP licenses provide the income stream that supports the R&D that improves the underlying technology. If the royalty rate is reasonable, and if unlicensed use is minimized, the required R&D can be funded. Absent a robust patent licensing model, vertical integration (closed innovation) is the model that technology implementers would be forced to adopt.

(ii) An implicit belief that standards development is little more than a matter of agreeing on interconnection protocols. It is quite different. In the case of mobile wireless, the SDO provides the platform for what is likely the largest cooperative R&D endeavor the world has ever seen – bigger and arguably even better than Bell Labs of yesteryear, the disappearance of which was unnecessary collateral damage from an antitrust-driven divestiture of AT&T, which paid too little attention to the future funding of breakthrough innovation.

(iii) A failure to understand that only four to five companies, most of them in the United States and Europe, provide more than 80% of the most important technology that gets embedded in standards, and that there are over 1,000 implementers, a number that is likely growing with 5G and IoT.

(iv) A failure to understand that unlicensed use of standards technology is common. Its presence threatens the technology licensing model and hence the open innovation business model that undergirds it.

(v) A failure to understand that patents are not self-enforcing. Nor is there a unified global enforcement mechanism for SEPs. As a consequence, there is not only unlicensed usage but also forum shopping. Implementers try to "divide and conquer."

These misunderstandings reflect a lack of appreciation that technological contributions to standards development (with the process managed under ETSI governance rules) require innovators to license their technology (and associated patent rights) to implementers, thereby giving up the right to sole use. This bargain works only if there is the expectation and the reality of royalty income sufficient to support past and future technology development activities. If this aspect is not understood and is not at the core of US public policy deliberations, then companies that seek to avoid paying market rates for the use of standards technology will likely succeed and, in doing so, undermine the long-term viability of the ecosystem. If policymakers and the courts allow even quasi free riding, the United States would be playing into the hands of those who have undermined US technological leadership, manufacturing capacity, and economic security.

A balanced approach is needed. If policy favors either side too much (implementers or upstream innovators), then the robust innovation ecosystem that has historically supported mobile wireless is put at risk.

In the early days of mobile wireless (that is, 2G and 3G), standardization activity was dominated or heavily influenced by vertically integrated firms. Today, the success of the open innovation global standards model has enabled nearly seamless wireless compatibility around the globe and allowed hundreds of new implementers to enter the mobile wireless ecosystem. These companies (for example, Apple, Samsung) typically do not contribute significant patented technology to assist in the creation of high-performance standards. They would prefer to use standards technology for free or for a nominal fee. Inasmuch as implementers can lobby government agencies, their sheer numerosity has tended to drown out the voice and perspective of technology developers. In 5G, for instance, the numbers of likely implementers are in the thousands, whereas the majority of the quality contribution comes from a small handful of companies (most notably, Qualcomm, Ericsson, Nokia, Interdigital, and Huawei) that spend heavily on R&D.[5]

In sum, for decades now interoperability standards have also incorporated technology covered by IP. However, this system is now at risk because the licensing landscape has changed somewhat: (1) The ratio of technology developers/contributors to implementers has diminished; and (2) US (and some foreign) antitrust agencies have injected antitrust issues[6] into FRAND[7] deliberations, creating uncertainty that has compromised the functioning of the market for technology. Some of this confusion was cleared up under Makan Delrahim's tenure as head of the DOJ Antitrust Division, but such progress is under threat of reversal under the Biden Administration.

## III. THE GEOPOLITICAL ENVIRONMENT

### A. *Complex System*

It is no longer acceptable – indeed, it is incredibly risky for Western democracies – for antitrust agencies to formulate policy without consideration of geopolitical consequences. Potential short-term domestic consumer welfare issues pale in comparison to many potentially existential geopolitical threats. Fortunately, dealing presciently with these issues will aid competition and innovation in the United States and elsewhere.

5 Huawei's success was aided by low-cost prior access to Western technology achieved by theft and antitrust actions. If the United States and other Western companies do not have the means to support ongoing R&D at the appropriate levels, then Huawei is likely to become the world's dominant technology provider in mobile wireless and related technologies.

6 At least in Europe, antitrust agencies intervened based on complaints submitted by implementers concerning licensor behavior.

7 FRAND stands for fair, reasonable, and nondiscriminatory (royalty rates).

There is reason for concern. In the past, antitrust enforcement actions (including by the Federal Trade Commission (FTC)) in the global technology marketplace have been misadventures in terms of their impact on not only competition but also US global competitiveness. Many are by now aware that China is endeavoring to stack international standards bodies with individuals who swear fealty to China. "China Standard 2035" lays out objectives for blockchain, quantum computing, AI, and other technologies. Impacting and controlling 5G standards development is also part of China's national mission. The Western democracies must be mindful of these activities and the underlying strategy. It is important to prevent the politicization of the standards development process.[8] The *14th Five-Year Plan for National Informatization* released in December 2021, for example, restates China's goals to "create a closed-loop innovation mechanism" to promote "standards building" in prioritized areas such as 5G, Big Data, AI, blockchain, industrial Internet, and so forth, and accelerate the completion and perfection of existing data sharing and data application standard systems. It also anticipates greater integration of China's information standards efforts into all sectors of the Chinese economy relying upon this "closed-loop" innovation system.[9] Policymakers and executives and members of standards bodies must be mindful of these activities and the underlying strategy.

The United States' antitrust policy, if the FTC is to be used as a guide, already has inadvertently strengthened the hand of China. The problems associated with standards technologies are not unlike the problems that the Western democracies are confronting with all emerging technologies. David Delpy, University College London, put it this way:

> Now, it's very difficult for countries to make sure that they get at least a fair share of the return on investment on emerging technologies .... If everybody's playing by the same rules, it's fine. But everybody isn't playing by the same rules. The issue is, value capture: how do liberal economies capture value in a world where not everybody is liberal?[10]

Western democracies must double down on R&D and strengthen the technological capabilities of business firms. That is a big – but necessary – task. The required R&D must, in the main, be private sector funded. For private sector R&D investment to occur, national policy must make sure that SEP owners receive fair compensation, sufficient to support the business models of those Western firms

[8] China may be somewhat delusional if it believes that providing monetary incentives for Chinese companies to put forward technologies will in and of itself sway professional bodies into not choosing the best technologies to incorporate in a standard.

[9] DigiChina, *Translation: 14th Five-Year Plan for National Informatization – Dec. 2021* (Jan. 24, 2021), https://digichina.stanford.edu/work/translation-14th-five-year-plan-for-national-informatization-dec-2021/.

[10] Richard Hudson, *New Year's Resolution: Research Group Aims to Fix the Way the World Collaborates on Technology*, Sci. Bus. (Jan. 4, 2022), https://sciencebusiness.net/news/new-years-resolution-research-group-aims-fix-way-world-collaborates-technology.

that generate the technology that ends up as part of the standards technology ensemble. Supporting the SEP licensing process so that it can, in turn, support the R&D necessary for technology development ought to be the critical policy objective of the DOJ, the United States Patent and Trademark Office, and the US National Institute of Standards and Technology. If achieved, it also will support subsidiary goals with respect to competition and consumer welfare.

### B. *The Changing Geopolitical Landscape*

"China Standards 2035," mentioned in Section III.A, was a galvanizing publication that can serve to remind executives and policy analysts that they need to develop a broader perspective with respect to standards development and standards setting. This initiative builds on "Made in China 2025" and heralds plans and financial and regulatory support for Chinese enterprises, public and private, to take control of the decentralized private (and substantially professionally driven) global standard development process. If successful, China will change the governance of global business, which in turn will augment China's geostrategic power. There are far-reaching consequences for international business, national security, and competition.

Chinese firms have already obtained substantial representation in the international standard-setting process. Unlike Western representatives, Chinese representatives, whether corporate or government, are held accountable to the nation-state. The Swedish Institute of International Affairs recently noted:

> For decades, and almost unnoticed by the general public and politicians, technical standards have been a driving engine behind globalization ... they [now] run the risk of turning into a core subject of great power competition over high technology ... Europe emphasizes its commitment to rules-based institutions in world affairs. Hence, it cannot simply adapt the new power approach to technical standards, since this undermines the existing institutional framework.[11]

The report further noted that China's state-directed approach to standards development "radically breaks with both the U.S. and European approaches that are both industry driven." Other sources draw attention to China, noting:

> The CCP has seized on the importance of these [standards development] bodies for the dual and mutually reinforcing objectives of increasing national competitiveness and building international influence on technology adoption.[12]

[11] Tim Nicholas Ruhling, *Technical Standardization, China and the Future International Order: A European Perspective*, Swedish Inst. Int'l Affs. (Feb. 2020), at 4–5, https://eu.boell.org/sites/default/files/2020-03/HBS-Techn%20Stand-A4%20web-030320.pdf.

[12] Lindsay Gorman, *The U.S. Needs to Get into the Standards Game – With Like-Minded Democracies*, Law Fare (Apr. 2, 2020), www.lawfareblog.com/us-needs-get-standards-game—-minded-democracies.

As a result, one might say that China is "inventing patents," in the sense that it is diligent about filing for both minor and major inventions. It also is very active at standards-setting meetings and on standards-setting governance.

Unfortunately, there is sometimes limited sophistication in understanding what is going on in the global technology marketplace. Some of this flows from the misreading of patent statistics.

Citing patent analytics company iPlytics, an article in the *Wall Street Journal* recently noted that companies from China own "36% of all 5G standard essential patents" and that "U.S. firms including Qualcomm and Intel hold just 14%." The article went on to state:

> Chinese companies own such a significant share of the patents [that] the Western companies need to pay to license from them, that is, the net royalty payments will be from Western companies to Chinese companies.[13]

This statement could be true only if the quality of Chinese patents is equivalent to or better than Western companies' patents, or if the infringing sales of Western firms are greater than that of Chinese firms. The licensing jurisdictions also need to be similar for such equivalences to be drawn.[14] The famous quote "not everything that can be counted counts, and not everything that counts can be counted" seems relevant in this context. However, it would be very imprudent to assume that all Chinese patents are valueless.

Patent statistics can be misleading. Regardless, the United States and its allies – including other liberal democracies, particularly Sweden, Finland, South Korea, and Japan – still maintain a fragile lead, even as the trends do not favor liberal democracies. Policy mistakes now could lead to the rapid dissipation of this fragile leadership by the liberal democracies, with very negative knock-on effects for the US economy and competition. In the next section, I step back and review the context in which SEPs need to be understood – at least with respect to mobile wireless.

## C. *The 5G Technological Ensemble*

The mobile wireless industry has a remarkable track record of developing continuously evolving and improving interoperable systems technology. GSM, Wideband Code Division Multiple Access (WCDMA), and, more recently, Long Term Evolution (LTE) are examples of successful technologies developed privately and separately, but combined by the 3rd Generation Partnership Project (3GPP), using consensus-driven governance, into a platform with massive economies of scale and scope.

Technology development for 5G occurs in a distributed manner with limited overall end-to-end supervision. A very few companies – such as Qualcomm, Nokia,

[13] Don Strumpf, *Where China Dominates in 5G Technology*, Wall St. J. (Feb. 26, 2019), www.wsj.com/articles/where-china-dominates-in-5g-technology-11551236701.

[14] *Id.*

and Ericsson – work hard to help ensure a high degree of end-to-end operability. Without these special efforts, 3GPP could fail. As an organization, 3GPP does not have its own resources to sponsor the development of "gap-filling" technology that, on a standalone basis, may not be financially viable. The real contributions of individual members are hard to calibrate and are not measured merely by counting the number of technical contributions made or patents declared by particular companies.

With 5G, 3GPP has the task of governing a collaborative effort among hundreds of different entities with different interests and incentives. Governance comes in at the time technologies are to be considered for inclusion in a standard. It oversees an iterative, nonlinear, consensus-based approach to technology selection and resulting standards development – systems engineering managed privately and in a decentralized manner. It has worked well, in part because the professionals involved are engineers. Historically, an engineering culture and commercial considerations dominated, and politics were held at bay. Members collectively (by vote) decided on the best technologies to go into a "standard" or new technological ensemble. This may change as Chinese national politics intervene.

Participating firms need confidence that each technology advanced for consideration is robust, has been or will be tested, and can be manufactured, and that the requisite software and applications support will be available. Sponsors of technology then are required to demonstrate that the technology is or can be commercially viable. Hence, by the time that patented technology becomes embedded in the standard, it already has undergone an early assessment as to commercial viability. Licensing executives need to understand this process, as it indicates that patents that are "truly essential" have in all probability passed a litmus test of commercial viability, and thus are likely to have value if indeed they are truly essential and not just "declared essential" by the patent owner.

Feedback from the validation and testing activities is critical and often leads to further development of the technology and/or changes in specification. This process is shown in Figure 1.1. Steps in validation include review, modeling, prototyping, and "plug tests/plug fests," where designers of equipment or software using the technology proposed for the standard test interoperability of products and designs with those of manufacturers. As standards go through revisions, multiple firms may submit proposals and work together toward final adoption of the standard.

The standard-setting for 5G is a continuous process. Updates are issued periodically. Licensing practices have evolved to support the open interoperable mobile wireless ecosystem, with royalties being set in the marketplace via negotiation at levels sufficient to encourage at least a few companies to make the large investment required to develop new 5G technologies.[15]

[15] Qualcomm alone spends over five billion dollars a year mainly on foundational wireless technologies.

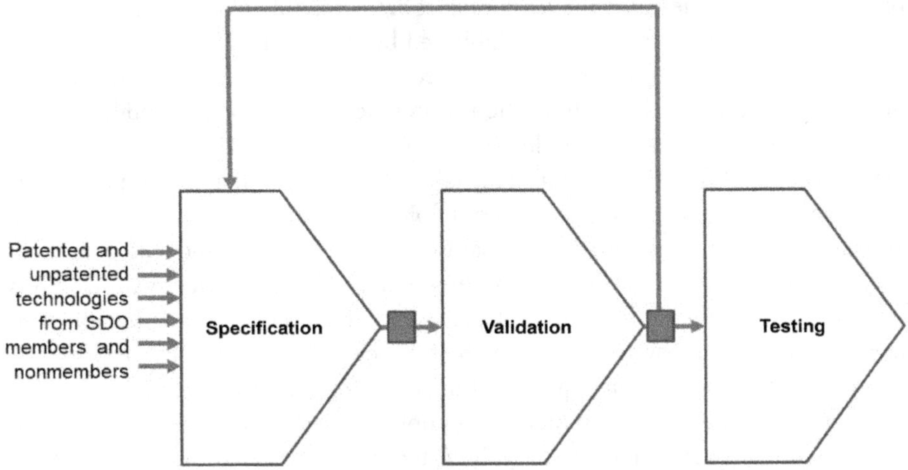


FIGURE 1.1. Some elements of standards development

## D. *Technological Ensembles and the Global Open Standardization Miracle*

As noted earlier, some observers have the naïve belief that standard-setting is just a matter of choosing compatibility/interoperability standards – like choosing between two- and three-pin electrical sockets and plugs. In reality, the situation is radically different.

Standard development is a distributed global R&D activity that takes place under loose 3GPP/ETSI governance.[16]

It is perhaps helpful to recognize that 3GPP is akin to a special kind of global technology development consortium that functions as the mobile wireless ecosystem's R&D arm for developing, assembling, and then standardizing foundational technologies.[17] Tens of thousands of engineers work on 5G foundational technologies. They are loosely coordinated by SDOs that work with 3GPP. This amazing technology development program – with billions of R&D dollars spent each year – is

[16] In the 1980s and 1990s, the International Telecommunications Union (ITU) played a major role with standard-setting for fixed-line telecom. Under the ITU, a new organization that orchestrated mobile wireless technology development was born: 3GPP. The 3GPP was established in 1998 when ETSI partnered with seven other SDOs around the world to develop technologies for 3G cooperatively. 3GPP is the de facto nexus for the remarkable combining of research efforts by the companies that are inventing 5G wireless technologies. Qualcomm, *Understanding 3GPP – Starting with the Basics* (Aug. 2, 2017), www.qualcomm.com/news/onq/2017/08/02/understanding-3gpp-starting-basics.

[17] *Id.*; David J. Teece, *Profiting from Innovation in the Digital Economy: Enabling Technologies, Standards, and Licensing Models in the Wireless World*, 47 RSCH. POL'Y 1367 (Oct. 2018).

mainly privately funded, and the technologies are combined in powerful ensembles by engineers operating under rules established by 3GPP/ETSI.

3GPP periodically releases documents incorporating important new advances in the foundational technologies from the research activities of global mobile wireless technology companies. For example, Release 16 was published in July 2020.[18] After a release document is published, it usually takes at least a year before cell phone and other subscribers have access to the fruits of the new technology. Infrastructure companies such as Ericsson, Nokia, Samsung, LG, Huawei, and Cisco have to design the technologies into their equipment; and chip and device manufacturers such as Samsung, Apple, LG, Motorola, ZTE, and Huawei have to embed them in new modems and in new generations of their devices. The network service providers also must install the requisite equipment upgrades for benefits to be realized. They do so, however, long after the standards technology is developed, whereas those that develop standards technology do so without the confidence that their technologies will ever be incorporated in the standards or have commercial payoff of any kind.

3GPP epitomizes cooperative global technology development at its apogee. It is a corollary of the US and European-led post–World War II liberal system of cooperative innovation and distributed economic organization. It is the de facto orchestrator of discoveries and inventions relevant to the mobile wireless ecosystem. With China's rise and expressed desire to dominate standards setting, this delicate organizational arrangement will be put at risk, particularly if the United States does one more antitrust policy zigzag.[19]

The type of governance that 3GPP affords is unique as to its global scale and scope. It is rule based and consensus driven. Constituencies include technology developers, systems operators, device makers, and governments around the world.

As described in Section IV, patents are important in the mobile wireless cooperative ecosystem, because patent licensing is how the global sharing and financing of new technology is achieved. Proprietary 2G, 3G, 4G, and now 5G technologies, many of them foundational, have been made available via nonexclusive patent licenses to the whole world, generation after generation. This is a major reason for the success of the mobile wireless ecosystem over the past 30 years. It has afforded countless benefits to billions of users, and fueled competition and economic growth in the United States and abroad. The licensing system promotes both interoperability and entry by device manufacturers by providing a necessary suite of technology input solutions.

[18] 3GPP, *Release 16*, www.3gpp.org/release-16.

[19] Examined more closely, 3GPP is a cluster of national and regional telecommunication standards organizations. It organizes its work into three different streams: radio access networks, services and system aspects, and core network and terminals for which 3GPP develops the technical specs. These are then converted into standards by seven regional standards-setting organizations (SSOs) that form the 3GPP partnership.

## IV. THE SALIENCE OF PATENTS AND PATENT LICENSING TO COMPETITION AND THE ROBUSTNESS OF THE INNOVATION ECOSYSTEM

While technology is adopted into the standards by vote of the members, developers who contribute patented technology do so only because of SDO rules requiring that patent owners who provide patents for inclusion in the standard "make licenses available" on the expectation that implementers/users take licenses and pay royalties, rather than infringing willy-nilly. That is, everyone can have access to 5G standardized technology (at least for a reasonable period of time) through patent licenses, but licensees must be willing to pay royalties. Commercial terms nevertheless must be FRAND. This two-way commitment somehow must be enforced for the open global R&D super consortia that is 3GPP to be viable.

As already noted, the open innovation mobile wireless technology model has become fragile. There are now thousands of downstream device manufacturers, but only a handful of firms provide 80% of the foundational technology that is incorporated into standards.[20] 5G connectivity would not be possible without decades of upstream R&D on foundational technologies by companies such as Ericsson, Nokia, Qualcomm, Samsung, and LG, as well as new players such as ZTE and Huawei. However, many downstream device manufacturers try to avoid paying royalties altogether.

A smoothly functioning market for patent rights cannot simply be assumed. A primary reason is that patents are not self-enforcing. When patents are issued, the invention is available for use worldwide through concurrent publication of the patented invention. The patented technology does not enjoy an automatic monetary collection mechanism. To some (unscrupulous or simply opportunistic) industry participants, the publication of patents and associated standards is an invitation for unlicensed use of the patented technology, because policing unlicensed usage is difficult and often costly. "Catch me if you can" is an all-too-common attitude that sometimes is aided and assisted by poorly designed competition policies.

Furthermore, the 2006 US Supreme Court decision, *eBay, Inc. v. MercExchange, LLC*,[21] had the effect, as a practical matter, of eliminating injunctive relief as a remedy for patent infringement in a wide range of circumstances. The decision put the US technology market and R&D spending to support standards at risk. I signed an amicus brief against *eBay*, because handicapping the right to enjoin would deeply compromise the licensing marketplace. Many of my fears have been realized, as "holdout" – discussed at length in the next part – is indeed a major problem today.

[20] In certain key industries (for example, wireless and automotive), the number of potential licensees is fairly small.

[21] 547 U.S. 388 (2006).

In sum, if the patent owner does its part and "makes licenses available," there is no guarantee that users will take a license and begin to pay royalties. The patent owner still needs to develop a licensing program and persuade unlicensed users to sign up and pay royalties. In the absence of injunctive relief and/or strong business ethics, that is a difficult and costly mission. Put differently, the patent owner is left with very limited means to bring putative licensees to the bargaining table unless courts or international trade regulators block market access for infringers.

## V. THE FRAND ROYALTY APPROACH

### A. *FRAND and the Open Innovation Ecosystem*

The mobile phone industry was in its infancy when ETSI was founded. The focus then and now is on what, in modern terms, we think of as creating a robust innovation ecosystem.

> ... the ETSI IPR POLICY seeks a balance between the needs of standardization for public use in the field of telecommunications and the rights of the owners of IPRs .... IPR holders whether members of ETSI and their AFFILIATES or third parties, should be adequately and fairly rewarded for the use of their IPRs in the implementation of STANDARDS and TECHNICAL SPECIFICATIONS.[22]

In what follows, I give consideration to these issues, with specific reference to ETSI policy. I also consider the cost of error and elaborate the point that under-rewarding the patent holder of an enabling technology[23] has very high societal costs and should be avoided.

From the outset, ETSI recognized the need for a forward-looking approach to technology development on mobile wireless. The original architects of ETSI's IP policies sought a "balancing of the interests" of technology contributors (patent owners) and implementers.

ETSI started as a European governmental initiative to assemble a broad set of actors committed to fairness and benefits to the broader telecommunications sector (ecosystem) and consumers. This broad constituency is still apparent today and includes chipset designers and fabricators, handset and base station makers, cellular service providers, app developers, and, of course, consumers.

The standards development system was not designed to favor one constituency over the others, or downstream over upstream. Indeed, initial versions of the ETSI IP policy that did not attract technology developers were rejected in favor of versions

[22] ETSI, *ETSI Directives – v44 – 21 December 2021, Rules of Procedure* at Annex 6 (Dec. 21, 2021), https://portal.etsi.org/directives/44_directives_dec_2021.pdf.

[23] An enabling technology is a generic or "platform" technology that has applications in many products/areas of application. It is a junior general purpose technology. *See, for example,* Teece, *supra* note 17.

that yielded "balance." When standards technology contributors enable so much of the subsequent downstream innovation, it is critical that technology developers not be shortchanged. This conclusion is not only consistent with ETSI IP policy, but also economically desirable and therefore entirely reasonable from a public policy perspective.

SDOs require that, before technologies are accepted into a standard, members that own patented technologies are "prepared to grant irrevocable licenses on fair, reasonable and non-discriminatory ('FRAND') terms and conditions under such IPR." What is "fair and reasonable" (FR) and what is "nondiscriminatory" (ND) often raise questions. In this chapter, I will address only the FR aspect of FRAND, in the context of ETSI.

There is little doubt – and Dr. Bertram Huber, an ETSI founder representative, confirms this view[24] – that ETSI was concerned with establishing a vigorous standards process to support the development of a robust telecommunication industry in Europe and around the world. ETSI requires FRAND commitments from its technology contributors, with the expectation that implementers would take a license under FRAND terms.

### B. *Patent Holdup: A Theory in Search of a Problem*

As noted, a vigorous mobile telecommunication industry requires a robust innovation ecosystem. Various parties and occasionally antitrust regulators have clumsily tried to undo ETSI's rules with respect to FRAND issues by advancing the concept of "patent holdup."[25] Patent infringement by implementers is excused and even encouraged by a "fig leaf" in the form of this antitrust theory of "patent holdup": "holdup" because the implementer supposedly only knows the royalties they must pay after they have committed capital. The theory chooses to disregard the R&D investments made by the technology developers.

By way of background, consider the sequencing of investment in the development and deployment of standardized technology. Figure 1.2 shows the sequencing of investments, which has important implications for licensing dynamics. Stage one investments are made having no guarantees that they will be successful; and even if it is technologically successful to some degree, the discoveries may not be good enough to go into the 5G standard. The fact that R&D to develop foundational technologies takes place before the equipment and device makers invest puts the licensor in a weak bargaining position with respect to the implementers/licensees. Moreover, there is no guarantee that the technology will be adopted over competing alternatives. This position is amplified by the twin facts that patents are not self-

24 Author conversations with Dr. Bertram Huber (2017).

25 ETSI leaves the rate to negotiations between the parties under the shadow of the FRAND commitment.

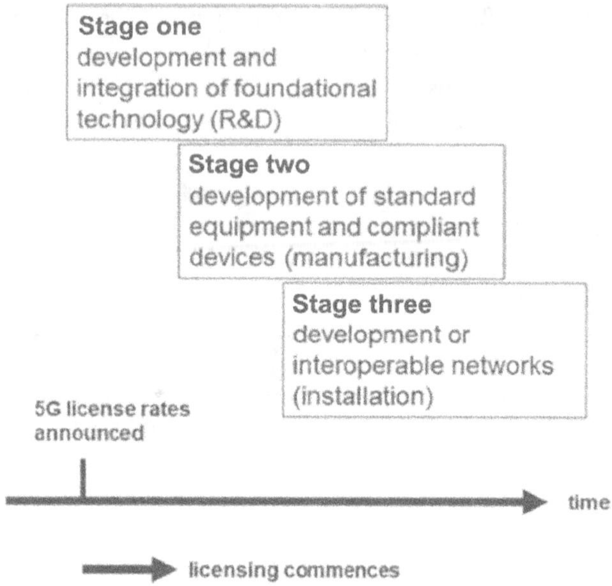


FIGURE 1.2. Sequencing of development and deployment of wireless infrastructure technology

enforcing and injunctions nearly impossible to come by in the United States post-*eBay*. Enforcement requires a court of law to back up the patent owner with an injunction or something similar. Otherwise "strategic" or "unscrupulous" putative licensees will engage in the similar sounding – but essentially inverse – practice of "holdout." As Makan Delrahim, then the head of the DOJ Antitrust Division, noted:

> If the implementers hold out, the innovator has no recourse, even if the innovation is successful. In contrast, the implementer has some buffer against the risk of hold-up because at least some of its investments occur after the royalty rates for new technology could have been determined. Because this asymmetry exists, under-investment by the innovator should be of greater concern than under-investment by the implementer.[26]

Put differently, implementers can "hold out," not take a license, and try to dodge paying royalties. To explore these issues further, one must examine in more detail FRAND issues

The mischief comes from implementers, aided and abetted by theoretically oriented academic economists whose models appear to have impacted antitrust enforcement agencies. Academic economic models of "patent holdup" ignore sunk

[26] Makan Delrahim, assistant attorney general, US Department of Justice, remarks as prepared for the USC Gould School of Transnational Law & Business Conference (Nov. 10, 2017), www.justice.gov/opa/speech/assistant-attorney-general-makan-delrahim-delivers-remarks-usc-gould-school-laws-center.

R&D investments by technology developers. They also ignore longer-term business model issues associated with funding R&D. These proclivities create a penchant for seeing royalties as SEP "taxes" and not precious fuel to support R&D.

The intellectual history of patent "holdup" theory is checkered. In my view, in the context of SEPs, it is a theory in search of a problem. It has been an economically damaging red herring that has compromised licensing activities and US technology leadership.

The first (mis)application of the holdup concept to the realm of patents was a paper published by Carl Shapiro in 2001.[27] Some IP scholars in the legal academy subsequently became aware of, and actively advanced, these patent holdup theories. Meanwhile, other scholars and practitioners actually close to the world of licensing understood this work to be theoretical musing and little else. There was considerable surprise when it was taken seriously by antitrust agencies and some courts.

The assumptions of rampant opportunism and guile by upstream technology providers lie at the heart of holdup theory.[28] Needless to say, there is no empirical support for either assumption in the context of patents. The theory also assumes that patents are self-enforcing (that is, there is an injunctive right appended to every patent exercisable at the discretion of the (SEP) patent owner), despite *eBay*. This constellation of assumptions is quite fanciful.

The theory – and theory is all it is – is sometimes articulated in terms of a patent owner promising to one or more implementers one rate, and specific investment is made by an implementer on the basis of that promise. The narrative is that patent owners subsequently, without good reason, strategically increase royalty demands once the implementer is locked in and committed to downstream investments. There is no evidence that patent holdup has ever occurred.

More commonly, the patent owner promises to make licenses available on FRAND terms, but without specifying in great detail until later on which rates it would seek to charge. Moreover, the right to enjoin, when it exists, requires the action of a court. There are, of course, reputational risks that technology developers would have to take should they act egregiously. Needless to say, there are contractual protections for implementers, which are, according to the law of many jurisdictions, third-party beneficiaries of the agreement between the SDO and the patent owner to set royalties that are FRAND.[29]

Technology developers, by contrast, can negotiate only after they have sunk their R&D investments and after their technology has survived a selection process to get

[27] Carl Shapiro, *Navigating the Patent Thicket: Cross Licenses, Patent Pools, and Standard Setting*, in Innovation Policy and the Economy 119–50 (Adam B. Jaffe, Josh Lerner, & Scott Stern, eds., 2001).

[28] Oliver Williamson, Markets and Hierarchies: Analysis and Antitrust Implications, A Study in the Economics of Internal Organization (1975); Oliver Williamson, The Economic Institutions of Capitalism: Firms, Markets, Relational Contracting (1985).

[29] China does not support third-party beneficiary theories.

into the standard. Thus, technology developers are at least as vulnerable to "reverse holdup" or, as it is more commonly known today, "holdout" by implementers, as implementers are vulnerable to hold up by developers.[30] Indeed, given that implementers could, but rarely do, seek licenses before they start making standards-compliant products, and the length of time that it takes licensors to conclude agreements with industry players, and the considerations noted below arising from the nature of the enforcement of patent rights, the dangers of holdout are quite real and substantially underestimated.[31] Although the theory of holdup historically has been advocated more vocally, it is holdout that is the greatest risk to licensing of ETSI standards.

Nevertheless, numerous implementers and some nation-states making standards-compliant products have found it convenient to claim to be victims of "holdup." China, for example, has also tried at the World Trade Organization (WTO) to have all SEPs declared barriers to trade.[32]

## C. *Patent Holdout*

Fortunately, in recent years, policy concern over holdout has received some attention. As noted earlier, Makan Delrahim, the former head of the DOJ Antitrust

[30] Luke Froeb & Mikhael Shor, *Innovators, Implementers and Two-Sided Hold Up*, ANTITRUST SOURCE (Aug. 2015). Froeb and Shor state that the "innovators' hold-up problem is more difficult to overcome" than any holdup problem facing the implementers. Under the leadership of Makan Delrahim, the DOJ Antitrust Division acknowledged that the holdup of innovators is a more serious problem than the holdup of implementers, discussed *infra*.

[31] This point is also made by Delrahim: "As a result [of hold-out], SEP holders either receive a below-FRAND payment, obtaining damages on the fraction of their portfolio that has been successfully litigated, or they need to file sequential litigation to obtain payment for all of their infringed SEPs . . . [P]atent hold-out can be a very attractive strategy for standards implementers." Delrahim, *supra* note 26, at 5. For similar sentiments, see Anne Layne-Farrar, *Why Patent Hold-Out Is Not Just a Fancy Name for Plain Old Patent Infringement*, COMPETITION POL'Y INT'L (Feb. 8, 2016); Michael Renaud, James Wodarski, & Sandra Badin, *Efficient Infringement and the Undervaluation of Standard-Essential Patents*, IAM (2016); Richard Epstein & Kayvan Noroozi, *Why Incentives for "Patent Hold-Out" Threaten to Dismantle FRAND, and Why It Matters*, 32 BERKELEY TECH. L.J. 1381 (2017).

[32] "China is of the view that, IPR issues in preparing and adopting international standards have become an obstacle for Members to adopt international standards and facilitate international trade. It is necessary for the WTO to consider negative impacts of this issue on multilateral trade and explore appropriate trade policies to resolve difficulties arising from this issue." WTO, *Intellectual Property Right (IPR) Issues in Standardization, Committee on Technical Barriers to Trade*, G/TBT/W/251 (May 25, 2005), https://docs.wto.org/dol2fe/Pages/FE_Search/FE_S_S009-DP.aspx?language=E&CatalogueIdList=84617,75721,51689,62632,61290,61833,52016,60561,72578,90497&CurrentCatalogueIdIndex=9&FullTextHash=1&HasEnglishRecord=True&HasFrenchRecord=True&HasSpanishRecord=True. Zhang Ping, a leading Chinese scholar who has trained a generation of Chinese professionals in IP, standards, and antitrust, similarly has relied heavily on Shapiro and Lemley in her influential book. ZHANG PING & MA XIAO, STANDARDIZATION AND INTELLECTUAL PROPERTY STRATEGY 39 (2nd ed. 2005).

Division, made considerable progress in helping develop the understanding that "collective holdout" behaviors in standard-setting are more pernicious and likely than unilateral holdup by SEP holders.[33] The recognition that holdup is a red herring is what may have led Contreras to argue that:

> To the extent that hold-up impedes the efficient operation of standard-setting processes, SDOs can, and have, adopted internal procedures, including disclosure and licensing requirements, to curtail that behaviour … it may thus be time to close the debate over the systemic prevalence of this form of behaviour.[34]

From an empirical perspective, Heiden and Petit note the emergence of a "long tail" of implementers or micro-vendors that are individually small but collectively account for a reasonable share of industry revenue, and that are not licensed.[35] Many of these implementers are based in China. They note in this context that "a systematic patent trespass effect can be deemed to occur when 30% or more of a relevant market is unlicensed." They relate this to a collective action problem: "why take a license if your competitors do not?" They note that the "systemic effect of patent trespass is primarily experienced through the impact on the technology market through the development of consensus-based standards."[36] Heiden, Padilla, and Peters note the presence of a similar "collective action" problem resulting in widespread holdout in the IoT sphere.[37]

These empirical observations echo the findings of Judge Essex of the US International Trade Commission (as summarized by Renaud, Wodarski, and Badin):

> [T]here is no evidence to support the notion that owners of SEPs have engaged in patent hold-up either in the investigations before him or in the telecommunications industry more generally. Rather, the evidence is all on the side of patent hold-out. The implementers of the standards are using the patented technology incorporated in the standards without authorization and without even engaging in licensing negotiations because they know that the worst that can happen is that they get sued, are found to infringe and are made to pay the same FRAND rate that they would have had to pay for using the patented technology in the first place.[38]

In summary, then, the very non-self-enforcing nature of patent rights directly indicates why holdout rather than holdup is a problem that we should expect to see

33 Delrahim, *supra* note 26.

34 Jorge L. Contreras, *Much Ado about Holdup*, 2019 U. Ill. L. Rev. 875, 904 (2019).

35 Bowman Heiden & Nicolas Petit, *Patent "Trespass" and the Royalty Gap: Exploring the Nature and Impact of Patent Holdout*, 34 Santa Clara High Tech. L.J. 179–249 (2018).

36 *Id.*

37 Bowman Heiden, Jorge Padilla, & Ruud Peters, *The Value of Standard Essential Patents and the Level of Licensing*, 49 AIPLA Q. J. 1 (2021).

38 Renaud et al., *supra* note 31. Judge Essex further concluded that this situation was "as unsettling to a fair solution as any patent hold up might be."

in licensing SEPs. My own experience with examining the smartphone licensing landscape in the context of litigation, and the empirical observations of other commentators, supports this. Royalty revenues are a small share of both the overall value-added from mobile telecommunications and smartphone implementers' revenues.[39] This calls into question the predictions of holdup theory and is consistent with the reality that holdout is an important characteristic of the licensing landscape today.

The reason that holdout is a present, and perhaps growing, danger is because it is a profitable strategy for implementers. It is profitable in part because of the weakness of injunctive relief, the fragmentation of the patent enforcement landscape at the global level, and the lack of corrective mechanisms in damages and license awards by courts.

### D. *Holdup versus Holdout: Time to Close the Debate*

Empirical studies have established the irrelevance of holdup theories. Notwithstanding this fact, implementers remain willing and eager to advance such theories. At the same time, it is now widely accepted that contractual mechanisms to redress holdup are available, should it ever occur.[40]

Galetovic, Haber, and Levine[41] provide a sophisticated empirical analysis with regard to the holdup issue. They find that "products that are SEP-reliant have experienced rapid and sustained price declines over the past 16 years," and that the "prices of SEP-reliant products have fallen at rates that are not only fast relative to a classic hold-up industry, they are fast relative to the patent-intensive products that are not SEP-reliant." Using a quasi-natural experiment to study the effect of the *eBay* decision on relative price declines in SEP-reliant versus non-SEP-reliant industries, they also do not find that prices in SEP-reliant industries were more affected by *eBay* than in non-SEP-reliant industries. If holdup were more of a problem in SEP-reliant industries, one would have expected to see a greater effect of *eBay* in these industries than in those not driven by SEPs.[42]

[39] Alexander Galetovic, Stephen Haber, & Lew Zaretzki, *An Estimate of the Average Cumulative Royalty Yield in the World Mobile Phone Industry: Theory, Measurement and Results*, 42 Telecomm. Pol'y 263 (2018). The authors estimate that, relative to smartphone manufacturer revenues of $425 billion in 2016, royalties were around $14 billion, or 3.3%.

[40] Further, it is not credible to think that sophisticated implementers in today's technology markets, with decades of SEP licensing and negotiating experience, are naïve with respect to what SEP holders might demand by way of royalties and non-price terms. Thus, what is being termed holdup is really just an assertion – that in circumstances where implementers plunge into making standards-compliant products before licensing the relevant SEPs, the threat of an injunction can confer substantial bargaining power upon the SEP holder and enable it to extract "too much" for a license.

[41] Alexander Galetovic, Stephen Haber, & Ross Levine, *An Empirical Examination of Patent Holdup*, 11 J. Competition L. & Econ. 549 (2015).

[42] *Id.*

The Galetovic et al. findings are unsurprising. The presence of the FRAND commitment, the lack of availability of injunctive relief (particularly in the United States), the repeat-game nature of standardization,[43] and the bargaining power of many implementers (for example, their ability to prolong litigation) all militate against holdup. Most fundamentally, holdup is unlikely in a setting where the implementer or prospective licensee can use the technology without paying for it, and injunctions are close to nonexistent, except perhaps in European courts. There is no way that the SEP holder or licensor by itself can prevent infringing use. Accordingly, there is a fundamental difference between "ordinary" goods and services, on the one hand, and IP, on the other – a point that Germany's Federal Court of Justice recently recognized:

> [U]nlike buyers of goods and services– standards implementers are in the favorable position to be able to access protected technology needed for producing standard compliant products, even without an agreement with the patent holder.[44]

The court's observation is another way of saying that patents are not self-enforcing. A patent holder cannot do what most suppliers of goods and services can do, which is simply to withhold supply to those customers who do not pay for the good or service.[45] Instead, patent holders must resort to costly and time-consuming litigation to enforce their rights. In such litigation, the risks to licensors and licensees are asymmetric. For example, a court decision that is substantially different from the licensors' position on the FRAND value of its portfolio potentially can have a significant adverse impact on that licensor's longer-term licensing strategy. Akemann, Blair, and Teece note:

> Intuitively, patent holders who face the prospect of having to litigate repeatedly against multiple infringers have to be concerned about what might be termed a "one-way ratchet" effect. If the patent holder wins one case against one infringer that does not mean that others will agree to take a license ... [H]owever, if the patent holder ever loses a case – especially on validity grounds – then there is likely to be a significant adverse effect on the patent holder's ability to gain license revenue from that patent in the future. In effect, the patent holder has to "win them all", while the infringers may only have to "win one." In this way ... risks

43 SEP holders who wish to continue participating in repeat rounds of standards-setting activities run the risk that other members will seek to exclude them from future standardization activities if they are seen to have violated their FRAND commitment.

44 English-language summary of Sisvel v. Haier, KZR 36/17 (Nov. 2020), https://caselaw.4ipcouncil.com/german-court-decisions/federal-court-of-justice-bgh/sisvel-v-haier.

45 While this problem of not being able to exclude infringers automatically also could apply to the circumstance of a non-SEP patent that a firm wishes to utilize exclusively in its own products, it may be more acute in the world of ETSI-related SEPs, where rights have to be enforced against multiple infringers, raising the costs of both detection and enforcement.

associated with a single loss . . . could lead to a set of rates in the marketplace that are significantly depressed relative to actual value.[46]

As noted earlier, the antitrust frameworks that employ holdup are static. They take the funding of (foundational) wireless technology for granted. This further undermines their relevance. It is hard not to agree with Barnett that the academy has led the judiciary and policymakers astray.[47]

Accordingly, it is time to shut down the antitrust ruse of patent holdup – not resurrect it, as the DOJ's draft policy statement on remedies for the infringement of SEPs (issued in December 2021 and withdrawn in June 2022) sought to do.[48] Should evidence of holdup ever emerge, the theory can be resuscitated; but it doesn't deserve time on the agenda now.

We nevertheless are left with the fallout of previous US policy misadventures in the form of: (a) some foreign competition agencies and bureaus using the holdup argument to support mercantilist policies and favor national champions;[49] and (b) a sense that the debate perhaps has come to a stalemate with legitimate arguments and evidence on both sides. The assessment in (b) is inappropriate. The weight of the evidence favors holdout as the problem. Holdup is merely a theoretical possibility remote from real-world situations. On the other hand, many SEP holders must wait years before they can achieve a license with implementers, or else must resort to litigation – a move that carries asymmetric risks for SEP holders, as discussed later – before they are able to obtain any payment. Moreover, the failure to enter license agreements with particular licensees almost always will have negative consequences for the SEP holder's broader licensing program.

Additionally, through delaying tactics, licensors may be able to extract significant discounts for past use, benefit from statutes of limitations on past damages, and benefit from potential expiry of patents that they have infringed for many years. If the worst outcome for an infringer is that ultimately it ends up paying a FRAND rate on only some portion of its infringing sales, it will have a great deal of bargaining power

[46] Michael Akemann, John Blair, & David Teece, *Patent Enforcement in an Uncertain World: Widespread Infringement and the Paradox of Value for Patented Technologies*, 1 Criterion J. Innovation 861 (2016).

[47] Jonathan M. Barnett, *Has the Academy Led Patent Law Astray?*, 32 Berkeley Tech. L.J. 1313 (2017).

[48] US Department of Justice. US Patent and Trademark Office and National Institute of Standards and Technology Withdraw 2019 Standards-Essential Patents (SEP) Policy Statement (June 8, 2022), www.justice.gov/opa/pr/justice-department-us-patent-and-trademark-office-and-national-institute-standards-and; US Department of Justice. Public Comments Welcome on Draft Policy Statement on Licensing Negotiations and Remedies for Standards-Essential Patents Subject to F/RAND Commitments (Dec. 6, 2021), www.justice.gov/opa/pr/public-comments-welcome-draft-policy-statement-licensing-negotiations-and-remedies-standards.

[49] It may be more precise to say that certain competition regulators are using patent holdup as theoretical justification for taking mercantilist actions to reduce royalty rates for local device manufacturers.

to bring to negotiations with SEP holders, meaning that infringement and holdout usually will be profitable strategies.

I consider there to be an urgent need in many jurisdictions to bring "balance" to SEP licensing in the cellular mobile space. While important new cases are moving to recognize the key role of technology developers, implementers still can engage in holdout with only distant sanctions to worry about.

One obvious lever is to strengthen the existing injunctive relief regimes to provide the licensee with stronger incentives to negotiate a license.

However, changes to injunctive relief may be practically difficult to achieve. In this context, FRAND royalty determinations in litigations may play an important restorative role too. Critical to this endeavor of bringing balance is the recognition that infringement and holdout must not be profitable. Recalcitrant licensees must not be put on the same footing as those who were (more) willing to sign up for licenses on FRAND terms. Nor must competition enforcement agencies assist renegades in their quest to avoid paying royalties or to crank down FRAND rates to below reasonable levels.

An important question that should be addressed is whether an "unwilling licensee" – one that is not prepared to accept any license terms other than those that it deems to be FRAND – should lose the benefit of the licensor's FRAND commitment. This means that not only should injunctive relief be immediate from the moment that the licensee's unwillingness is established – as the UK courts recognized in a recent decision involving Apple and Optis, and as has been common in mainland Europe – but also potentially that royalty rates in damages awards need not be based on assuming that the FRAND constraint applies.

Inherently, a FRAND royalty rate is reasonably linked to the (likely) value contribution of the technology to the product. However, as discussed earlier (and in more detail later), holdout exerts other costs on the licensor, especially on its overall licensing program. Damage awards based solely on the value contributed to the product may not suffice to restore the licensor's economic position to where it would have been absent the infringing behavior. Nor would they truly address the harm to the licensing marketplace that holdout behavior inflicts.[50]

[50] David Teece, Peter Grindley, Edward Sherry, & Keith Mallinson, *Maintaining Ecosystem Innovation by Rewarding Technology Developers: FRAND, Ex Ante Rates and Inherent Value* (Working Paper Series No. 21, Tusher Center on Intellectual Capital, Apr. 24, 2017). However, even if punitive or exemplary damages and enhanced injunctive relief are ambitious and unlikely steps, courts can and should take measures to recognize the distinction between willing and unwilling licensees. One important and constructive step concerns the use of the licensor's existing licenses or other parameters of a licensor's existing licensing program in evaluating FRAND rates to apply to infringers who have compelled litigation. The key consideration is that, even in any given licensing situation between a given licensor and a given licensee, a range of rates (here, "rates" is used as a shorthand for terms of agreement, including non-price terms) may be consistent with meeting the "balance" envisioned in FRAND. The theoretical upper bound for this FRAND range is, as I explain, the value

In practice, many licensees will obtain rates that are well within the FRAND range (and perhaps even below it), because the licensor will be prepared to accept low rates to achieve (relatively) quick settlement and avoid litigation. It would be wrong for courts to base rates for infringers on these negotiated rates without recognizing the context in which these rates were achieved. To do so would put the unwilling licensee on the same footing as the willing licensee. Instead, court-awarded rates (whether applied to licenses or past-use damages) at a minimum should be based on the FRAND (benchmark) rate. However, even this may be too lenient a corrective for the problem of holdout.

Further, the nondiscrimination (ND) prong of FRAND should not be invoked as a reason to base awards either on "best prices" or even averages across licensees – the ND prong cannot be interpreted in such a way that nondiscrimination trumps the fundamental idea of balance. The comparison of royalty rates achieved by different licensees is at most relevant for an ND analysis to the extent that differences in royalty rates distort competition.[51] If two licensees operate very different business models and aim at very different market segments, then unwarranted distortions to competition are unlikely. Royalty rates paid to individual licensors are a small sliver of the implementer's overall cost stack, and so differences in these rates paid are unlikely to distort competition.[52]

As noted earlier in Section IV (especially Figure 1.2), the commonplace situation that exists is one in which an implementer begins manufacturing devices and implementing SEPs long before taking licenses to any of them. Many licensors issue notice letters to implementers years after use of the SEPs actually began. It is often at least a couple of years before licenses are agreed. Reasonably often, there is

contribution that the technology makes to the product, which I refer to as a "FRAND benchmark rate." This value contribution should be allowed to reflect the value that the technology offers as part of a standard. In practice, this value contribution will be difficult or impossible to measure. It could be conservatively proxied, however, by using the upper end of negotiated licenses or "standard program rates." (Such proxies are likely to understate the FRAND benchmark rate because, as explained herein (and *id.*), all licensees will have bargaining power sufficient for them to extract a significant share of the value contribution of the technology for themselves.)

51 Of course, the issue here is what is meant by "ND" in ETSI's conception of FRAND, as opposed to what sort of behavior is sanctionable under antitrust law. So-called secondary-line price discrimination in the provision of licenses would not be sanctionable in the United States under antitrust laws, but it might be in Europe. But even under the latter, more expansive, price discrimination regime, a mere difference in royalty rates would not in itself be seen to distort competition. Given this backdrop of antitrust law and given the complete lack of evidence that ETSI has ever seriously contemplated a hard-edged MFN (most favored nation)-style nondiscrimination policy, the persistent advocacy of such an approach by some implementers is unmerited. David J. Teece, Edward F. Sherry, & Peter C. Grindley, *On the "Non-Discrimination" Aspect of FRAND Licensing: A Response to the Indian Competition Commission's Recent Orders*, 30 IIMB Mgmt. Rev. 10 (2018), https://doi.org/10.1016/j.iimb.2017.09.002.

52 Further, many negotiated licenses involve lump-sum payments, which do not affect marginal pricing and production decisions.

no agreement, and unlicensed use continues unabated. In many cases, perhaps something close to a decade will go by before litigation is launched, and then perhaps one or two further years will pass before any decision – and before any court-awarded damages or court-determined license are made available. In all, a decade or more might pass before the infringer pays anything for its use, by which point a new generation of the standard has replaced the old standard upon which the litigated decision was based!

Even then, the licensee may pay no more than a relatively low rate that is deemed to be FRAND – often on the basis of rates derived from comparable licenses without necessary adjustments to account for the difference between willing licensees and those who force the licensor into litigation. Alternatively, this rate may be based on inherently licensee-friendly formulations such as the "top-down" method of allocating some aggregate reasonable royalty among the different licensors. Either way, a licensee that delays or strings out discussions has little incentive to take a license – at worst, it will have to pay the same FRAND rate that it would have paid anyway. At best, it secures an advantage over its licensed rivals by remaining under the licensor's radar.

Growing diversity in the nature and geographic loci of implementers and the emergence of new use cases such as those associated with IoT exacerbate the problem. The share of implementers with licenses likely is falling, and at least in some environments – such as the licensing of IoT implementers – a collective action problem is emerging in which widespread infringement may be self-perpetuating.[53] The source of the collective action problem is simple: Licensed users will see themselves as being competitively disadvantaged relative to unlicensed users.

That such pervasive problems exist is not surprising. The perverse focus on holdup and the introduction of antitrust lawsuits has made these problems worse. Attention to the specifics of licensing negotiations and actual business behavior always has suggested that holdout rather than holdup is the bigger problem.

## VI. RESTORING AND REVITALIZING TECHNOLOGY MARKETS

### A. *Why Corralling SEP Infringers Is Difficult*

The primary issue that should animate competition policymakers and the courts is the threat posed to open consensus-based standards development by holdout behavior. At its root, a holdout licensee will do no worse than a willing one; and it might well do better. That is, it could end up paying nothing or, by negotiating a license

[53] Heiden & Petit (2018), *supra* note 35, provide statistics that suggest that licensing coverage fell from roughly 73% of "implementing firms that are potential licensees" in 2006 to 39% of such firms in 2016. Their results appear to be based on interviews that they conducted with licensing experts and firms that are participants in the licensing marketplace. They attribute this "patent trespass" to a long tail of unlicensed implementers.

late in the day (when most of its sales and profits from using the technology are safely in the past), pay heavily discounted rates.

Similarly, consider a situation in which the licensor offered a license on FRAND terms, and the licensee declined such a license and made counteroffers that were not FRAND or otherwise indicated its unwillingness to take a license on FRAND terms. In this case, even if the licensee were enjoined, it can still avail itself of a FRAND license at some point in the future.

The perennial availability of a FRAND license weakens even the power of any injunctive relief remedy – where such is available. It encourages the licensee to try its luck in the courts rather than negotiate for a license. If the licensee escapes an injunction, it can continue to enjoy the benefit of infringement perpetually or until it forces the licensor into conceding terms that it likes.[54] Alternatively, if the licensee is informed, it may still have the option to have the injunction lifted by accepting a FRAND license – possibly the very same FRAND license it had been offered and turned down before. Nonetheless, the threat of injunction can still have a powerful effect on some recalcitrant licensees.

The present standardization and licensing systems lack sufficient corrective or countervailing forces to prevent this problem from not just entrenching itself but getting worse – as the unlicensed share of the industry increases, the stronger will be the incentives for other firms to resist taking licenses too. Short of some form of enhanced damages, there do not seem to be innate solutions to this problem of potential mass infringement – at least not for implementers who are not major contributors to standards and therefore do not face consequences for past opportunistic conduct.

Worse still, there is some indication that holdout is now something of a norm – for example, very few implementers sign a license anywhere close to their date of first infringement, and most limit the period of past use for which they pay royalties and/or are able to negotiate steep discounts for past infringement. Thus, as delay and discounts related to that delay become a norm across all implementers, the harder it becomes to sanction such behavior through the standard-setting process.

Instead, it seems that the best restraint on holdout is likely to arise from a willingness on the part of the courts (and the competition agencies) to engage with the intent of the FRAND commitment (that is, to recognize fully the centrality of balanced incentives to ETSI's standardization activities). It also is important that

[54] In Apple v. Optis, 2021 EWHC 2564 (Pat.), there was another possibility, which arises in the context of the "FRAND injunction" paradigm used in the United Kingdom. Under this paradigm, an injunction is an alternative to accepting a license on FRAND terms. In that proceeding, Apple had communicated to the court its unwillingness to accept FRAND terms as determined by the court. This put Apple into the category of an "unwilling licensee," and the court ruled that this meant the injunction could be applied even before the court determined FRAND terms. In the alternative scenario, Apple would have continued to infringe Optis's SEPs at least until the court handed down a decision on FRAND terms, which Apple then could elect to refuse.

antitrust enforcement agencies keep their distance and don't provide a helping hand to infringing holdouts.

### B. *Strengthening Injunctive Relief*

The most obvious factor encouraging holdout is the difficulty of obtaining injunctive relief. No US court has granted an injunction in a SEP-related case, at least not since *eBay*. The situation in Europe is somewhat better than that in the United States, and there is a well-developed framework (Huawei-ZTE) for assessing when injunctive relief is an appropriate remedy.[55] However, even that framework arguably provides too much leeway to the putative licensee or infringer. This can be understood usefully by contrasting the Huawei-ZTE framework to Germany's "Orange Book" standard.[56]

Under Germany's Orange Book standard, an infringer could avail itself of a competition law defense against an injunction only if it had made an unconditional offer to license either on terms that the SEP holder could not reasonably refuse or at a rate proposed by the SEP holder but subject to review and modification by courts. Under Huawei-ZTE, the licensee is not required to make such commitments. Importantly, the "unconditional" commitment to license under Orange Book means that the acceptance of a license cannot be delayed through challenges to selected patents on validity and infringement grounds. The Orange Book standard was seen as very generous to patentees and was not designed in the context of ETSI SEPs, where the SEP holder has made an explicit FRAND commitment. Nonetheless, some of its provisions may have merit in the context of cellular SEPs. As a practical matter, the licensing of ETSI-related SEPs is almost always at the portfolio level. Portfolios that have been licensed and/or litigated frequently almost certainly contain at least some valid and infringed patents. In this context, the leeway of licensees to resist taking portfolio licenses by challenging specific patents

55 Judgement of the Court of Justice of the European Union, Case C-170/13, Huawei Technologies Co. v. ZTE Corp and ZTE Deutschland GMBh. This decision spelled out the sequence of steps that licensor and licensee must undertake in order to show that they are negotiating in good faith toward an agreement on FRAND terms. If the licensee is unable to show evidence of its willingness to accept a license on FRAND terms (for example, by making counteroffers based on FRAND, or putting money into escrow to contribute toward payment for past infringement, or by not making timely responses to the licensor's offers), it may be appropriate for the licensor to seek injunctive relief under this framework.

56 The Orange Book was a document setting out technical standards relating to recordable compact discs. In this case, Philips – which had been a key contributor to the standards – sought an injunction against several manufacturers on the grounds of infringement of its SEPs related to the Orange Book. One of the defendants then attempted to mount a competition law defense on the grounds that Philips had a dominant position in relation to one of its SEPs, and was seeking to abuse this dominant position by seeking an injunction. The German Federal Court's decision in this case spelled out the circumstances in which a competition law defense could be used to avoid an injunction in relation to SEP infringement.

on validity and infringement grounds (as allowed under Huawei-ZTE) is an avenue for delay and another mechanism by which licensees are able to raise SEP holders' enforcement costs. The practical upshot is that under the Huawei-ZTE regime, holdout has become easier, and SEP-related injunctions have become harder to obtain.

The UK Supreme Court in *Unwired Planet International Ltd v. Huawei Technologies (UK) Co. Ltd* devised the concept of a "FRAND injunction."[57] Injunctions were available only at the point that the licensee turned down a FRAND determination from the court. In practice, this allows licensees to go all the way up to the "FRAND trial" without any penalty for infringement – in the United Kingdom, the trial to determine FRAND terms would occur after a typically lengthy and costly process of validity and infringement trials, as well as trials on separate, discrete issues. Again, given this potential for delay and given that the licensee's worst-case scenario is a court-determined FRAND license, the threat of an injunction in the UK court may not be a powerful motivation for implementers to negotiate a license with urgency. Indeed, some implementers may be quite happy to accept an injunction in the relatively small UK market if the alternative is to avoid a global license agreement on terms determined by a UK court.[58]

### C. *Adjusting Damages to Disincentivize Holdout*

Courts should recognize that holdout creates significant economic harms for the SEP holder in question, to the innovation ecosystem, and for the licensing marketplace. The problem in individual litigations is typically that a given SEP holder has failed to secure a license after a prolonged period of infringement by the implementer. As noted, the harm in that particular case is not just the cost of delayed payment to the patent owner, but the indirect harm to the SEP holder's licensing program.

There are clear externalities in licensing – securing a given license can confer credibility and momentum for the SEP holder's broader licensing efforts. Conversely, failure to secure a license can damage progress with other would-be

57 Unwired Planet International Ltd. v. Huawei Technologies (UK) Co. Ltd. [2020] UKSC 37.

58 Of course, this option is not straightforward. The margins from UK sales relative to the cost of making those sales may be substantial (as much of the implementer's cost base is incurred to support global rather than UK sales, it will still incur most of these costs even if it ceased its UK operations). Nonetheless, the implementer will consider the fact that in the alternative – where it took a license on terms decided by the court – it would pay royalties on all its global sales. Further, it also might worry that other licensors will use the UK courts to pursue global royalty claims. In this situation, the additional royalties it could pay would cumulate to a potentially substantial amount – whereas the benefit of remaining in the UK market would not change. Accepting a UK injunction will prevent other claimants from pursuing claims in the United Kingdom. Clearly, the smaller the implementer's anticipated UK sales, the easier it will be to accept an injunction. Further, if global FRAND rates were applied just to UK past sales, this also might be conducive to the strategy of accepting an injunction.

licensees. In particular, some implementers – especially those in product segments such as IoT or among implementers focused on some emerging markets and on China – will seek to tie either the terms of a license (for example, payment) or the very signing of a license to the SEP holder's success in signing on other implementers that they perceive as being in the same segment. There may be a broader impact on the licensing marketplace as a whole. Akemann, Blair, and Teece explain how widespread infringement begets more infringement, thus creating a "bandwagon effect":

> [W]hen there are many infringers, each infringer might believe there is a perceived safety in numbers, as each infringer might believe that the chance it will be pursued is low . . . the patent holder cannot refuse to supply the technology in the way that suppliers of tangible goods can . . . . We would therefore expect to see that royalties negotiated in a marketplace with widespread infringement will typically be lower than those negotiated in circumstances where infringement was less common . . . . In this regard it is worth emphasizing the significant and asymmetric risks that a patent holder faces as it tries to enforce its patent rights against a long line of potential infringers.[59]

How courts handle issues such as damages or the terms of FRAND licenses can thus have an impact in terms of not just alleviating economic harm arising from prolonged infringement, but also correcting the distortions in the wider licensing marketplace that arise from allowing such conduct to persist. The actions that courts take today will dictate not just future litigation outcomes, but future negotiations in the marketplace – which always happen in the "shadow of the court."

### D. *Limiting the Availability of FRAND*

Consider an SEP holder that has been attempting to negotiate a license with an implementer for several years and finally has brought the matter to litigation. If the evidence suggests that the putative licensee essentially had no interest in negotiating a license on FRAND terms, is it adequate that the redress available is a license on FRAND terms, and damages for past infringement on FRAND terms? This is problematic for two reasons: (a) It potentially puts the litigious infringer on the same footing as more willing licensees; and (b) it actually does not restore the SEP holder into the position it would have been had the infringement never taken place, as the SEP holder will not be compensated adequately for the harm it has suffered as a result of the infringement. While in principle the harm caused by the delay in taking the license might be quantifiable, the harm to the wider licensing program may not be so readily quantifiable. With respect to addressing the harm caused just by the delay to taking the license, courts should be prepared to address this harm and

[59] Akemann et al., *supra* note 46, at 873–75 (citations omitted).

to do so in an economically robust way – for example, instead of using statutory interest rates in damages awards, courts could use the licensor's cost of capital or some other measure of economic opportunity cost to address the issue.

More importantly, instead of giving the licensee the choice of eventually taking a FRAND license, once a licensee has been found unwilling, the most obvious corrective is to withdraw the option of a FRAND license. The threat of being found unwilling and thus losing an entitlement to a FRAND license is likely more potent than the threat of being enjoined (which will only apply in a single jurisdiction anyway) and then being able to claim a FRAND license to lift the injunction. (If courts across different jurisdictions consistently applied this logic, then it would also prevent the situation wherein the implementer can swallow an injunction in a less important jurisdiction and then prevail upon another court to award it a FRAND license.)[60]

These proposals might seem radical to some and might push courts into territory that seems controversial. After all, there is nothing explicit in the ETSI IP policy that suggests that the FRAND commitment applies regardless of the licensee's willingness; but nor is there an explicit provision that limits its application in the case of an unwilling licensee. To the extent that above-FRAND awards might contain a deterrent element, they may be seen by some as legally difficult to justify, but they are economically very easy to justify.[61] Nonetheless, at least the broader concept of restoring the licensor's economic position by recognizing the harm it suffers from infringement fits in with the idea of restorative or equitable damages, rather than being punitive.[62]

A minimalist measure that would at least somewhat restore the balance is that courts should not give unwilling licensees the same terms as willing ones. Quite often, the determination of FRAND rates is based (at least partly) on a review of the licensor's "comparable" licenses. In many instances, the rates that the licensor negotiated with the licensee might reflect concessions that were made in the context of meaningful negotiations. The licensor has an interest not just in securing the best "rates" but in ensuring that its licensing program maintains momentum (given that concluding license agreements lends impetus to concluding other licensing

60 In this case, the injunction may be lifted if the parties negotiate a license, but that negotiation does not have to be bound by FRAND terms. To the extent that the SEP holder may be able to extract value based on excluding the implementer from the marketplace altogether, a part of the supra-FRAND value extracted is to compensate the SEP holder for the broader harm that the licensee's conduct has inflicted on it, while the remainder serves as a deterrent that could sharply correct incentives in the marketplace.

61 For example, in the United Kingdom, punitive or exemplary damages are rarely available, although perhaps deliberate or misleading conduct by the licensee to avoid taking a license might conceivably qualify.

62 In non-SEP-patent cases, courts sometimes award damages on a "lost profits" basis, rather than a reasonable royalty basis. This is conceptually similar to recognizing the economic loss created by infringement that I refer to previously.

agreements – that is, there are externalities at work). Clearly, such a calculus does not apply in the case of the unwilling licensee, and equally clearly, there is a compelling policy rationale for not putting unwilling and willing licensees on the same footing. Put another way, while a range of rates may be consistent with FRAND, the unwilling licensee ought not be entitled to the "best" rate in that range, or even the average rate in that range. In fact, it would still be quite accommodating to give it a rate that is at the top end of the FRAND range.

### E. *The Role of Competition Policy*

With their endorsement of holdup theories, US competition agencies sometimes have eschewed an evidence-based approach to antitrust and wittingly or unwittingly aided and abetted patent holdout, thereby compromising the development of foundational technology that supports innovation and drives dynamic competition. The primary beneficiaries of this strategy are new entrants in China that end up receiving a discount, and/or escape royalty payments for many years – and often permanently.[63]

Competition policy needs to favor the future and embrace innovation and business models that support innovation in the key enabling technologies. They must support innovation both upstream and downstream. Avoiding the temptation to resurrect patent holdup arguments is the most concrete step that can be taken to help innovation and competition. Challenging the open innovation model simply plays into the hands of those that might prefer vertical integration of upstream wireless technology and downstream devices.

## VII. CONCLUSION

Markets for technology don't function well without strong IP.[64] Technology still may get developed without IP protection, but such developments will be confined to vertically integrated enterprises. Technology needs to be embedded in, and priced into, goods and services supplied via integrated enterprises. That is how consumers and producers usually pay for technology licensing.[65] The same is true for many intermediate products, such as automotive parts. Only in unusual cases is the division of labor between technology developer and product maker nearly complete, at which point the technology is made available through licensing.

[63] Even if they ultimately pay royalty payments, they do so on terms that have been adjusted downward artificially under the threat of antitrust intervention.

[64] David J. Teece, *Profiting from Technological Innovation: Implications for Integration, Collaboration, Licensing and Public Policy*, 15 RSCH. POL'Y 285 (1986).

[65] David J. Teece, *Business Models, Business Strategy and Innovation*, 43 LONG RANGE PLAN. 172 (2010).

Mobile wireless is different. 3G, 4G, and (hopefully) 5G wireless telecommunications are cases in point. Competition agencies should rejoice and support rather than undermine the (licensing) business models. Paradoxically, competition policy advocates admire open innovation models. "Consumers benefit from open innovation strategies," according to the FTC,[66] in part because it allows specialization to flourish. Open innovation models require that licensing regimes be supported with royalties that are paid in a timely fashion at levels sufficient to draw forth the investment needed to make the ecosystem robust.

Unlocking the full potential of 5G, 6G, and other frontier technologies will require robust patent protection to ensure rewards sufficient to induce investment in new technologies. Each generation of wireless technology – 3G, 4G, now 5G – has taken more than five years (10 years for 5G) to define and many more years to perfect. While initial 5G wireless standards have been set, there will be many updates and improvements.

If competition agencies fail to recognize holdout issues and use antitrust to shield unwilling licensees, such decisions will stimulate the emergence of a vertically integrated "Big Tech" business model in mobile wireless and transfer wealth to "net user" economies that primarily specialize in the adoption and imitation of new technologies, rather than into critical R&D for next-generation wireless. Not much in this scenario is appealing from a competition policy perspective. Should this scenario play out, it will be necessary to declare that policy mistakes helped destroy the greatest model of technological cooperation and innovation that Western civilization ever created. The poorest members of global society, who have benefited enormously from mobile technology, are likely to suffer disproportionately. Adopting a broader intellectual framework that recognizes the unique requirements of open innovation along with geopolitical realities should help avoid such a calamity.

[66] Federal Trade Commission, The Evolving IP Marketplace: Aligning Patent Notice and Remedies with Competition 7 (Mar. 2011), www.ftc.gov/sites/default/files/documents/reports/evolving-ip-marketplace-aligning-patent-notice-and-remedies-competition-report-federal-trade/110307patentreport.pdf.

2

# Antitrust Convergence on Substantive Norms for SEP Licensing Negotiations

## *Should and Could It Be?*

*Maureen K. Ohlhausen and Jana I. Seidl*

## I. INTRODUCTION

Two truths must guide antitrust agency policy and enforcement with respect to intellectual property (IP). First, strong patent rights foster innovation. Second, licensing is a cornerstone of a strong system of IP. With the advent of 5G and the proliferation of the Internet of Things (IoT), it is critically important that the US antitrust agencies calibrate policy and enforcement priorities with respect to IP in a manner that ensures efficient licensing – in turn, maintaining strong patent rights. The agencies can achieve this by striking the appropriate balance between the rights of innovators and those of implementers. They should take a cautious and clear approach to wielding antitrust as a tool to address licensing disputes lest they inadvertently exacerbate bargaining frictions resulting from legal standards that are ambiguous. European courts have gone further than US courts and agencies in some of these areas to date. Recent developments signal that the United States may be taking cues from the European approach going forward where courts have begun articulating guardrails surrounding the interplay of IP and antitrust with respect to licensing negotiations.

## II. THE EVOLUTION OF IP/ANTITRUST AND SEP LICENSING IN THE UNITED STATES

Antitrust regulators have long sought to strike the right balance and tone in approaching and evaluating the exercise of IP. After all, the antitrust laws prohibit monopolies, while the patent laws confer exclusive rights on an IP holder. It comes as no surprise then that the evolution of the interplay between antitrust and IP – and specifically, whether and how antitrust should be brought to bear on situations involving IP – has taken some twists and turns over the decades.

This is especially the case where standard-essential patents (SEPs) are involved. Open standardization and healthy competition on the merits when technologies vie for inclusion in a standard carry tremendous consumer benefits, for example in the form of interoperability, safety, or energy consumption. And collaborative technical standards have been critically important to global growth. It bears emphasizing in this context that IP and antitrust are "two bodies of law [that] are actually complementary, as both are aimed at encouraging innovation, industry and competition."[1] But innovation spurred by technical standards and progress toward new and improved standards such as 5G and new environments such as the IoT can only come about when innovators are assured that their contributions will secure them the appropriate return.

On the flip side, in order to realize these standards, implementers must be assured access to patented technologies incorporated into a standard once the standard-development process is complete. Voluntary commitments by innovators to make SEPs available to implementers on a fair, reasonable, and non-discriminatory (FRAND) basis emerged as a means to promote access after a standard is adopted. These FRAND commitments are contractual obligations between a SEP holder and standards-development organizations (SDOs), to which implementers are a third party and meant to facilitate and guide bilateral negotiations between SEP holders and implementers. Of course, the devil is in the details and disputes can arise during these bilateral licensing negotiations over what exactly constitutes FRAND rates and terms.

This is where industry participants have called for guidance from the antitrust agencies as to when and how antitrust law will step in. On one side of the debate are those who view most IP/antitrust issues to be a matter for contract law. Others call for a more expansive role for antitrust law in enforcing companies' practices with respect to wielding their IP portfolios in what may be perceived to be an anticompetitive manner. As a result, depending on who you ask, US antitrust agency guidance over the years has been viewed as either too implementer friendly or too innovator friendly. As the agencies embark on what may be viewed as yet another shift in policy, it is critical that they be careful to shape policy in such a manner that bargaining frictions attendant to SEPs not devalue the contribution of patents to standards so much that innovators are incentivized to instead create walled garden technologies with closed standards. Historical shifts – and constants – can be instructive here.

## A. *An About-Face on Package Licensing*

In the 1970s – long before the current disputes over SEP licensing – the then Deputy Assistant Attorney General of the Antitrust Division of the United States

[1] Atari Games Corp. v. Nintendo of Am., Inc., 897 F.2d 1572, 1576 (Fed. Cir. 1990).

Department of Justice (Division) articulated a list of "Nine No-Nos" – patent licensing practices that the Division would likely view as presumptively unlawful.[2] One of these "No-Nos" was "requiring mandatory package licensing."[3] Package licensing is a license on a bundle or portfolio of patents, which can be charged at a single royalty rate or a formula that does not take into account the specific subset of patents used by the licensee. That approach was informed by the concern that aggregating licenses in such a manner may be a form of a tying arrangement that in certain circumstances violates antitrust law. Today, of course, package licenses – and global portfolio licenses – are often the norm as it pertains to standard-essential technology, including 5G. And for very good reason.

In 1979, the Supreme Court weighed in on package licensing when it decided *Broadcast Music, Inc. v. Columbia Broadcasting System* (*BMI Decision*).[4] The Court unequivocally removed package licensing from the universe of per se prohibitions, announcing that these licensing arrangements should instead be evaluated under the rule of reason framework, a case-by-case, fact-based analysis. In its decision, the Court extolled the procompetitive virtues of such licensing arrangements: They provide for "unplanned, rapid, and indemnified access to any and all of the repertory of [works], and [provides rights owners] a reliable method of collecting for the use of their [intellectual property]."[5] Indemnification and lowering monitoring costs promote patent peace. And aggregating licenses to IP into portfolios, or packages, lowers transaction costs to negotiating access to those rights. As such, the Court changed the trajectory of the IP/antitrust interplay in an important manner. Informed in large part by the *BMI Decision*, the Division characterized the "No-Nos" by 1981 as "contain[ing] more error than accuracy" when viewed through the lens of "rational economic policy."[6]

In 2020, the Division issued a business review letter (BRL) to Avanci regarding its platform for joint licensing of SEPs for 5G telecommunications technologies for use in vehicles and other IoT devices.[7] The Division reaffirmed the principles of the *BMI Decision*. By acting as a centralized agent for licensing a large percentage of 5G SEPs, the BRL notes that Avanci can facilitate licensing and help integrate emerging 5G technologies into vehicles faster, with less infringement risk, and at

2 Bruce B. Wilson, Deputy Assistant Attorney General, Antitrust Division, Address before the Michigan State Bar Antitrust Section and the Patent, Trademark and Copyright Section (Sept. 21, 1972), *partial text reprinted in* 4 Trade Reg. Rep. (CCH) ¶ 13,126.

3 *Id.*

4 441 U.S. 1 (1979).

5 *Id.* at 20.

6 Abbott B. Lipsky, Jr., Deputy Assistant Attorney General, *Current Antitrust Division Views on Patent Licensing Practices*, 50 Antitrust L.J. 515, 517 (1982) (text of remarks before the American Bar Association Antitrust Section, Washington, DC (Nov. 5, 1981)).

7 US Department of Justice, Antitrust Division, Response to the Avanci LLC's Request for a Business Review Letter (July 28, 2020), www.justice.gov/atr/page/file/1298626/download.

reduced transaction costs.[8] And given Avanci's scale, it could also reduce other transaction costs such as those associated with monitoring and compliance.[9]

### B. *Steadfast Adherence to the Principle that the Antitrust Laws Require Harm to Competition*

In April 1995, more than a decade after jettisoning the "No-Nos," the Division, together with the Federal Trade Commission (FTC), set out their first formal guidance on enforcement policies with respect to IP issues in their *Antitrust Guidelines for the Licensing of Intellectual Property* (1995 IP Guidelines).[10] The two agencies (Agencies) then issued a joint report in 2007, *Antitrust Enforcement and Intellectual Property Rights: Promoting Innovation and Competition*,[11] that affirmed the principles of the 1995 IP Guidelines and applied them to conduct beyond licensing. The 2007 publication was bookended by two reports issued by the FTC: one in 2003, *To Promote Innovation: The Proper Balance of Competition and Patent Law and Policy*;[12] the other in 2011, *The Evolving IP Marketplace: Aligning Patent Notice and Remedies with Competition*.[13] Both reports followed extensive hearings with industry participants to inform observations and recommendations.

The Agencies then modernized their 1995 IP Guidelines in 2017.[14] The 2017 IP Guidelines largely reaffirmed the Agencies' core enforcement philosophy first announced in 1995. Both the 1995 and 2017 IP Guidelines embrace the Agencies' stance that recognized the procompetitive and welfare-enhancing effects of licensing IP. The 2017 Guidelines state:

> Licensing, cross-licensing, or otherwise transferring intellectual property ... can facilitate integration of the licensed property with complementary factors of production. This integration can lead to more efficient exploitation of the intellectual property, benefiting consumers through the reduction of costs and the introduction

8 *Id.* at 9.

9 *Id.* at 10.

10 US Dep't of Just. & Fed. Trade Comm'n, Antitrust Guidelines for the Licensing of Intellectual Property (Apr. 6, 1995), www.justice.gov/sites/default/files/atr/legacy/2006/04/27/0558.pdf.

11 US Dep't of Just. & Fed. Trade Comm'n, Antitrust Enforcement and Intellectual Property Rights: Promoting Innovation and Competition (Apr. 2007), www.justice.gov/sites/default/files/atr/legacy/2007/07/11/222655.pdf.

12 Fed. Trade Comm'n, To Promote Innovation: The Proper Balance of Competition and Patent Law and Policy (Oct. 2003), http://ftc.gov/os/2003/10/innovationrpt.pdf.

13 Fed. Trade Comm'n, The Evolving IP Marketplace: Aligning Patent Notice and Remedies with Competition (Mar. 2011), www.ftc.gov/sites/default/files/documents/reports/evolving-ip-marketplace-aligning-patent-notice-and-remedies-competition-report-federal-trade/110307patentreport.pdf.

14 US Dep't of Just. & Fed. Trade Comm'n, Antitrust Guidelines for the Licensing of Intellectual Property (Jan. 12, 2017), www.justice.gov/atr/IPguidelines/download.

of new products. Such arrangements increase the value of intellectual property to consumers and owners.[15]

When the Agencies announced the proposed updates to the 1995 IP Guidelines, then Chairwoman Ramirez stressed that "U.S. antitrust law leaves licensing decisions to IP owners, licensees, private negotiations, and market forces unless there is evidence that the arrangement likely harms competition."[16] It is important to note that this principle applies to all patent licensing negotiations, including those over SEPs, subject to voluntary FRAND royalty rate commitments. Simply put, as courts have held repeatedly over the years in agreement with the Agencies' approach, a breach of FRAND by itself cannot be a violation of the Sherman Act.[17]

Even when the Agencies seemed to have taken a divergent path on some IP/antitrust approaches during the Trump Administration (as further discussed), then Chairman Simons could not have been more clear that the FTC and the Division saw eye to eye on this fundamental principle: "We agree ... that a breach of a FRAND commitment, standing alone, is not sufficient to support a Sherman Act case, and ... the breach, fraud or deception must also contribute to the acquisition or maintenance of monopoly power ... or involve an agreement that unreasonably restrains trade."[18]

The Agencies have not wavered in their conviction that efficient licensing boosts innovation and that antitrust laws should stay out of the way until and unless there is cognizable harm to the competitive process and thus cause to intervene to preserve consumer welfare. In fact, the Division reaffirmed this principle in several statements of interest filed during the Trump Administration. For example, in *Lenovo v. Interdigital*, the Division emphasized that alleged violations of FRAND commitments are not cognizable under Section 1 or Section 2 of the Sherman Act.[19] In *Continental v. Avanci*, the Division again submitted a statement of interest

[15] *Id.* at 5.

[16] Press Release, Fed. Trade Comm'n, FTC and DOJ Seek Views on Proposed Update of the Antitrust Guidelines for Licensing of Intellectual Property (Aug. 12, 2016), www.ftc.gov/news-events/press-releases/2016/08/ftc-doj-seek-views-proposed-update-antitrust-guidelines-licensing.

[17] Rambus Inc. v. FTC, 522 F.3d 456, 464 (D.C. Cir. 2008); FTC v. Qualcomm Inc., 969 F.3d 974 (9th Cir. 2020); Cont'l Auto. Sys. v. Avanci, LLC, 485 F. Supp. 3d 712 (N.D. Tex. 2020), *appeal docketed*, No. 20-11032 (5th Cir. Oct. 15, 2020).

[18] Joseph Simons, Chairman, Fed. Trade Comm'n, Prepared Remarks of Chairman Joseph Simons before the Georgetown Law Global Antitrust Enforcement Symposium 5–6 (Sept. 25, 2018), www.ftc.gov/system/files/documents/public_statements/1413340/simons_georgetown_lunch_address_9-25-18.pdf.

[19] Statement of Interest of the United States of America, Lenovo v. Interdigital, No. 20-493-LPS (D. Del. July 17, 2020), ECF 13, www.justice.gov/atr/case-document/file/1295526/download. The court in this case held the deception claim to be cognizable.

arguing that alleged violations of FRAND commitments are not cognizable under Section 2.[20] In that case, the district court agreed and dismissed the claims.[21]

### C. *The Perceived Back-and-Forth Regarding Remedies Available for SEPs*

In addition to the IP Guidelines, the Division also put forth in 2013, in collaboration with the US Patent and Trademark Office (USPTO), the *Policy Statement on Remedies for Standards-Essential Patents Subject to Voluntary F/Rand Commitments* (2013 Remedies Statement).[22] Issues at the forefront of the ongoing IP/antitrust debate, especially with respect to SEP licensing negotiations, include the availability of injunctions for infringement of SEPs, the related issue of holdup versus holdout, and the essential facilities doctrine – that is, whether a duty to deal should apply to SEPs. Holdup refers to bad faith behavior by innovators, which is typically a threat of exclusion from the market to extract unreasonably high royalty rates or licensing terms that are unreasonably favorable to the SEP holder. Of course, effectuating such an exclusion requires the SEP holder to seek, and then be granted, a court order. Holdout, on the other hand, refers to conduct by implementers to drag out licensing negotiations and legal maneuvers such as anti-injunction suits that in effect prolong their SEP infringement and are meant to pressure innovators to accept unreasonably low royalty rates or unreasonable licensing terms in the implementer's favor. The US International Trade Commission (ITC) has summed up holdout as follows:

> [A]n implementer utilizes declared-essential technology without compensation to the patent owner under the guise that the patent owner's offers to license were not fair or reasonable. The patent owner is therefore forced to defend its rights through expensive litigation. In the meantime, the patent owner is deprived of the exclusionary remedy that should normally flow when a party refuses to pay for the use of a patented invention.[23]

[20] Statement of Interest, Cont'l Auto. Sys. v. Avanci, LLC, No. 3:19-cv-02933 (N.D. Tex. Feb. 27, 2020), ECF 278, www.justice.gov/atr/case-document/file/1253361/download.

[21] Cont'l Auto. Sys. v. Avanci, LLC, 485 F. Supp. 3d 712, 734 (N.D. Tex. 2020). The Fifth Circuit vacated and remanded with instructions to dismiss the case for lack of Article III standing but did not opine on antitrust injury or the merits. Opinion, Cont'l Auto. Sys. v. Avanci, LLC, 27 F.4th 326, 336 (5th Cir. 2022). The court found that (1) Continental's claim of injury was too speculative – it had not established that OEMs accepted non-FRAND licenses and then invoked indemnification rights against Continental, and (2) Continental could not establish that it was a third-party beneficiary of the FRAND commitment at issue – it was not a member of the relevant SDO and does not need the SEPs at issue to operate. *Id.* at 332–34.

[22] US Dep't of Just. & US Patent & Trademark Off., Policy Statement on Remedies for Standards-Essential Patents Subject to Voluntary F/RAND Commitments (Jan. 8, 2013) [hereinafter 2013 Remedies Statement], www.justice.gov/atr/page/file/1118381/download.

[23] Certain Elec. Devices, Including Wireless Commc'n Devices, Portable Music & Data Processing Devices, & Tablet Computs., Inv. No. 337-TA-794, at *38 (USITC July 5, 2013) (Comm'n Op.).

There are divergent schools of thought in the United States on whether a SEP holder's breach of FRAND or an implementer's holdout should be considered an antitrust concern rather than a dispute to be left strictly to contract law. And there are disagreements over whether an implementer should be able to seek an injunction against an infringing potential licensee – mainly centering on whether that infringer is a willing or unwilling licensee.

The 2013 Remedies Statement aimed to address the availability of injunctive relief in ITC investigations under Section 337 of the Tariff Act of 1930.[24] Importantly, the Statement took the position in no uncertain terms that injunctive relief (in the form of an exclusion order by the ITC) may be an appropriate remedy in certain circumstances involving an unwilling licensee, including, for example, where a "putative licensee refuses to pay what has been determined to be a F/RAND royalty, or refuses to engage in a negotiation to determine F/RAND terms."[25]

Some industry participants, however, read the 2013 Remedies Statement, coupled with prior Division statements and speeches, as advancing an anti-injunction stance for SEPs.[26] For example, in 2012, the Division's then deputy assistant attorney general, Renata Hesse, gave a speech calling on SDOs to clarify FRAND commitments, limit injunctions, create guidelines or arbitration provisions governing determinations of FRAND rates, and the like.[27] That same year, the Division, in its statement in connection with Google's acquisition of Motorola Mobility (including its SEP portfolio), had also lauded "clear commitments" by rights holders to license on FRAND terms and "not to seek injunctions in disputes involving SEPs."[28] In addition, the 2013 Remedies Statement was expressly invoked by the US Trade Representative in vetoing an ITC exclusion order in the high-profile dispute between Samsung and Apple over Apple's infringement of cellular SEPs, in which Apple had failed to show that Samsung violated FRAND commitments. The US Trade Representative wrote, "[E]xclusionary relief … based on FRAND

24 2013 Remedies Statement, *supra* note 22, at 1 ("[The agencies] provide the following perspectives on … whether injunctive relief in judicial proceedings or exclusion orders in investigations under section 337 of the Tariff Act of 1930 are properly issued when a patent holder seeking such a remedy asserts standards-essential patents that are encumbered by a RAND or FRAND licensing commitment." (citations omitted)).

25 *Id.* at 7.

26 *See, for example*, Paul H. Saint-Antoine, *IP, Antitrust, and the Limits of First Amendment Immunity: Shouting "Injunction" in a Crowded Courthouse*, ANTITRUST MAG. (Summer 2013), at 41, 43; J. Gregory Sidak, *The Antitrust Division's Devaluation of Standard-Essential Patents*, 104 GEO. L.J. ONLINE 48 (2015).

27 Renata B. Hesse, Deputy Assistant Att'y Gen., Antitrust Div., US Dep't of Just., Six "Small" Proposals for SSOs before Lunch, Address before the ITU-T Patent Roundtable 9–10 (Oct. 10, 2012), www.justice.gov/atr/file/518951/download.

28 Press Release, US Dep't of Just. Antitrust Div., Statement of the Department of Justice's Antitrust Division on Its Decision to Close Its Investigations of Google Inc.'s Acquisition of Motorola Mobility Holdings Inc. (Feb. 13, 2012), www.justice.gov/opa/pr/statement-department-justice-s-antitrust-division-its-decision-close-its-investigations.

encumbered SEPs should be available based only on the relevant factors described in the [2013 Remedies] Statement."[29] This was despite the fact that the 2013 Remedies Statement made clear the examples of factual scenarios in which an exclusion order may be appropriate "is not an exhaustive one."[30]

By 2014, the United States' top specialized patent court, the US Court of Appeals for the Federal Circuit (Federal Circuit), clearly articulated in *Apple v. Motorola* that claims involving infringement of SEPs subject to FRAND commitments were to be treated as any other patent case would be in an analysis as to whether an injunction should issue in federal court.[31] The Federal Circuit was explicit – the Supreme Court's *eBay* framework for injunction standards, grounded in the traditional principles of equity, applies to SEPs:

> To the extent that the district court applied a *per se* rule that injunctions are unavailable for SEPs, it erred. While Motorola's FRAND commitments are certainly criteria relevant to its entitlement to an injunction, we see no reason to create, as some *amici* urge, a separate rule or analytical framework for addressing injunctions for FRAND-committed patents.[32]

Similarly, with respect to damages, the Federal Circuit explained in *Ericsson v. D-Link*: "We believe it unwise to create a new set of *Georgia-Pacific*-like factors for all cases involving RAND-encumbered patents."[33] The court thus made clear that cases involving SEPs do not warrant special rules.

Against this backdrop, during the Trump Administration, the Division took to heart calls for more clarity on its enforcement policy in the IP/antitrust space. The Division announced a policy change – the "New Madison" approach.[34] That new approach included several important points, namely that: (1) holdup is not an antitrust problem; (2) holdout is a danger to incentives to innovate; (3) injunctions for SEP infringement should be protected rather than persecuted; and (4) innovators have no duty to deal, for example, to license a valid patent. The FTC agreed with some of this approach but did not go so far as to minimize the antitrust risks from holdup. The Division then filed several statements of interest in cases involving

[29] Letter from Ambassador Michael B. G. Froman, US Trade Rep., to The Hon. Irving A. Williamson, Chairman, US Int'l Trade Comm'n, Vetoing ITC-794 Exclusion Order 2 (Aug. 3, 2013), www.ustr.gov/sites/default/files/08032013%20Letter_1.PDF (citing the policy statement and instructing the ITC to make findings regarding the potential for patent holdup).

[30] 2013 Remedies Statement, *supra* note 22, at 7.

[31] Apple Inc. v. Motorola Inc., 757 F.3d 1286, 1331 (Fed. Cir. 2014), *overruled on other grounds by* Williamson v. Citrix Online, LLC, 792 F.3d 1339 (Fed. Cir. 2015) (en banc).

[32] *Id.* at 1331–32.

[33] Ericsson v. D-Link Sys., Inc., 773 F.3d 1201, 1232 (Fed. Cir. 2014).

[34] Makan Delrahim, Assistant Att'y Gen., Antitrust Div., US Dep't of Just., The "New Madison" Approach to Antitrust and Intellectual Property Law, Address before the University of Pennsylvania Law School (Mar. 16, 2018), www.justice.gov/opa/speech/file/1044316/download.

issues at the core of the IP/antitrust debate as discussed earlier – all as part of its "multi-pronged effort to help educate and modernize the approach to antitrust and intellectual property law."[35]

The Division also explicitly disavowed prior guidance where it felt that developments showed that the intended message had been misunderstood. For example, in 2019, the Division withdrew the 2013 Remedies Statement over concerns it was misconstrued as calling for a different set of rules for licensing SEPs than nonessential patents. At the time, the Division issued a new remedies statement in conjunction with the USPTO and the National Institute of Standards Technology (NIST), *Policy Statement on Remedies for Standard-Essential Patents Subject to F/RAND Commitments* (2019 Remedies Statement).[36] The press release accompanying the 2019 Remedies Statement elaborated: "A previous statement on the matter issued in 2013 had been misinterpreted .... Today's joint statement seeks to ensure that US patent law is appropriately calibrated ... [and] sets a positive example for other jurisdictions that have sought to diminish the value of SEPs."[37]

The 2019 Remedies Statement made clear – in line with prevailing case law and pointing to the Federal Circuit's *Apple v. Motorola* decision – that SEPs and non-standard-essential patents are subject to the same remedies, including injunctions, and that the same framework applies for any analysis as to the availability of remedies.[38] Specifically, the 2019 Remedies Statement advanced the position that "[a]ll remedies available under national law, including injunctive relief and adequate damages, should be available for infringement of standards-essential patents subject to a F/RAND commitment, if the facts of the case warrant them."[39] The 2019 Remedies Statement also cited to examples of both holdout and holdup when discussing conduct of negotiating parties that would be relevant to remedies determinations.[40] Ultimately, the 2019 Remedies Statement pointed confidently to "courts – and other relevant neutral decision makers – [to] continue to determine remedies for infringement of standards-essential patents subject to F/RAND

35 US Dep't of Just., New Heights for the New Madison Approach (June 23, 2020), www.justice.gov/atr/division-operations/antitrust-division-update-2020/new-heights-new-madison-approach.

36 US Patent & Trademark Off., Nat'l Inst. of Standards & Tech. & US Dep't of Just., Policy Statement on Remedies for Standard-Essential Patents Subject to F/RAND Commitments (Dec. 19, 2019) [hereinafter 2019 Remedies Statement], www.justice.gov/atr/page/file/1228016/download.

37 Press Release, US Patent & Trademark Off., US Patent and Trademark Office Releases Policy Statement on Standards-Essential Patents Subject to Voluntary F/RAND Commitments (Dec. 19, 2019), www.uspto.gov/about-us/news-updates/us-patent-and-trademark-office-releases-policy-statement-standards-essential#:~:text=Today's%20joint%20statement%20seeks%20to, diminish%20the%20value%20of%20SEPs.

38 2019 Remedies Statement, *supra* note 36, at 6 (quoting Apple Inc. v. Motorola, Inc., 757 F.3d 1286, 1331–32 (Fed. Cir. 2014)).

39 *Id.* at 4–5.

40 *Id.* at 5.

licensing commitments pursuant to the general laws" that would preserve competition and incentives to innovate.[41]

Similarly, the Division took "the extraordinary step to supplement"[42] a 2015 BRL to an SDO, the Institute of Electrical and Electronics Engineers (IEEE), explaining that recent developments had "proven [the 2015 letter] outdated and [the Division] fear[ed] that reliance on its analysis, both in the United States and abroad, could actually harm competition and chill innovation."[43] The Division pointed to three primary ways that the IEEE's policy "may undercut current U.S. law and policy": (1) by limiting the scope of rights available to a SEP owner, including that of seeking injunctive relief against an infringer (here the Division went so far as to suggest the IEEE consider changing its policy to make is easier for SEP holders to pursue injunctive relief); (2) by not dedicating sufficient attention to holdout, conduct that would undermine the bargaining position of innovators; and (3) by possibly limiting the scope of royalties.[44]

### D. *There Is No Special Duty to Deal for SEPs in US Antitrust Law*

The FTC's highest profile case during the Trump Administration that brought antitrust law to bear in an IP dispute was its monopolization case against Qualcomm over licensing practices related to modem chips.[45] In that case, the FTC had actually filed its complaint in the last days of the Obama Administration and eventually took the case all the way to a request for rehearing en banc at the US Court of Appeals for the Ninth Circuit (Ninth Circuit). While an outlier, the district court's decision in the case threw into flux well-settled antitrust law on the essential facilities doctrine when it ruled in favor of the FTC.[46] That decision inappropriately expanded a company's antitrust duty to deal beyond any prior course of conduct by extrapolating from a prior course of licensing certain patents to certain limited parties a duty to deal across all patents and with all allegedly similarly situated parties.

But decades of precedent establish that US antitrust law does not support a broad duty to deal. The Sherman Act imposes a duty to deal with or continue dealing with rivals only in the rarest circumstances because "once you start, the Sherman Act may

[41] *Id.* at 7.

[42] Letter from Makan Delrahim, Assistant Attorney General, US Dept. of Justice, Antitrust Div., to Sophia Muirhead, General Counsel and Chief Compliance Officer, IEEE (Sept. 10, 2020) [hereinafter 2020 IEEE BRL], www.justice.gov/atr/page/file/1315291/download.

[43] Press Release, US Dep't of Just., Antitrust Div., Justice Department Updates 2015 Business Review Letter to the Institute of Electrical and Electronics Engineers (Sept. 10, 2020), www.justice.gov/opa/pr/justice-department-updates-2015-business-review-letter-institute-electrical-and-electronics.

[44] 2020 IEEE BRL, *supra* note 42, at 4–9.

[45] Fed. Trade Comm'n, *Qualcomm, Inc.*, www.ftc.gov/enforcement/cases-proceedings/141–0199/qualcomm-inc.

[46] FTC v. Qualcomm Inc., 411 F. Supp. 3d 658, 820–24 (N.D. Cal. 2019), *rev'd and vacated*, 969 F.3d 974 (9th Cir. 2020).

be read as an antidivorce statute."[47] The extremely limited circumstances include, for example, the unilateral termination of a voluntary (and thus presumably profitable) prior course of dealing that suggests a willingness to forsake short-term profits to achieve an anticompetitive end.[48]

The Ninth Circuit reversed the district court's *Qualcomm* decision,[49] making clear that the long-standing precedent of *Aspen Skiing Co. v. Aspen Highlands Skiing Corp.*[50] continues to be "at or near the outer boundary of § 2 liability"[51] – the Sherman Act simply does not impose a duty to deal with or continue to deal with competitors absent the rarest exceptions.[52] The FTC petitioned for rehearing en banc,[53] which the Ninth Circuit denied. *FTC v. Qualcomm* also brought to light a rift between the Division and the FTC as the Division filed a statement of interest at the district court level asking Judge Koh to schedule a hearing on a remedy should she find for the FTC, followed by an amicus curiae brief in which the Division sided with Qualcomm at the Ninth Circuit.[54] The Division's amicus brief specifically addressed that antitrust law did not require Qualcomm to deal on specific terms with component-level manufacturers even if it was part of the FRAND commitment.[55]

## III. INTERNATIONAL IP/ANTITRUST DEVELOPMENTS GO FURTHER THAN THE UNITED STATES HAS TO DATE

Courts in the United States, Europe, and China have repeatedly found that where an SEP holder is seeking to license a worldwide portfolio of cellular SEPs, and the implementer's operations are worldwide, a FRAND license is a global portfolio

47 Philip Areeda, *Essential Facilities: An Epithet in Need of Limiting Principles*, 58 ANTITRUST L.J. 841, 850 (1990); see also Aspen Skiing Co. v. Aspen Highlands Skiing Corp, 472 U.S. 585, 600–01 (1985).

48 Verizon Commc'ns v. Law Offices of Curtis V. Trinko, LLP, 540 U.S. 398, 409 (2004). See also Brief of Dr. Janusz A. Ordover as Amicus Curiae in Support of Petitioner, Swisher Int'l Inc. v. Trendsettah USA, Inc., 140 S. Ct. 443 (2019) (No. 19-349), www.supremecourt.gov/DocketPDF/19/19–349/118554/20191009124442125_19-349acDrJanuszAOrdover.pdf.

49 FTC v. Qualcomm Inc., 969 F.3d 974 (9th Cir. 2020).

50 472 U.S. 585 (1985).

51 Trinko, 540 U.S. at 399.

52 Aspen Skiing, 472 U.S. at 600–01.

53 Petition of the Fed. Trade Comm'n for Rehearing En Banc, FTC v. Qualcomm Inc., No. 19-16122 (9th Cir. Sept. 25, 2020), www.ftc.gov/system/files/documents/cases/1410199qualcommrehearingpetition.pdf.

54 *See, for example*, John D. McKinnon & James D. Grimaldi, *Justice Department, FTC Skirmish over Antitrust Turf*, WALL ST. J. (Aug. 5, 2019), www.wsj.com/articles/justice-department-ftc-skirmish-over-antitrust-turf-11564997402.

55 FTC v. Qualcomm, Inc., Brief of the United States of America as Amicus Curiae in Support of Appellant and Vacatur, No. 19-16122 (9th Cir. Aug. 30, 2019), www.justice.gov/atr/case-document/file/1199191/download. The Division also argued that the district court failed to identify harm to competition where it observed only that Qualcomm's power to demand high royalties was a function of its patents and did not find that Qualcomm was pricing below cost in the chip market.

license.[56] In fact, a UK court in *Unwired Planet v. Huawei* specifically found that where a portfolio is "sufficiently large and has sufficiently wide geographical scope that a licensor and licensee acting reasonably and on a willing basis would agree on a worldwide licen[s]e. They would regard country by country licensing as madness. A worldwide licen[s]e would be far more efficient."[57]

In contrast to the United States, however, courts in Europe and the United Kingdom have provided industry participants more guidance in terms of FRAND licensing and the negotiation process by giving more examples and commentary around the contours of what is considered good faith negotiations and circumstances pertaining to the availability of injunctive relief. For example, in Europe, an innovator who does not provide notice of infringement and does not explain why the license terms and rates sought are FRAND – considered the proper negotiation process – risks losing its right to injunctive relief in case of a finding that it abused its dominant market position.[58] On the other hand, when an implementer is unwilling to take a license on FRAND terms or unduly delays negotiations, this conduct can open the path to injunctive relief for the innovator.[59] Some courts have found a delay of several months (for example, five months) to signal an unwilling licensee.[60] And European courts have found that a steadfast refusal to pay any royalties whatsoever to the SEP holder classifies as a case of holdout.[61] In the United

56 Unwired Planet Int'l Ltd. v. Huawei Tech. (UK) Co. Ltd. et al. [2020] UKSC 37 [15] (upholding trial court's reasoning that a global portfolio license is the appropriate and efficient approach); BGH, May 5, 2020, KZR 36/17 (¶ 78) (Ger.) https://juris.bundesgerichtshof.de/cgi-bin/rechtsprechung/document.py?Gericht=bgh&Art=en&sid=3abd1ba29fc1a5b129c0360985553448&nr=107755&pos=0&anz=1; Landgericht Mannheim [LG] [Mannheim Regional Court] Jan. 8, 2016, 7 O. 96/14 (¶ 63) (Ger.) (in light of usual industry practice, offer of a worldwide license is FRAND); see also Britain Eakin, *China's Top Court Affirms Right to Set Global FRAND Rates*, LAW 360 (Sept. 10, 2021), www.law360.com/articles/1419376/china-s-top-court-affirms-right-to-set-global-frand-rates (discussing Oppo v. Sharp (2020) Yue 03 Min Chu 689 (Shenzhen Interm. People's Ct. 2020)).

57 Unwired Planet Int'l Ltd. v. Huawei Tech. (UK) Co. Ltd. et al. [2017] EWHC 2988 [543] (Pat).

58 BGH May 5, 2020, KZR 36/17 (¶¶ 69–72).

59 *Id.* ¶¶ 69–70; Unwired Planet Int'l Ltd. v. Huawei Tech. (UK) Co. Ltd. et al. [2020] UKSC 37, [145–47, 158]; Landgericht Düsseldorf [LG] [Düsseldorf Regional Court] July 16, 2015, C-170/13 (¶ 74) (Ger.) https://eur-lex.europa.eu/legal-content/EN/TXT/?uri=CELEX:62013CJ0170.

60 BGH May 5, 2020, KZR 36/17 (¶ 92) (implementer taking several months to respond to a notification of infringement indicative of unwilling licensee); Landgericht Düsseldorf [LG] [Düsseldorf Regional Court] May 9, 2016, I-15 U 36/15 (¶¶ 2, b, bb) (Ger.) (taking five months to respond to an infringement letter is a delay tactic).

61 TQ Delta LLC v. Zyxel Communications Ltd. & Zyxel Communications A/S, [2019] EWHC 745 [12] (Pat) ("I accept that this is a case of 'hold-out' by ZyXEL. They have not paid any royalties to TQ Delta (or any other patent holder) in respect of any standards essential patent. Of the two patents from TQ Delta's portfolio which have now been litigated in this jurisdiction, infringement of the '268 Patent has been established, and has been continuing for many years. ZyXEL have blown hot and cold as to whether they will accept whatever licence is considered by the Court to be RAND. They have refused to 'agree to submit to the outcome of an appropriate [RAND] determination' and yet have claimed the benefit of the RAND undertaking.").

Kingdom, a "willing licensee" is "one willing to take a FRAND licen[s]e on whatever terms are in fact FRAND."[62]

Additionally, the case law in China has shifted on injunctive relief in line with European developments. In an encouraging development, in 2018, the Beijing IP Court handed down a landmark decision in *Iwncomm v. Sony*, upholding the first injunction related to a dispute over FRAND licensing terms for a SEP in that country.[63] The court made clear the circumstances under which a SEP holder may secure an injunction. It held that SEP holders may obtain an injunction where a potential licensee negotiated in bad faith (for example, procrastinated as a tactic to draw out discussions and avoid paying royalties).[64]

European courts have also recognized that FRAND licensing negotiations are context-specific in assessing both the process and the terms offered.[65] As such, they have weighed in on specific issues such as making non-FRAND offers,[66] substantiating infringement claims,[67] considering comparable licenses,[68] requiring confidentiality and nondisclosure agreements,[69] licensing downstream users,[70] and selective licensing.[71] And while some European courts have avoided wading into this particular area, the United Kingdom and at least one court in China have asserted

62 Unwired Planet Int'l Ltd. v. Huawei Tech. Co. Ltd. et al. [2017] EWHC 711 [708] (Pat).

63 Hui Zhang, Mengling Liu, & James Yang, *Beijing High Court Upholds China's First-Ever SEP Injunction in Iwncomm v. Sony*, Kluwer Patent Blog (May 29, 2018), http://patentblog.kluweriplaw.com/2018/05/29/beijing-high-court-upholds-chinas-first-ever-sep-injunction-iwncomm-v-sony/.

64 Jacob Zhang & Li Yang, *New Lessons From a Milestone SEP-Based Infringement Litigation*, Managing IP (Mar. 21, 2018), www.managingip.com/article/b1kbppqbxg3j5s/new-lessons-from-a-milestone-sep-based-infringement-litigation.

65 BGH May 5, 2020, KZR 36/17 (¶ 79).

66 Landgericht Mannheim [LG] [Mannheim Regional Court] Nov. 17, 2016, 7 o. 96/14 (§ IV.I) (Ger.).

67 *Id.*

68 Unwired Planet Int'l Ltd. v. Huawei Tech. (UK) Co. Ltd. et al. [2020] UKSC 37 [105–19] (assessing discrimination by comparison to comparable licenses as part of a "single, unitary obligation" to license on FRAND terms).

69 Landgericht Düsseldorf [LG] [Düsseldorf Regional Court] July 13, 2017, 4a O 16/16; Landgericht Düsseldorf [LG] [Düsseldorf Regional Court] Mar. 22, 2019, I-2 U 31/16 (assessing the effect of "contractual nondisclosure agreements with its licensees" in light of a licensor's FRAND commitment).

70 *See, for example*, Landgericht Düsseldorf [LG] [Düsseldorf Regional Court] Nov. 26, 2020, 4c O 17/19 (¶¶ 17–29) ("[W]hile an SEP holder may preferentially solicit end-manufacturers of a particular product to take a licence, it may not ignore or reject legitimate licence requests/offers from a supplier."); Landgericht Düsseldorf [LG] [Düsseldorf Regional Court] July 11, 2018, 4c O 81/17 ("[A] patentee must be allowed to freely choose the distribution level at which he intends to conclude license agreements.").

71 Landgericht Düsseldorf [LG] [Düsseldorf Regional Court] July 11, 2018, 4c O 81/17 (stating that "[a] difference of treatment shall be permissible where objectively justified," but finding discrimination in selective enforcement).

jurisdiction to set global FRAND rates,[72] offering an alternative to what is typically left for determination by a jury in the United States.[73]

On the international front, it is concerning that, as some have remarked, China has both misunderstood or misapplied the essential facilities doctrine[74] and recently announced that it has a national policy to advance its own companies' interests in standards organizations.[75] Neither of these developments serves to uphold strong patent rights and maximizes incentives to innovate. The United States should continue to lead by example here and avail itself of potential avenues for engagement to share its experience on these fronts.

## IV. WHERE DO WE GO NEXT: IS THE UNITED STATES MOVING TOWARD SUBSTANTIVE CONVERGENCE WITH EUROPE?

We have come a long way in refining the interplay of antitrust and IP. And US courts have certainly made some headway in clarifying in what instances antitrust can and should be used to address IP disputes. The Division and the FTC have sought to clarify their policy approaches but at times took divergent paths on some critical issues, creating uncertainty both within the United States and with respect to its global leadership on substantive convergence regarding principles impacting innovation incentives and technological progress.

Against this backdrop, when the American Bar Association's Antitrust Law Section (Section) submitted its 2021 Presidential Transition Taskforce Report to the Biden Administration, it called for additional guidance on licensing practices and obligations associated with SEPs.[76] The Section's "Presidential Transition Taskforce" reports – the tradition of which goes back to special reports first compiled

[72] Unwired Planet Int'l Ltd. v. Huawei Tech. (UK) Co. Ltd. et al. [2020] UKSC 37; Britain Eakin, *China's Top Court Affirms Right to Set Global FRAND Rates*, Law 360 (Sept. 10, 2021), www.law360.com/articles/1419376/china-s-top-court-affirms-right-to-set-global-frand-rates (discussing Oppo v. Sharp, (2020) Yue 03 Min Chu 689 (Shenzhen Interm. People's Ct. 2020)).

[73] TCL Commc'n Tech. Holdings v. Telefonaktiebolaget LM Ericsson, 943 F.3d 1360, 1375 (Fed. Cir. 2019), *cert. denied*, 141 S. Ct. 239 (2020).

[74] *See, for example,* Maureen K. Ohlhausen, Comm'r, Fed. Trade Comm'n, Hearing on the Foreign Investment Climate in China: U.S. Administration Perspectives on the Foreign Investment Climate in China, Before the U.S.-China Econ. & Sec. Review Comm'n (Jan. 28, 2015), www.uscc.gov/sites/default/files/Maureen%20Ohlhausen_Testimony.pdf.

[75] *See, for example,* Arjun Gargeyas, *China's 'Standards 2035' Project Could Result in a Technological Cold War*, The Diplomat (Sept. 18, 2021), https://thediplomat.com/2021/09/chinas-standards-2035-project-could-result-in-a-technological-cold-war/.

[76] Am. Bar Ass'n, Antitrust Law Sec., Presidential Transition Report: The State of Antitrust Enforcement 17–19 (Feb. 2021) [hereinafter 2021 Presidential Transition Report], www.americanbar.org/content/dam/aba/administrative/antitrust_law/lp-files/presidential-transition-report.pdf.

with the election of President George H. W. Bush in 1988[77] – are prepared every presidential election year.[78] They are meant to educate the incoming administration on the then current state of antitrust and suggest areas of focus going forward.

The task force, chosen anew every four years, includes "attorneys in private practice, in-house counsel, and antitrust law and economics scholars."[79] Its members typically also represent a cross section of "political, ideological, and professional views, ... [leading to] often vibrant and spirited debate among the Members" who reach consensus on the recommendations in the task force report.[80] The intersection of IP and antitrust has featured in these transition reports since at least 2001, and the report's observations and recommendations present a timely look at industry participants' understanding of the state of agency enforcement policy and case law. Notably, the Section did not endorse a specific policy view for IP/antitrust in general but rather requested that the Agencies provide transparency and additional, more detailed guidance, for example, "on what may constitute exclusionary conduct where a breach of a FRAND commitment is involved" and "when seeking an injunction related to FRAND-encumbered patents might raise antitrust concerns."[81]

Since then, President Biden issued Executive Order No. 14036 on Promoting Competition in the American Economy (Competition EO), which in part encourages the attorney general and the secretary of commerce to consider reevaluating their positions on the intersection of IP and antitrust to safeguard the standard-development process and potential harm to competition from industry participants leveraging their IP in anticompetitive ways.[82] Specifically, the Competition EO questioned whether the Division's 2019 Remedies Statement should again be revised. It did not take long for the Division to heed the Administration's call – about five months later, the Division, in conjunction with the USPTO and NIST, released a new *Draft Policy Statement on Licensing Negotiations and Remedies for*

77 The current format of the reports was adopted with the 1993 transition report. Prior to this, the Section produced the 1989 Report of the ABA Antitrust Section Special Committee to Study the Role of the Federal Trade Commission, the 1989 Report of the ABA Antitrust Section Task Force on the Antitrust Division of the US Department of Justice, and the 1991 Report of the ABA Antitrust Section Special Committee on International Antitrust. *See* AM. BAR ASS'N, SEC. OF ANTITRUST LAW, 1993 REPORT OF THE ABA ANTITRUST SECTION SPECIAL TASK FORCE ON COMPETITION POLICY 1 (Feb. 23, 1993), www.americanbar.org/content/dam/aba/administrative/antitrust_law/v12/report_1993-comp-policy.pdf.

78 AM. BAR ASS'N, SEC. OF ANTITRUST LAW, THE STATE OF FEDERAL ANTITRUST ENFORCEMENT 9 (2001), www.americanbar.org/content/dam/aba/administrative/antitrust_law/v12/report_antitrustenforcement.pdf.

79 2021 Presidential Transition Report, *supra* note 76, at 6.

80 *Id.*

81 *Id.* at 18.

82 Exec. Order No. 14036, 86 Fed. Reg. 36987 (July 9, 2021).

*Standards-Essential Patents Subject to Voluntary F/RAND Commitments*[83] and solicited public comments (2021 Draft Remedies Statement).[84]

The 2021 Draft Remedies Statement followed a speech by the Division's Economics Director of Enforcement, Jeffrey Wilder, that already walked back some of the statements contained in the 2019 Remedies Statement.[85] Comments from the Division's Assistant Attorney General Jonathan Kanter during his confirmation hearings before the Senate Judiciary Committee were largely consistent with Mr. Wilder's speech.[86] That speech previewed some significant potential shifts, including seemingly suggesting that a breach of FRAND may amount to deception under relevant IP/antitrust case law and as such present a cognizable antitrust claim,[87] while also promising "clearer guidance on what good-faith [licensing] negotiation looks like and how bad-faith conduct can hinder competition."[88] Related to the latter, Mr. Wilder also seemed to indicate that the Division would favor IP policies that prescribe what licensing negotiations should look like.[89]

The 2021 Draft Remedies Statement correctly described the purpose of the FRAND commitment as one to "facilitat[e] access on F/RAND terms to the technology needed to implement a standard and help[] to ensure that the rights of patent holders whose technology is used are appropriately respected."[90] While the 2021 Draft Remedies Statement retained the central point of the 2019 Statement and developed case law that there is not "a unique set of legal rules for SEPs subject to F/RAND commitments,"[91] other aspects were concerning. For example, the

[83] US Patent & Trademark Off., Nat'l Inst. of Standards & Tech. & US Dep't of Just., Draft Policy Statement on Licensing Negotiations and Remedies for Standards-Essential Patents Subject to Voluntary F/RAND Commitments (Dec. 6, 2021) [hereinafter 2021 Draft Remedies Statement], www.justice.gov/opa/press-release/file/1453826/download.

[84] Press Release, US Dep't of Just., Public Comments Welcome on Draft Policy Statement on Licensing Negotiations and Remedies for Standards-Essential Patents Subject to F/RAND Commitments (Dec. 6, 2021), www.justice.gov/opa/pr/public-comments-welcome-draft-policy-statement-licensing-negotiations-and-remedies-standards.

[85] Jeffrey Wilder, Econ. Dir. of Enf't, Leveling the Playing Field in the Standards Ecosystem: Principles for a Balanced Antitrust Enforcement Approach to Standards-Essential Patents (Sept. 24, 2021), www.justice.gov/opa/speech/file/1437421/download.

[86] Gene Quinn, *Jonathan Kanter Responses to Senate Provide Insight on Approach to Antitrust-IP Nexus*, IPWATCHDOG (Oct. 24, 2021), www.ipwatchdog.com/2021/10/24/jonathan-kanter-responses-senate-provide-insight-approach-antitrust-ip-nexus/id=139124/.

[87] Wilder, *supra* note 85, at 3 ("Consequently, this commitment assures standards implementers that they will have access to SEPs on reasonable terms .... While SDO IPR policies should facilitate efficient licensing, there are often disputes and unsavory negotiation tactics that make reaching a licensing agreement difficult."); *id.* at 6 (stepping back from recent position that "a patent owner's breach of a FRAND commitment can *never* constitute an antitrust violation").

[88] *Id.* at 5.

[89] *Id.* at 3 ("While SDO IPR policies should facilitate efficient licensing, there are often disputes and unsavory negotiation tactics that make reaching a licensing agreement difficult. In these circumstances, standardized products can be delayed and consumers suffer.").

[90] 2021 Draft Remedies Statement, *supra* note 83, at 3.

[91] *Id.* at 8.

2021 Draft Remedies Statement contained various legally unsupported suggestions of antitrust liability and vague references to what negotiators "should" do to act in good faith.

After a review of the many public comments received, the Division, USPTO, and NIST announced in June 2022 that they were withdrawing the 2019 Remedies Statement rather than revising it.[92] Such a move, they concluded, "is the best course of action for promoting both competition and innovation in the standards ecosystem."[93] USPTO and NIST spokespersons highlighted that the decision was informed by the importance of ensuring American companies' continued global leadership in research and development as well as engagement by those stakeholders in international standards development.[94] The Division revealed its plan to use a case-by-case approach in evaluating conduct by SEP holders and implementers – with a focus on scenarios involving small- and medium-sized businesses or highly concentrated markets.[95]

In implementing this case-by-case enforcement approach, the Division (and FTC) should be careful to heed clear case law that antitrust liability does not attach where a SEP holder merely seeks an injunction as a remedy to an infringing implementer or supra-FRAND rates or terms in SEP licensing negotiations. More is required for a cognizable antitrust claim. And with respect to lesser explored IP/antitrust issues in US case law to date, including, for example, factual scenarios that could indicate a party is either a willing or unwilling licensing negotiation participant, the Agencies would do well to look to European case law developments to benefit from lessons learned by European and UK courts that have already grappled with these issues in more detail than their US counterparts have done. This approach would also serve to foster convergence substantively on IP/antitrust principles and inject certainty for innovators and implementers alike who must negotiate global portfolio licenses. The important tenet that must remain front and center is that Agency guidance, even through case-by-case developments, must be clear, and it cannot stand settled case law on its head – lest the Agencies undermine efficient licensing negotiations and thereby undermine stability and certainty for industry participants.

## V. CONCLUSION

A reliable IP system – one that maintains strong patent rights – is essential for enabling US companies to compete on a level playing field and maintain their

[92] Press Release, US Dep't of Just., Justice Department, US Patent and Trademark Office and National Institute of Standards and Technology Withdraw 2019 Standards-Essential Patents (SEP) Policy Statement (June 8, 2022), www.justice.gov/opa/pr/justice-department-us-patent-and-trademark-office-and-national-institute-standards-and.

[93] *Id.*

[94] *Id.*

[95] *Id.*

leadership position. Case law developments worldwide have moved the needle to clarify legal rules and create an environment with at least some guardrails for SEP licensing that helps shape conduct by industry participants during negotiations. In the United States specifically, there is legal consensus that a breach of FRAND alone is not a cognizable antitrust claim under the Sherman Act, and injunctions for SEP infringement are properly sought and issued against an unwilling licensee (but not a willing licensee). Any next steps in terms of agency guidance, including its newly announced case-by-case enforcement approach, should be careful not to devalue technical contributions to standards by innovators or depart from the lessons learned to date not only in the United States but also abroad.

PART II

# Patent Holdup, Royalty Stacking, and the FRAND Standard

# 3

# Cellular SEP Royalties and 5G

## *What Should Competition Policy Be?*

*Alexander Galetovic, Stephen Haber, and Lew Zaretzki*

## I. INTRODUCTION

For the past four decades, technological progress in semiconductors and wireless technology has been driven by specialized firms that develop technologies and license them to downstream chip manufacturers and implementers in exchange for royalties. Specialization and trade in technology along the vertical chain has been one of the hallmarks of fast technological progress, and in wireless communications this pattern has repeated itself across multiple rounds of change in standards. Indeed, 5G is being developed in significant part by the same specialized technology firms that previously helped to develop 3G and 4G. As with previous wireless generations, standardization, patents, and licensing support vertical specialization.

As in all markets, 5G market demand determines the amount of economic surplus produced. Over the past 20 years, technology developers and implementers have battled over the surplus and have sought to leverage policy, regulation, legislation, jurisprudence, popular sentiment, and other dynamics to capture the largest possible share for their stakeholders. Their duty to capture maximum surplus for their stakeholders discourages concern for unintended or unexpected consequences for the market as a whole and for other critical institutions in society.

Meanwhile, however, the battle over surplus has taken place in a market where technology developers compete. We have shown in previous work (which we discuss later) that an equilibrium emerged in wireless 3G and 4G licensing. In this equilibrium, innovators license standard-essential patents (SEPs) and implementers (chiefly, smartphone Original Equipment Manufacturers (OEMs)) pay royalties under a licensing regime in conformity with "fair, reasonable and nondiscriminatory" (FRAND) principles. We estimated that for more than 10 years the cumulative royalty yield paid by the entire phone manufacturing value chain hovered between a market average of 3% and 3.5% of the average selling price of a phone, which seems to be the agreed market price for the technologies that make phones work. We have also shown that the transition between 3G and 4G licensing occurred smoothly, as licensing practices and aggregate royalty levels did not change, despite substantial changes in products and the manufacturers involved.

5G will allow wireless technologies to become part of a broader array of products. When compared with current 4G networks, 5G promises to provide between 10 times and 100 times faster data rates, at latencies up to 10 times smaller. Faster data rates and lower latencies, in turn, may enable new technologies for automated manufacturing, private mobile networks, and fixed wireless access (FWA). These new technologies will create new markets and attract new entrants. For example, Multi-Access Edge Computing (MEC) will invite new participants into the 5G system, which may include firms in semiconductors (for example, Intel), computing (for example, Dell, VMWare), cloud services (for example, AWS) and more.[1] These new participants may take advantage of opportunities for specialization to create and patent innovations, and may implement external innovations to create new products and services. As such, they may become part of the 5G licensing landscape, whether as licensor, licensee, or both.

What role should competition policy play in emerging 5G markets? The seeming uncertainty about the practices and royalties that will emerge in new licensing markets may tempt antitrust authorities to intervene to regulate licensing or fix and apportion aggregate royalties. Indeed, for many years the Federal Trade Commission, the Department of Justice, the European Commission, and several academics and industry consultants have argued that the market prices that emerge from royalty agreements are the result of "excessive royalties" wrought by the additional monopoly power conferred by standardization. The central point of this chapter, however, is that economic theory and actual experience with royalty setting in 3G and 4G suggest that the revenues that patent holders obtain are not monopoly rents. Indeed, the level of royalties indicate that rents enjoyed by patent holders are Ricardian – that is, these rents reflect the fact that the selected technology creates more value per unit of input than the alternatives, not the exercise of economic market power.[2] Thus, the rents enjoyed by technology developers are the result of competition among technologies.

[1] Multi-access Edge Computing (MEC) deploys cloud-computing capabilities "at the edge of the network" – that is, the equivalent of the base station – thus obtaining ultra-low latency and high bandwidth. In addition, it allows real-time access to radio network information that can be used by applications. Operators can open their Radio Access Network (RAN), enabling service providers to offer applications and services to mobile subscribers, households, and firms. *Multi-access Edge Computing* (MEC), ETSI, www.etsi.org/technologies/multi-access-edge-computing (last visited May 18, 2022).

[2] A Ricardian rent arises from differential productivity or costs per unit among factors of production on the types of rent; see Roger G. Noll, *"Buyer Power" and Economic Policy*, 72 Antitrust L.J. 589, 592–94 (2005). In the business literature, this is usually called a competitive advantage or a differential firm capability.

Indeed, 5G technology developers and implementers continue to collaborate to develop, produce, and deploy 5G products and services. Further, as with prior cellular technologies, consumers, enterprises, and implementers can choose among many alternatives to 5G for various tasks. Thus, 5G is being deployed under the same conditions that characterize the well-functioning cellular SEP licensing market. This market has achieved a long-term equilibrium across the period of our research (2007–2021), spanning the deployment and use of 2G, 3G, 4G, and 5G technologies. In this equilibrium, cumulative royalties have converged to market values, and the market has apportioned them according to the incremental value generated by each holder's intellectual property (IP) assets. 5G and prior cellular technologies earn Ricardian rents determined by the differential value that those technologies create over alternatives.

We are, therefore, not aware of a prima facie argument to justify the intervention of competition authorities to regulate 5G royalties. Competition authorities should instead be watchful over the equilibrium that has existed in the market for cellular SEPs over the past decade or more, ensuring that parties do not undermine it through tactics employed to maximize their own share of economic surplus. If parties undermine the equilibrium in this manner, the results could be tragic for the associated technology, product, and service markets.

Further, competition policy should anticipate the arrival of new market participants, and the important role that 5G technology developers will play in the emergence of new markets. It should also anticipate that these new participants and markets may differ from present 5G-enabled participants and markets in terms of structures, behavioral norms, and other salient features. These new participants may lack a history of cellular SEP licensing. Such markets may need to find new solutions to adopt 5G or may need to adopt or adapt approaches from current 5G-enabled cellular SEP markets. Based on the experience of the past two decades across multiple cellular technology generations, it is our view that these markets will find their way to remunerate investments in R&D, so long as SEPs and SEP licensing transactions can be reliably enforced as elsewhere and as appropriate.

The rest of the chapter is organized as follows. In Section II, we briefly describe 5G and how its technological characteristics expand capabilities and alter network architecture. In Section III, we argue that the incremental value that the technology creates at the margin is a rent, and the source of this rent is the factor that informs competition policy. Section IV reviews substitutes that perform part of the functions that 5G can execute. Section V concludes.

## II. WHAT IS 5G?

5G provides an integrated cellular system performing the useful and relevant functions typical of prior cellular networks. Like 3G and 4G, 5G also provides a new radio technology (5GNR) and expands technical scope and functionality.

In addition, 5G provides revolutionary new options for deployment architecture and opens the door to a broad range of entirely new applications.

### A. *New 5G Technologies and Capabilities*

Table 3.1 shows that 5G will provide data rates potentially up to 20 gigabytes per second (gbps) (v. 4G: 20 megabytes per second (mbps)) and latencies as low as ~1 millisecond (ms) (v. 4G: 20–30 ms), and allow for considerably denser networks, of more than one million connections per square kilometer, which can support massive Internet of Things (IoT) deployments. In addition, 5G networks facilitate private networks (a network built for a specific organization – for example, on a university or corporate campus) and network slicing – that is, reserving part of the network for tailor-made applications for one or more specific clients.

5G also provides greater integration of useful capabilities that were formerly typically located outside cellular systems. For example, MEC (Mobile Access Edge Computing) incorporates processing capabilities at the edges of the network, in part to achieve broader system goals such as low latency. Integration also serves new applications through new specialized subsystems such as C-V2X (Cellular Vehicle to Everything) for safety coordination among vehicles, infrastructure, and other connected devices as well as UAS (Unmanned Aerial Systems) for 5G control of aerial drones.

These new subsystems stand to create value, which mobile operators, infrastructure vendors, users, and suppliers of the attendant new technologies may appropriate. Table 3.2 contains an exemplary list of markets and suppliers that could benefit.

TABLE 3.1. *New 5G technologies and capabilities*

| Capability | Description |
|---|---|
| eMBB | Enhanced Mobile Broadband, providing data rates potentially up to 20 gbps. Enabled in part by 5G improvements such as New Radio (NR) and Millimeter Wave (24–40 GHz frequency band). |
| URLLC | Ultra-Reliable Low-Latency Communication, providing latency as low as ~1 ms, for highly sensitive applications such as manufacturing automation. |
| mMTC | Massive Machine Type Communication (>1 m device connections/ km), supporting colossal IoT deployments. |
| Private Networks | Independent 5G networks, providing organizations with greater control of deployment, performance levels, and security. This may be useful for the most security conscious organizations, as well as for those most reliant upon network performance (for example, for mission critical industrial operations). |
| Network Slicing | Virtual 5G networks, enabling network operators to tailor services to specific users or uses on their general network. For example, a carrier could offer an IoT-optimized slice to IoT users, or a custom slice to one organization in lieu of building and operating its own private network. |

TABLE 3.2. *Potential beneficiary suppliers*

| System/ feature | Description |
|---|---|
| MEC | • Compute hardware vendors (Dell, HPE, Supermicro)<br>• Compute software vendors (VMWare, RedHat, etc.)<br>• Compute processor vendors (Intel, NVIDIA, AMD)<br>• Cloud service providers (AWS, Google Cloud, Microsoft Azure) |
| C-V2X | • Automotive OEMs (Toyota, Volkswagen)<br>• Automotive tier 1 suppliers (Denso, Bosch, Continental)<br>• Automotive other suppliers (embedded, module, TCU, drive computer) |
| UAS | • Drone OEMs (DJI, Yuneec, 3D Robotics, Parrot) |
| Private Networks | • Enterprises (perhaps especially large/sophisticated organizations) |
| Network Slicing | • Enterprises (perhaps especially SMEs or specialist firms) |

## B. *New 5G Distributed Architecture Options*

The providers of radio access networks (RANs) for 2G, 3G, and 4G have long provided deeply integrated solutions comprising network equipment, software, and services. The situation resembles that of IBM in the mainframe era, in that a small number of incumbents provide end-to-end solutions to mobile operators. These incumbents, along with their consolidated predecessors such as Nortel and Motorola Networks, helped create the mobile industry. At present, four firms dominate the RAN market. Of these, Huawei and ZTE dominate the market in the People's Republic of China (PRC) but are excluded from several markets, including the United States, where Nokia and Ericsson enjoy an effective duopoly, as seen in their market revenue shares depicted in Figure 3.1.[3]

As we see in Figure 3.2, however, 5G enables new underlying architecture options, including vRAN and Open RAN, which provide network operators with additional supplier options to help build their networks. This architectural opening resembles the move from IBM mainframe to Client/Server, which launched a computing revolution. This change potentially provides opportunities for highly competitive vendors to participate in the 5G system, making it more efficient, less expensive, and more innovative. Leading industry analyst Dell'Oro forecasts Open RAN as 15% of the market by 2026, while vRAN will be 5–10%, and combined these will represent 20–25% of the market within four years.[4]

[3] *Base Station Market Poised for Strong Year Thanks to 5G and China According to Omdia*, MICROWAVE J. (Dec. 7, 2020), www.microwavejournal.com/articles/35089-base-station-market-poised-for-strong-year-thanks-to-5g-and-china-according-to-omdia.

[4] Press Release, Dell'Oro Grp., Open RAN on Track Comprise 15 Percent of RAN by 2026 (Jan. 21, 2022), www.delloro.com/news/open-ran-on-track-comprise-15-percent-of-ran-by-2026/.

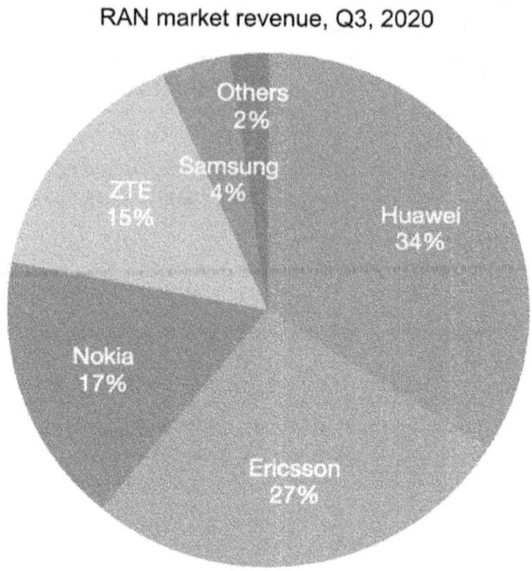


FIGURE 3.1. RAN market revenue by OEM
*Source: Base Station Market Poised for Strong Year Thanks to 5G and China According to Omdia*, Microwave J. (Dec. 7, 2020)

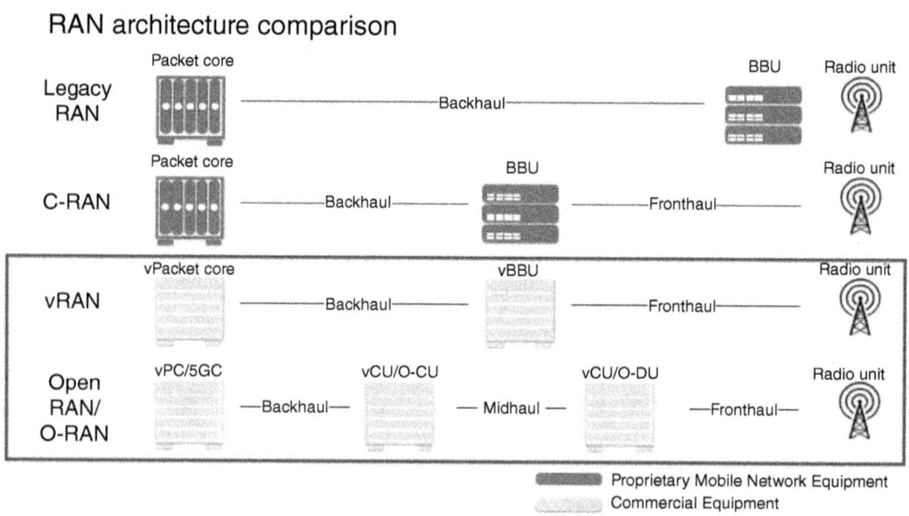


FIGURE 3.2. Comparison of RAN architecture options

Open RAN and vRAN enable network operators to move workloads to Commercial-Off-The-Shelf (COTS) equipment from leading computing vendors. Providers can base their equipment on world-class silicon, hardware, and software from some of the most competitive firms in the world, as well as from innovative startups, providing further diversification in the mobile value chain. Operators could

TABLE 3.3. *Firms that potentially can participate as decentralized providers*

| Element | Potential new beneficiaries |
|---|---|
| Centralized Unit<br>Distributed Unit | • Compute hardware vendors (Dell, HPE, Supermicro)<br>• Compute software vendors (VMWare, RedHat, open source)<br>• Compute processor vendors (Intel, NVIDIA, AMD)<br>• Cloud compute providers (AWS, Google Cloud, Microsoft Azure) |
| NFV Software | • OpenRAN startups (Mavenir, Rakuten Symphony, Airspan) |
| Radio Unit | • RU providers (Airspan, Casa Systems)<br>• Modem providers (Qualcomm, for small cells) |

use cloud service providers who can further enhance efficiency and handle these workloads as a service. Table 3.3 lists firms that potentially can participate as decentralized providers.

In the special case of Open RAN, operators can obtain radio units from competing vendors beyond the incumbent leaders such as Ericsson and Nokia. Qualcomm's 5G RAN Platform for Small Cells and FSM200xx processors promises a ready avenue for small-cell entrant providers to provide competitive radio unit products. Enhanced competition in this area appears particularly desirable given the national security imperative to exclude Huawei and ZTE from many key markets.

These new architecture and vendor options promise to enable new players to enter the market for cellular services. Examples include Rakuten in Japan and, following their example, DISH in the United States. New entrants may be very helpful in the United States given the merger of T-Mobile and Sprint, which leaves just three major wireless carriers to serve the US market. The US market suffers from higher wireless prices than many other countries.[5]

The nature of the new system enables new and superior applications. For example, the combination of 5G's wireless nature and the improved performance through eMBB make 5G FWA a practical alternative to fixed broadband services based on FTTx and DOCSIS. This should benefit wireless carriers, FWA vendors, and broadband consumers.

In short, 5G brings an array of new technologies, scope improvements, features, and applications. While each may create value, the value may accrue to different market segments and participants in each case based on the nature of each improvement. Most improvements may chiefly benefit carriers, while others may chiefly benefit enterprises, vendors of particular types of equipment, software or services, or others. Interestingly, the expanded scope of 5G means that many mobile device vendors may be agnostic to many of these improvements since they are targeted at other markets and other parts of the mobile value chain.

[5] *Worldwide Mobile Data Pricing 2021*, CABLE.CO.UK, www.cable.co.uk/mobiles/worldwide-data-pricing/ (last visited Apr. 9, 2021).

### C. *The 5G Value Chain*

A notable feature of the 5G value chain is that it involves multiple, specialized firms that act in a decentralized fashion, coordinated by standards and market interactions. In that respect, the 5G value chain is like the value chain of previous technological generations, only more advanced and perhaps more complex.

As is well known, the theory of patent holdup and royalty stacking predicts that a market characterized by multiple agents contributing to a standardized technology will be able to exploit monopoly power. According to the theory, the existence of multiple monopolies strangles markets and most of the price paid by consumers will redound to the profit margins of the technology development firms.[6]

A testable implication of the theory is that implementers "see down" the game tree, and therefore refrain from making investments. It is therefore curious that implementers in 5G do not seem to be concerned about this possibility. They are making sizable investments. We think that they are likely drawing on the history of 3G and 4G in making their investment decisions.

## III. WHAT DETERMINES ROYALTIES? LESSONS FROM 3G AND 4G[7]

In this section, we argue that the appropriate framework to think about royalties in new markets is standard price theory, which explains where value comes from and how it is distributed among the factors of production that create value in any given market. In 5G, IP is one of these factors of production, and its market price – the royalties paid by implementers – will indicate the incremental value that the technology creates at the margin. The main point is that this incremental value is a rent, and the source of this rent informs competition policy.

### A. *Value and Distribution in Wireless Mobile Markets*

Price theory observes that in equilibrium the price paid by consumers equals the value created by the entire production chain at the margin, as determined by the good's demand curve. This is true whether the good or service is a pound of bread, a bundle of pins, or a phone.

Price theory also shows that the total revenues generated by sales in a market are distributed among the factors of production – those inputs that were involved in the production of the final good – based on the value each adds to total revenues at the

6 Mark A. Lemley & Carl Shapiro, *Patent Holdup and Royalty Stacking*, 85 Tex. L. Rev. 1991 (2007).

7 This section is based on Alexander Galetovic & Stephen Haber, *SEP Royalties: What Theory of Value and Distribution Should Courts Apply?*, 17 Ohio St. Tech. L.J. 189 (2021).

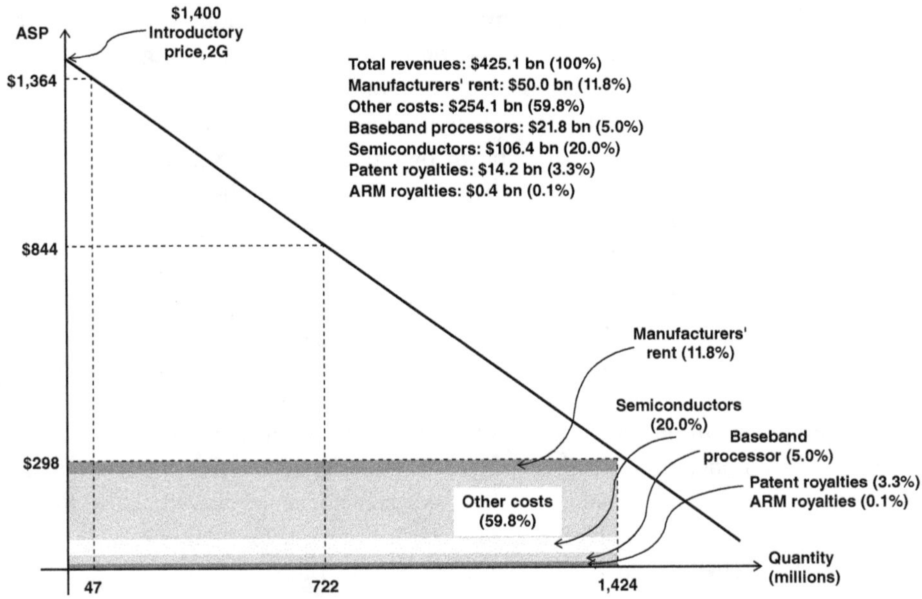


FIGURE 3.3. Value and distribution in the smartphone value chain

margin. It follows that revenues are exhausted by the payments made to the input suppliers along the value chain. These revenues remunerate the opportunity cost of the inputs used in the production of the final good and any rent that factors of production receive.

We follow our 2021 paper[8] and illustrate these insights with a simple supply and demand graph. Figure 3.3 depicts the observed equilibrium in the smartphone market in 2016. For simplicity, we parameterize the world market with a single linear demand curve and assume that all consumers paid the average selling price of a smartphone.[9]

In 2016, phone manufacturers sold 1.42 billion units for $425.1 billion, at an average selling price of $298.[10] Because consumers are free to buy a phone, the

[8] Galetovic & Haber, *supra* note 7, at 197.

[9] To draw the intercept of the linear demand curve on the price axis, we followed Alexander Galetovic, Stephen Haber, & Lew Zaretzki, *Is There an Anticommons Tragedy in the World Smartphone Industry?*, 32 BERKELEY TECH. L.J. 1527, 1550 (2017), who use the fact that 2G phones, which were considerably inferior devices compared with a 2016 smartphone, were introduced at $1,400 in 1992. It is likely that this underestimates consumers' willingness to pay and hence biases our calculation toward obtaining smaller market power rents. Consumers buy different models and brands. Nevertheless, they are free to choose among models and brands, and in equilibrium, marginal consumers are indifferent among them. Consequently, one could build a quality-adjusted average phone and work with this demand curve.

[10] The August 2017 update of the database, which shows the sources and calculations in detail, is available in an Excel workbook available upon request. Alexander Galetovic, Stephen Haber,

demand curve shows how much consumers value a smartphone at the margin. On average, $298 is what the least willing consumer in 2016 was willing to pay for a smartphone. Figure 3.3 also shows that most consumers valued their phones at more than $298 and obtained a net surplus when they bought a phone, the difference between their willingness to pay, as shown by the demand curve, and the market price. It follows that the total consumer surplus was equal to the area between the demand curve and the market price for phones. According to the demand curve depicted in Figure 3.3, consumer surplus in 2016 was equal to $784 billion.[11]

Figure 3.3 also shows how the revenues generated by the sale of smartphones were distributed among phone manufacturers and input suppliers. Roughly 20% of the revenues accrued to semiconductor manufacturers ($85 billion; $60 per smartphone, on average); 5% accrued to manufacturers of baseband processors ($22 billion; $15 per smartphone, on average); and 60% of the revenues ($254.1 billion; $178 per smartphone, on average) accrued to the producers of other inputs, such as the firms that made the cameras, gorilla glass, and housings, as well as the firms, such as Foxconn, that assembled the phones. Around 12% ($50 billion; or $35 per smartphone) reached the firms that sold the phones in the form of profits, most of which accrued to Apple.

IP is an asset, and royalties are the revenues that this asset generates. As can be seen in Figure 3.3, 3.4% of the revenue generated by the smartphone market reached the owners of patents ($14.2 billion, or roughly $10 per smartphone). Most of this ($12.4 billion) was earned by SEP owners. The remainder was largely earned by non-SEP patents, held by firms such as Microsoft (which earned royalties mainly on the patents on its operating system and software technologies), the patent pools that license audio and video codecs, and patent assertion entities that own the patents necessary to manufacture semiconductors.

The distribution of the $425.1 billion in revenues among input providers reflects the choices that firms in the production chain made to substitute away from more expensive inputs toward less expensive inputs. Thus, firms at the end of the production chain, which designed and marketed the phones (for example, Samsung, Apple), combined inputs from many suppliers to minimize the costs of producing the smartphones that consumers valued. Similarly, the firms that produced the intermediate inputs and IP for those smartphones (for example, Corning, Ericsson) also combined inputs from many suppliers to minimize costs. Those suppliers, in turn, purchased the necessary inputs from firms even further up the production chain, and so on. Each input in the production chain had its own demand curve. That is, the demand curve each producer faced was derived from the

& Lew Zaretzki, *An Estimate of the Average Cumulative Royalty Yield in the World Mobile Phone Industry: Theory, Measurement and Results*, 42 TELECOMM. POL'Y 263, 266 (2018).

[11] This is equal to about 1% of world GDP.

demand for smartphones, and the elasticities of each demand curve depended in part on the possibilities for substituting away from that input. Consequently, firms along the production chain equalized the value created by each input at the margin with the input's market price.

The share of each input in the $425.1 billion in revenues in the smartphone market was therefore the equilibrium outcome of a complex process of cost minimization. Because the output of an upstream firm is the input of firms further downstream, and all value stems ultimately from consumers' willingness to pay, no stage of the production chain is independent of, and separable from, the others – prices are determined simultaneously in all of them.

What does price theory tell us about how to value the IP necessary to produce a smartphone? The royalty is the rental price of IP and is a function of the value that consumers were willing to pay for the capabilities created by those patented technologies, at the margin, and the possibilities that producers had to substitute away from using those IP assets toward alternative technologies. The finding that the patent holders earned 3.4% of the value of the average smartphone in 2016 has three complementary interpretations. First, the purchaser with the lowest willingness to pay for the average smartphone valued those technologies at the equivalent of just 3.4% of the price she paid for her smartphone. Second, there must have been alternative technologies toward which producers could eventually substitute. If that had not been the case, then the owners of the IP property would have operated as monopolists and charged far more than 3.4% of the value of a phone, a point to which we return later. Third, and importantly, IP owners did not enjoy market power; they could not constrain output to raise prices.

### B *Economic Rent and the Distribution of Value across the Stages of the Production Chain*

Should competition authorities care about the level of royalties? Some argue that patent holders can charge royalties because they enjoy monopoly power, even beyond that granted by the patent. According to this argument, the profits that patent owners obtain are rents wrought by market power, and there is a natural role for competition policy. A different explanation, however, is that the revenues that patent holders obtain are the fair remuneration of their investment, given the risks they took when researching and developing the technology. In this view, the rents enjoyed by patent holders are Ricardian – their origin is that the selected technology creates more value per unit of input than the alternatives, not the exercise of market power. According to this argument, rents are the result of competition among technologies, and there is little, if any, role for competition policy.

Thus, the nature and origin of the rents made by patent holders are central to assessing whether competition policy must do something, if anything, about

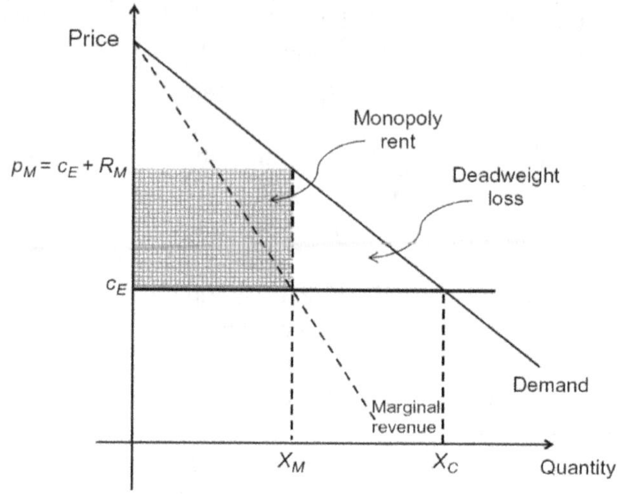


FIGURE 3.4. Monopoly licensor sets the royalty rate

royalties. To appreciate the difference between a monopoly rent and a Ricardian rent, a few diagrams help.

Consider first a royalty set by a patent holder who holds a monopoly because nobody can produce without infringing the monopolist's patent. To keep the argument simple, we assume that the monopolist licenses the technology to competitive manufacturers that can produce phones at constant cost *c*. As Figure 3.4 shows, when fixing the royalty rate, the monopolist patent holder confronts the market demand for the final good, and reasons that if she sets the royalty equal to *R*, competitive manufacturers will pass on the royalty and set a price $p = c + R$ for a phone.

Thus, by fixing *R*, the monopolist patent holder controls the final price *p*. Also, by setting *R*, the monopolist patent holder controls the per-unit rent. Standard economic theory says, and Figure 3.4 shows, that the monopolist patent holder will increase *R* to contract output until the marginal revenue from selling phones is equal to the manufacturing cost *c*. In equilibrium, the price of a phone will be equal to $p_M = c + R_M$ and then

$$\frac{p_M - c}{p_M} = \frac{R_M}{p_M} = \frac{1}{\eta}, \tag{3.1}$$

where $\eta$ is the elasticity of the demand for phones. It follows that, just as for a standard monopolist, pricing is determined by the classic Lerner formula.

A Ricardian rent, by contrast, emerges when a firm produces more or higher quality output per unit of input than its competitors. Because a Ricardian rent remunerates a competitive advantage, it can emerge in a competitive market. Observed royalties have nothing to do with market power.

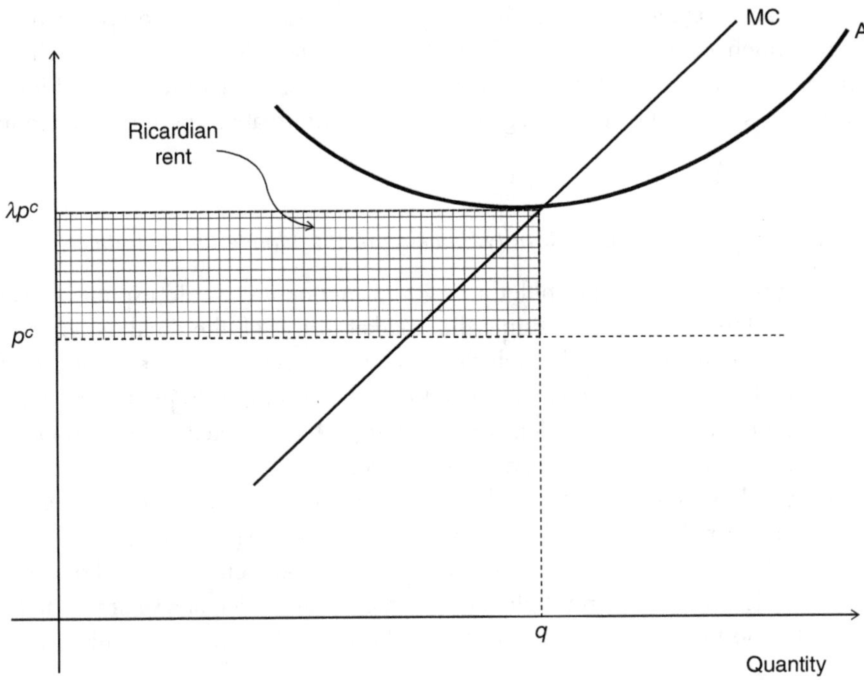


FIGURE 3.5. Patents and Ricardian rents

To appreciate this point, Figure 3.5 draws the standard average and marginal cost curves of a competitive licensee. The figure shows an innovation that increases quality and consumers' willingness to pay by a factor $\lambda > 1$ over the alternative, and in equilibrium, products that use the technology command a market premium equal to $(\lambda - 1)p^c$. Thus, the licensee obtains additional revenues equal to the shaded rectangle. This rectangle is the Ricardian rent wrought by the technology, which the patent holder can appropriate through licensing and charging royalties. But its origin is the market premium $(\lambda - 1)p^c$, and its total amount is capped by the incremental value created by the technology.

As can be seen in Figure 3.5, the royalty is part of the licensee's average cost. The licensee covers all her costs, and the good's market price equals its long-run marginal cost. But because a better technology commands a price premium, which is equal to consumers' differential willingness to pay, well-functioning markets naturally create the rents and rewards that incentivize investments in R&D. This rent may be transferred to the patent holder via per-unit royalties, a lump-sum payment, or a combination of both. Whatever the means whereby the Ricardian rents are transferred, they remunerate those technologies that are more productive and deliver more output or value per unit of input.

It follows that competition policy may have a role if the source of the rents is the exercise of market power – that is, if technology developers are able to raise the running royalty to create scarcity at the margin. By contrast, competition policy does not have much of a role if technology developers earn Ricardian rents, which emerge in competitive markets. The point we make next is that the observed level of the royalties charged by technology developers is informative about the source of the rents.

### C. *Monopoly Power and Royalty Stacking in the Mobile Phone Industry*

As Figure 3.6 shows, a patent holder acting as a monopolist would exploit market power by restraining output, raising the market price of the final good, and extracting the monopoly rent through the royalty. A direct test for the existence of a monopoly therefore uses the Lerner margin, as shown in (3.1) to predict the level of the royalties that patent holders would charge if they acted as a monopoly. We apply this reasoning to the smartphone market.

Figure 3.6 shows the same demand curve as Figure 3.3 but assumes that patent holders act as a single profit-maximizing monopolist. The main result is that patent holders would have earned 66% of all revenues of the value chain, instead of 3.3%. Higher royalties would have multiplied the average selling price of a smartphone by a factor of almost three – from $298 to $844. Consequently, only 722 million units

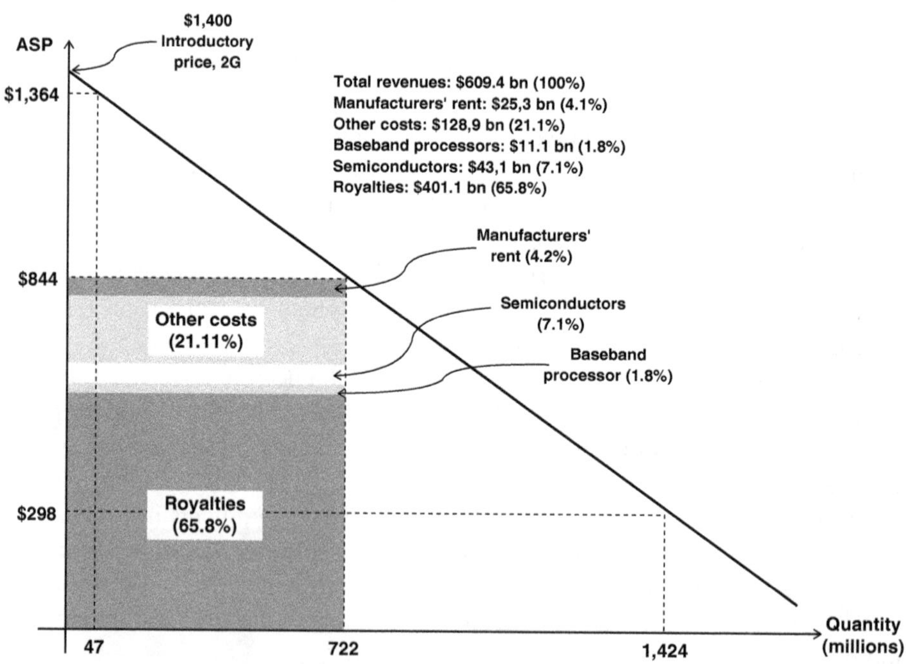


FIGURE 3.6. Value and distribution in the smartphone value chain with a single Monopoly Patent Holder (2016)

would have been sold, instead of 1.42 billion. Despite the decline in unit sales, however, total revenues would have risen from $425.1 billion to $609.4 billion. More than two-thirds of those revenues (about $400 billion) would have accrued to the patent holders. Those revenues would have been pure rent, as they exceeded the long-run cost of the inputs used to produce the patented technologies.[12]

An influential literature, known as patent holdup and royalty stacking, implies that the situation should have been even worse than outlined in the prior paragraph. The theory claims that because multiple firms own the IP that is essential to make a phone interoperable and compatible, each can exploit its monopoly independently.[13] That is, the magnitude of the cumulative royalty in the smartphone value chain predicted by the theory of royalty stacking grows with the number of patent holders.[14] Monopoly will stack upon monopoly, strangling the industry. Formally, if the number of patent holders is $n$, then the Lerner margin is:

$$\frac{p_s - c}{p_s} = \frac{R_s}{p_s} = \frac{n}{\eta} > \frac{1}{\eta} \tag{3.2}$$

In our 2017 paper,[15] we parameterize a standard royalty stacking model. We observe that in 2016, there were 21 identified patent licensors who received royalty revenue, and that the cumulative royalty yield predicted by the theory of patent holdup and royalty stacking is 79%. That is, if patent holders were each exploiting their monopoly power independently, they would receive four out of every five dollars paid for a smartphone.

A predicted royalty of 67% (a single monopolist) or 79% (21 monopolists) compares with the observed average cumulative royalty yield from the 21 identified patent licensors of 3.4% in 2016.[16] That is to say, the actual yield is more than 20 times lower than either the yield predicted by the theory of patent holdup and royalty stacking or by the predicted royalty that would be charged by a single profit-maximizing monopolist.

The implication is that patent holders in the smartphone value chain do not exercise any meaningful monopoly power. On the contrary, the remuneration that patent holders receive is a Ricardian rent. Thus, the evidence from 3G and 4G is that there is no evidence that standardization creates market power. Because of this, there is little ground to claim that competition authorities have any meaningful role to play in cellular SEP licensing as traditionally practiced within the smartphone industry.

[12] Indeed, the profits of patent licensors would have been of the order of 0.6 % of world GDP.

[13] Lemley & Shapiro, *supra* note 6.

[14] We use the term royalty "yield" rather than royalty "rate." "Rate" refers to the actual royalty paid by a licensee to a licensor as a percentage of the licensee's sales. "Yield" is the sum of patent royalty payments divided by the total value of mobile phones shipped, the latter of which might include the production of those who evade patent licenses.

[15] Galetovic et al. (2018), *supra* note 10.

[16] *Id.*

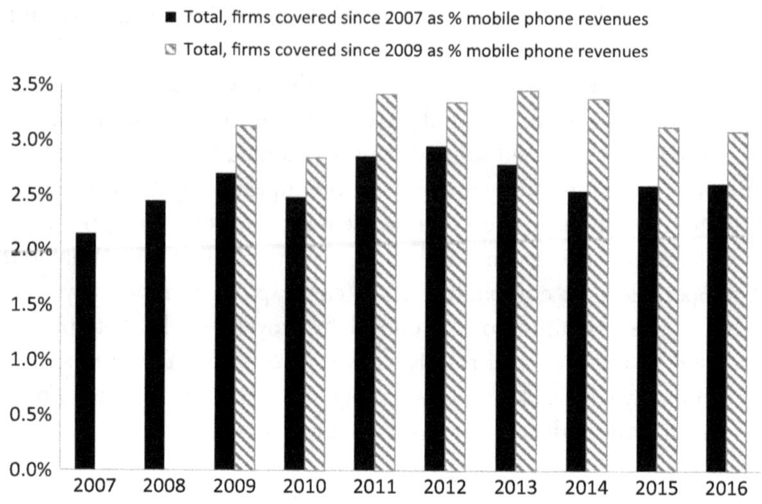


FIGURE 3.7. Patent royalties as percentage of the value of mobile (smart and feature) phones shipped (2007–2016)

## D. 3G *and* 4G: *A Functioning Licensing Market*

As we have seen, in 2016, the average royalty yield for a smartphone was 3.4% of the average selling price of a smartphone. In our 2018 paper,[17] we estimated royalty yields from 2007 through 2016 and found that they showed remarkable stability. Figure 3.7 shows the average royalty yield since 2007 for 16 licensors, which accounted for 78.2% of all royalty revenues in 2016. Since 2009, we have data for 21 licensors, which accounted for 92.5% of all royalty revenues in 2016. As Figure 3.7 shows, both series are remarkably stable. The average cumulative royalty yield of firms with data since 2007 hovers between 2.1% and 3%; the average cumulative royalty yield of firms with data since 2009 hovers between 3% and 3.5%, falling only marginally during the last three years. Note that, as can be seen in Figure 3.8, the composition of sales between feature and smartphones changed significantly during the period, and the value of sales roughly doubled, and yet the average cumulative royalty yield remained stable.

## E. *New Estimates of the Cumulative Royalty Yield*

Since our research described above, we have continued to monitor the market for cellular SEP licenses, although we have modified our methodology so that we now focus on cellular SEP licensing specifically.[18] The leading cellular SEP licensors are

[17] *Id.*

[18] This means that we now omit licensors focused on non-SEP licensing as well as those focused on licensing SEPs not related to cellular technologies, for example those related to digital

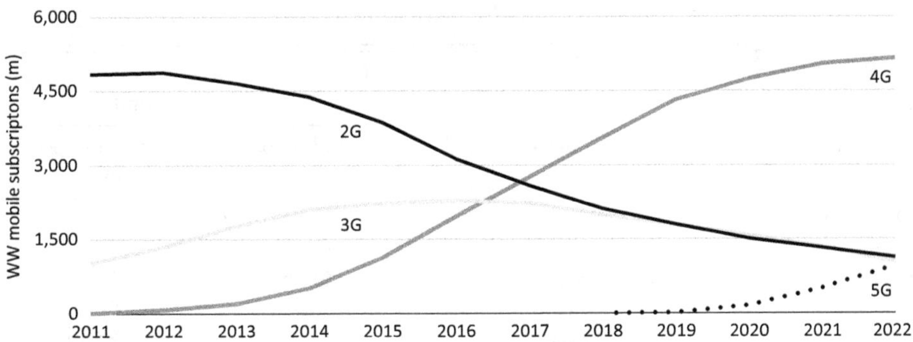


FIGURE 3.8. WW mobile subscriptions by cellular generation
*Source*: Ericsson Mobility Visualizer, Sept. 16, 2023

Qualcomm, Nokia, Huawei, Ericsson, and Interdigital. These constituted 84% of royalties in 2016, the last year covered in our prior studies.[19]

As can be seen in Table 3.4, these five licensors obtained $82 billion in royalty revenues during 2015–2022, on average about $10.2 billion per year. On average, royalties were $9.05 per smartphone. If we make room for other licensors, average royalties per mobile device are about $9.79, in line with the orders of magnitude in our previous research. Basically stated, the data are consistent with our earlier estimates. In conclusion, little seems to have changed since the period covered in our previous research – a profound observation given the passage of so many years and the corresponding changes in markets, technologies, products, and companies.

## IV. LESSONS FOR 5G LICENSING

We have seen that the R&D investments that yielded the previous technological generations have been remunerated by Ricardian rents, which emerge in competitive markets and reflect that the selected technology creates more value per unit of input than the alternatives. What does this suggest about the licensing equilibrium that will emerge in 5G?

So far, the evolution of 5G suggests that technology developers and implementers have and will continue to collaborate to develop, produce, and deploy 5G products

audio and video technologies and Wi-Fi. This results in our removing licensors such as IBM, Rambus, and Xperi from our study, as well as patent pools such as Via Licensing's AAC program and MPEG-LA's H.264 and MPEG-4 programs.

19 For the methodology behind this calculation and the calculations in the subsequent paragraph, see the Appendix.

TABLE 3.4. *Leading cellular SEP licensing businesses (2015–2022)*

| Cellular SEP licensor | Total royalties identified ($b) | Implied effective royalties/cellular mobile device unit | |
|---|---|---|---|
| | | Excluding past units | Including past units |
| Qualcomm | $55,601 | $5.07 | $5.05 |
| Nokia | $10,806 | $1.47 | $1.30 |
| Huawei | $3,216 | $0.69 | $0.59 |
| Ericsson | $8,865 | $1.28 | $1.13 |
| Interdigital | $3,238 | $0.54 | $0.46 |
| Subtotal (5 licensors) | $81,726 | $9.05 | $8.54 |
| Others (est. 10%) | $8,173 | $0.74 | $0.74 |
| Total | $89,899 | $9.79 | $9.29 |

*Sources*: IDC, Mobile Phone, Tablet, and Wearables, 2015–2022; Ericsson, Investor Relations materials, 2015–2022; Interdigital, Investor Relations materials, 2015–2022; Nokia, Investor Relations materials, 2015–2022; Qualcomm, Investor Relations materials, 2015–2022; Qualcomm QTL Licensee Search; CNBC, "Huawei to start charging royalties to smartphone makers using its patented 5G tech," Arjun Kharpal, Mar. 16, 2021, "Huawei announces royalty rates for its patent license programs," Huawei, July 13, 2023

and services. Thus, 5G replicates the conditions that yielded the functioning licensing market for 2G, 3G, and 4G. An additional reason to think that investments in 5G will be remunerated by Ricardian rents is that consumers, enterprises, and implementers can choose among alternatives to perform many of the tasks that 5G will perform. In a competitive market, therefore, 5G will earn a rent, but it will be determined by the differential value that the technology creates over the alternatives.

Indeed, parties in each market are likely to consider the appropriateness of 5G for their specific situations in light of alternatives and substitutes. Different jobs for different customers may be accomplished with different devices, on different networks with different underlying network technologies.[20] This may depend upon the required data rates, latencies, ranges, power consumption limitations, mobility characteristics, and costs. These factors may determine whether and when a substitute or alternative displaces 5G and when it cannot. In what follows, we briefly review several potential alternative technologies that perform at least some of the tasks that 5G can accomplish, effectively acting as substitutes.

[20] Clayton M. Christensen, Taddy Hall, Karen Dillon, & David S. Duncan, *Know Your Customer's "Jobs to Be Done,"* HARVARD BUSINESS REVIEW (Sept. 2016), https://hbr.org/2016/09/know-your-customers-jobs-to-be-done.

## A. 5G *Alternatives and Substitutes*

### 1. Competition from Predecessors

5G must contend with its predecessors, 4G/LTE and 2G/ GSM. GSM remains highly useful and popular in specific geographic regions, powering roughly the same number of devices shipped in 2020 as in 2007. What does this mean? We can look at this from the perspective of the "Jobs to be Done"[21] framework developed by Clayton Christensen and colleagues. In this framework, it is critical to understand the "progress a consumer is trying to make in particular circumstances." The answer can indicate which products consumers will "hire" for a job. The decisions by consumers here suggest that 2G continues to sufficiently address the job-to-be-done for some customer segments. Interestingly, as Figure 3.8 shows, consumers quickly moved on from 2G's immediate successors, 2.5G and 3G, to 4G. Time will tell whether such segments move on to 5G, relegating 4G to the historical record, or whether 4G can, like 2G, retain a segment of consumers over the longer term.

Similarly, 4G has successfully addressed smartphone-based jobs such as getting to a meeting, arranging a dinner, or planning a vacation. Some 5G-capable smartphones continue to enable such tasks over 4G networks to save power, moving users to 5G only for select tasks where higher bandwidth justifies higher power consumption. Such tasks might include the much-discussed use case of preparing for a long trip by instantly downloading entire movies, but more likely will involve fundamentally new jobs most of us cannot currently foresee.

### 2. IoT Applications

From an IoT use case standpoint, 5G provides support through its mMTC capabilities. However, some IoT use cases may continue to employ traditional proprietary wireless technologies such as DJI OccuSync for UAV control, or standardized alternatives such as Wi-Fi, Bluetooth LE, Wireless HART, LoRaWAN, or other options. In each case, markets can consider performance, cost, and suitability for the use case at hand. For example, within the home or enterprise, many IoT devices do not require a mobile network to connect, and so Wi-Fi or Bluetooth may be satisfactory and lower in cost. For some IoT deployments in the field, LoRaWAN may remain appropriate, while in others 5G may provide advantages. At a minimum, 5G's mMTC capabilities should provide value in massive field deployments.

### 3. Fixed Wireless Access Applications

5G delivers a legitimate mobile broadband service, which should enable it to deliver a highly competitive FWA service that its predecessors could not deliver in many

[21] *Id.*

geographic areas. This may enhance competition in areas with fixed broadband build-outs, such as suburbs, and also provide a first true broadband option in areas presently lacking fixed broadband services. FWA can also serve vehicles such as buses, RVs, and mobile duty trailers. Even here, 5G should face competition, however, for example, from SpaceX's Starlink service (see Section 5, SpaceX Starlink).

### 4. Network Combinations and Offload

Users operating from homes, workplaces, and well-equipped "third places" such as airports, libraries, university campuses, and chain restaurants often enjoy Wi-Fi connectivity with fixed broadband backhaul for both better coverage and bandwidth. Historically, the worldwide Wi-Fi mobile data offload[22] from 4G was 59%, and forecasts call for 70% offload from 5G. This may be desirable, as it reduces the burden on cellular networks in use cases where cellular network attributes such as mobility do not add value. The Wi-Fi plus FTTx combination may continue to address many jobs to be done for mobile device users.

Note that many jobs may leverage such network combinations. For example, the major Japanese cellular carrier KDDI will use Starlink backhaul for rural base stations. This may enable 5G or other cellular coverage to serve customers in remote locations, but only through combination with Starlink.

Different geographic locations and even regions may present different preferred combinations. For example, outstanding fixed broadband offerings in Pacific Rim metro areas may enable Wi-Fi substitutes and derivatives such as Amazon Sidewalk networks. Rural areas within the United States may lend themselves to combinations involving cellular networks plus SpaceX's Starlink backhaul. When performing a job requires multiple networks working together, the different networks and technologies involved create value commensurate with their roles in completing the entire job.

## B. *Future Developments*

During the lifespan of 5G, we can expect much improvement in 5G itself, as well as in some of these substitute and alternative technologies. Cisco's Annual Internet Report observes the following trajectories for some key 5G alternatives and substitutes:[23] Fixed broadband speeds will more than double by 2023, to 110 mbps

[22] Cisco VNI Global Mobile Data Traffic Forecast, 2017–2022, Cisco (Feb. 2019), https://twiki.cern.ch/twiki/pub/HEPIX/TechwatchNetwork/HtwNetworkDocuments/white-paper-c11-741490.pdf.

[23] *New Cisco Annual Internet Report Forecasts 5G to Support More Than 10% of Global Mobile Connections by 2023*, Cisco (Feb. 18, 2020), www.cisco.com/c/en/us/solutions/collateral/executive-perspectives/annual-internet-report/white-paper-c11-741490.html.

(v. 46 mbps in 2018); Wi-Fi speeds from mobile devices will triple by 2023, to 92 mbps (v. 30 mbps in 2018); Wi-Fi hotspots will grow four times, to 628 million public hotspots (v. 169 million in 2018). Cisco forecasts these improvements in the context of rapidly growing quantities of users, connected devices, and changing use cases, including for UHD video and IoT devices.

For nascent technologies, the trajectory is more uncertain and speculative, but perhaps has greater potential upside. We look at three interesting cases to consider.

## 5. SpaceX Starlink

This is a satellite constellation of 3,185 operating satellites,[24] with FCC approval obtained for 12,000 and plans filed with ITU regulators for an additional 30,000,[25] to reach a total of 42,000. It would provide fixed wireless access for 25 countries, and there are pending applications to serve many more. Starlink can serve vehicles such as RVs and mobile duty trailers and promises to soon serve vehicles in motion, including planes, boats, trucks, and cars.[26]

## 6. Amazon Sidewalk

This is a LoRaWAN network technology embedded into Amazon devices such as Echo (Alexa voice assistants) and Ring (surveillance camera, doorbell, and flood light) devices, as well as Sidewalk network devices. Amazon has distributed, and continues to distribute, millions of these devices, creating a rapidly growing global Sidewalk network. There is no end-user charge for Sidewalk technology or usage. Ultimately, this network could become highly useful, at least in particular geographic areas and, for certain use, cases such as IoT deployments in those areas. Tile and similar tracker tags provide an early use case. Amazon could expand the capabilities of Sidewalk in the future to do more jobs, perhaps including some jobs presently performed by Wi-Fi or cellular networks such as messaging.

## 7. Helium

Another LoRaWAN network is Helium. While Sidewalk is centralized under Amazon control, Helium is decentralized. As with Sidewalk, any party can deploy one or more network nodes at their convenience and personal cost on their site. For

[24] Starlink Statistics (Oct. 22, 2022, 11:48 PM), https://planet4589.org/space/stats/star/starstats.html.

[25] Caleb Henry, *SpaceX Submits Paperwork for 30,000 More Starlink Satellites*, Space News (Oct. 15, 2019), https://spacenews.com/spacex-submits-paperwork-for-30000-more-starlink-satellites/.

[26] Michael Kan, *SpaceX Preps 'Ruggedized' Starlink Dish for Cars, Boats, and Planes*, PC Mag (Aug. 4, 2021), www.pcmag.com/news/spacex-preps-ruggedized-starlink-dish-for-cars-boats-and-planes.

TABLE 3.5. *Comparison of 5G and substitutes, alternatives, and complements*[a]

| Category | Technology | Max rate (mbps) | Latency (ms) | Range (km) |
|---|---|---|---|---|
| Cellular | 5G mm Wave | 20,000 (down)[b] | 1 (URLLC)[c] | <0.5[c] |
| | 5G Sub 6 | 900 (down)[d] | <10[e] | 30[e] |
| | 4G/ LTE (excl. advanced) | 150 (down)[f] | <35[f] | 200[g] |
| WiFi | Wi-Fi 7[h] | 46,000 (down) | <5 | n/a |
| | Wi-Fi 6E[h] | 9,600 (down) | <10 | n/a |
| PAN | Bluetooth Classic (v1.0-3.0)[i] | <3 | <100 | 0.01 |
| Wireless | Bluetooth LE (v4.0-5.x)[i] | <1 | 6 | 1.0[j] |
| IoT | Zigbee[k] | 0.250 | n/a | 0.1 |
| | LoRaWAN [k] | 0.006 | n/a | 10 |
| Fixed Broadband | Starlink (satellite constellation)[l] | <500 | 25–50 | n/a |
| | DOCSIS 4.0 (cable) | 10,000 (down)[m] | <1[n] | n/a |
| | DOCSIS 3.1 (cable) | 10,000 (down)[o] | <5 (LLD)[p] | n/a |
| | 50G-PON (fiber)[q] | 50,000 (down) | n/a | n/a |
| Proprietary | DJI OccuSync 3.0 (Drone)[r] | 15 (down) | 130 | 15 |

[a] These figures provide a basic perspective on these technologies' capabilities and position to compete to do specific jobs-to-be-done. Some of these figures may be theoretical maximums, while others may be viewed as practical maximums. Many may represent estimates of varying rigor and quality. In either case, a host of assumptions may apply; for example, one might achieve a maximum rate at a minimal range. Future advances in technology, or perhaps in implementation techniques, may improve upon these figures.

[b] Report ITU-R M.2410-0 (11/2017), Minimum requirements related to technical performance for IMT-2020 radio interface(s), www.itu.int/dms_pub/itu-r/opb/rep/R-REP-M.2410-2017-PDF-E.pdf.

[c] Ronan McLaughlin, "5G low latency requirements," Broadband Library, https://broadbandlibrary.com/5g-low-latency-requirements/.

[d] Jeremy Horwitz, "The definitive guide to 5G low, mid, and high band speeds" (Dec. 10, 2019), https://venturebeat.com/mobile/the-definitive-guide-to-5g-low-mid-and-high-band-speeds/.

[e] Ronan McLaughlin, "5G low latency requirements," Broadband Library.

[f] "How fast are 4G and 5G?," 4G.co.uk, www.4g.co.uk/how-fast-is-4g/.

[g] "Ericsson and Telstra extend reach on an LTE network up to 200km," Ericsson (Feb. 27, 2020), www.ericsson.com/en/press-releases/2/2020/2/ericsson-and-telstra-extend-reach-on-an-lte-network-up-to-200km.

[h] Rajiv, "Wi-Fi 7 Specifications and Features," RF Page (May 8, 2023), www.rfpage.com/wi-fi-7-specifications-and-features/#:~:text=It%20can%20support%20data%20rates,of%20less%20than%2010%20milliseconds

[i] "Bluetooth low energy basics: classic Bluetooth vs. Bluetooth LE," How to Electronics (May 20, 2023), https://how2electronics.com/classic-bluetooth-vs-bluetooth-low-energy-comparison/

[j] Jon Gunnar Sponås, "Things You Should Know About Bluetooth Range," Nordic Semiconductor (Jan. 25, 2023), https://blog.nordicsemi.com/getconnected/things-you-should-know-about-bluetooth-range.

[k] "Comparison of Wireless Technologies: LoRaWAN and Zigbee, WiFi, NB-IoT," DFRobot (Apr. 12, 2023), www.dfrobot.com/blog-1646.html.

[l] Anthony Spadafora, "Starlink internet coverage, cost, speeds and the latest news," Tom's Guide, (June 16, 2023), www.tomsguide.com/news/starlink-internet-coverage-speed-cost-satellites-ipo-and-latest-news.

[m] "DOCSIS 4.0 Technology," CableLabs (July 3, 2023), www.cablelabs.com/technologies/docsis-4-0-technology.

[n] Vivian Susko, "DOCSIS 4.0, Explained," Minim (Apr. 21, 2021), www.minim.com/blog/docsis-4.0-explained.

[o] "DOCSIS 4.0 Technology," CableLabs (July 3, 2023).

[p] Vivian Susko, "DOCSIS 4.0, Explained," Minim (Apr. 21, 2021).

[q] Jeff Heynen, "50G PON moves closer to deployment," Dell'Oro (Apr. 19, 2023), www.delloro.com/50g-pon-moves-closer-to-deployment/.

[r] "Consumer Drones Comparison" (July 5, 2023), DJI.com, www.dji.com/products/comparison-consumer-drones.

example, a homeowner could deploy in their neighborhood, while a business property owner could deploy across its locations. In both cases, the parties can create a new revenue stream while also serving their own users. Helium coordinates these actors using a crypto-based system where node operators earn HNT crypto tokens, and network users can "burn" such tokens to obtain service. Helium has deployed over 966,000 nodes across over 180 countries and 74,000 cities.[27]

Helium is now deploying a 5G network with plans for over 40,000 small cells (6,300 deployed to date) and has formed a partnership with DISH to supplement Dish's new network build-out. It could also provide its node operators with improved or additional wireless technologies to deploy in the future under this model.

In general, we should expect these alternatives and substitutes to do various wireless networking jobs in competition with 5G technologies, and for that competition to continue over years to come (Table 3.5).[28]

## V. CONCLUSION

We started this chapter by asking what role competition policy should play in emerging 5G markets. Our analysis suggests that the R&D investments that brought about previous technological generations have been remunerated by Ricardian rents, which emerge in competitive markets and reflect that the selected technology creates more value per unit of input than the alternatives. Moreover, the evidence shows that royalty yields are an order of magnitude smaller in prior wireless generations (through 4G) than those that would obtain if technology developers enjoyed market power. In short, there is evidence that royalty rates are determined in a competitive market, and there is no evidence that technology holders operate as monopolists.

So, royalty rates have been determined in a competitive market, and as we have observed, the market for mobile device cellular SEPs appears to be in a long-term equilibrium. What might enable such an equilibrium? Cellular SEP licensing typically relies upon long-term agreements, often on the order of five years but occasionally as long as 10 years and often involving one or more parties with long-term involvement in the industry. These agreements also tend to span periods of time during which cellular technologies arrive and depart, and during which their relative utility changes. These license agreements often cover several or even all such technologies simultaneously (for example, 5/4/3/2G and even "future

[27] HELIUM.COM, www.helium.com/ (last visited Oct. 23, 2022).

[28] These figures provide a basic perspective on these technologies' capabilities and position to compete to do specific jobs-to-be-done. Some of these figures may be theoretical maximums, while others may be viewed as practical maximums. Many may represent estimates of varying rigor and quality. In either case, a host of assumptions may apply; for example, one might achieve a maximum rate at a minimal range. Future advances in technology, or perhaps in implementation techniques, may improve upon these figures.

generation" rights), in the form of license grants and/or in complex combinations with other rights such as covenants. Some license agreements provide this coverage for a fixed amount of royalties paid out in a structured fashion within the term of the agreement.

Because of this, cellular SEP royalties appear "sticky" in the aggregate, and do not change quickly as might prices for commodities traded in spot markets. Many implementers may have entered the 5G market while operating under agreements established years before. Of those who entered the market with 5G licenses, some may have done so through serendipity when prior licenses happened to expire in a timely fashion to parallel their product roadmap transition to 5G. Note that individual parties may conduct their licensing activities differently on one or more of the deal parameters described earlier.

There are indications that Ricardian rents drive 5G as well. 5G has been developed in decentralized fashion with collaboration among specialists and others, as has been the norm. 5G confronts competition for the jobs-to-be-done from alternative technologies, including from 4G/LTE and 2G/GSM. This suggests that the equilibrium will continue, with royalties converging to market values and apportionment among licensors according to the incremental value generated by their respective intellectual property assets.

For these reasons, competition authorities should remain vigilant as always in observing the market but should not presume that monopolistic market power or other problems will distort the market for 5G cellular SEPs. Instead, competition authorities should be watchful over the equilibrium that has determined cellular SEP royalties for many years lest market participants employ tactics that undermine it and produce tragic consequences for its corresponding product and service markets.

## APPENDIX: ESTIMATING ROYALTIES 2015–2021

We continue to gather publicly stated licensing revenues as before, but we also attempt to gather public information regarding the identities of licensees each year. We then use market analyst data to understand the shipment volumes for these licensees to arrive at implied royalty rates per unit for each leading cellular SEP licensor. Our method, as with most "outside-in" methods, is imperfect but nevertheless provides insight.

We note that our implied royalty rates cover cellular-enabled mobile devices such as smartphones, tablets, and watches, but we omit feature phones, dongles, hotspots, modules, embedded solutions, automotive TCUs, and automobiles and other vehicles in general. This will tend to reduce shipment volumes and hence increase implied mobile device royalty rates. However, we believe that the undercounted volumes will be small in comparison to the covered markets, and hence the impact will be relatively small.

Several of the largest mobile devices makers such as the BBK entities OPPO, Vivo, and realme are privately held and do not share revenue information. Therefore, we prefer to calculate implied royalty rates on a royalty per unit basis rather than a percentage basis.

We compute implied royalty rates over the entire period observed, covering 2015–2021. We do this because royalties often arrive in nonuniform streams owing to agreements featuring significant upfront payments provided as a "pay down" of future royalties, a payment for past unlicensed shipments, or for other reasons. In some cases, this period may span agreements – for example, Apple and Nokia signed a new license in May 2017 but had a prior license that they had signed in June 2011. Our observation period incorporates portions of both licenses but the entirety of neither.

We overestimate cellular SEP royalties to the extent that these licensors (i) perform any other patent licensing, (ii) ascribe any patent licensing value to their noncellular SEP patents, and (iii) perform any other business in their licensing business. For example, during our observation period, Nokia Technologies included its brand licensing business (for example, licensing the Nokia brand to HMD), as well as a small digital health business (Withings, divested in 2018), and a small digital media business (Ozo VR camera, terminated in 2018). Similarly, all but Huawei participated in Avanci (a patent licensing platform), earning royalties for licensing automotive OEMs. This will tend to inflate our implied mobile device royalty rates, but we expect that the effects will be minor because these selected licensors overwhelmingly focus on cellular SEP licensing in the mobile device markets.

Furthermore, we do not know the identities of all licensees. Consequently, we will tend to underestimate the quantity of licensed shipments. This will also tend to raise implied royalty rates. However, we commonly find evidence for licenses among the largest players, and so again, the effects may be minor or principally impacting the least transparent licensor (Huawei).

While we can often identify that firms have established licenses with each other, we cannot count on reliable information regarding the start date and termination date of each agreement. We make a simplifying assumption that licenses typically span a full calendar year, and so if a license is present in say 2018, then we assume that it covers all units shipped in 2018. We know this is not always correct, as some licenses expire midyear. This will tend to depress rates by increasing the quantity of units covered. However, renewals will tend to obviate this overcounting if they occur within our lengthy observation period. We believe this simplifying assumption will again have minor impact.

When a new licensee arrives, we do not include the collective past units shipped prior to the year of the first known license. For example, in mid-2021, Interdigital and Xiaomi established their first patent license. Xiaomi had shipped over 600 million smartphones between 2015 and 2020, which we do not include, although we expect that the license agreement addressed these in some manner, such as with a release.

Incorporating these units would lower our implied royalty rates. This is likely the largest distortion caused by our methodology, although it appears small in the context of the total volumes licensed in the period. Consequently, we provide an alternative view in which we make an opposite assumption under which all licenses include coverage of past sales, whether by license, release, covenant, or any other approach. This scenario may be the most likely to replicate reality.

We note that the implied royalty rates may also reflect other information aside from productivity benefits. For example, licensors and licensees may exchange other sources of value in license agreements or other deals and may also enjoy different relationships, leading to different results.

We should directly address several unique points about Huawei. Huawei is the only licensor in our study that is a private company and does not provide audited financial statements under the auspices of rigorous securities regulation, as do the others. Instead, we have Huawei's public representations regarding the size of its licensing business over a period of years. Furthermore, Huawei does not tend to publicize the identities of its licensees in press releases or elsewhere to our knowledge. Consequently, we both take on faith Huawei's royalty figures in a manner unlike the treatment of the other licensors, and we also make assumptions about Huawei's licensees that we do not make for others. Huawei's business is the least transparent of those considered and so we leave it to the audience to consider the utility of the figures derived for it.

In addition, it is generally understood that the PRC restricts Huawei's licensing program such that it focuses on non-PRC licensees. If Huawei could and did license PRC firms such as OPPO, Vivo, Xiaomi and realme, it might grow demonstrably larger, and it might also arrive at a different implied royalty rate. To the extent that these companies presently enjoy an effective zero royalty rate, one might consider an alternative approach to calculating its effective market rate accounting for those "unlicensed" volumes.

Our implied royalty rates are market averages, and so any individual license might involve a rate that is higher or lower for a host of reasons, which might include shipment volumes, geographic areas of operation, and many other factors. This approach may in part indicate the productivity impact of each portfolio's inventions and hence an average Ricardian rent for the corresponding licensor's portfolio.

# 4

# The Fair Division of Surplus from a FRAND License Negotiated in Good Faith

*J. Gregory Sidak*

## I. NEGOTIATING FRAND LICENSES IN GOOD FAITH

Government agencies in Japan, China, the European Union, the United States, and other countries have issued guidelines to facilitate private negotiation to license the use of SEPs that a patent holder has voluntarily committed to a standard-setting organization (SSO) to offer to license on FRAND terms to a third party seeking to implement the standard.[1] In 2022, government agencies renewed their efforts to issue or reissue such guidelines.[2] Although those guidelines differ in several respects, a

[1] Guanyu Shenli Biaozhun Biyao Zhuanli Jiufen Anjian De Gongzuo Zhiyin (Shixing) (关于审理标准必要专利纠纷案件的工作指引 (试行)) [Working Guidelines on the Trial of Standard-Essential Patent Dispute Cases (for Trial Implementation)] (Apr. 26, 2018); *Guide to Licensing Negotiations Involving Standard Essential Patents*, Japan Patent Office (June 5, 2018), www.jpo.go.jp/e/support/general/sep_portal/document/index/guide-seps-en.pdf; Communication from the Com. to the Eur. Parliament, the Council, & the Eur. Econ. and Soc. Comm., COM (2017) 712 final, at 2 (Nov. 29, 2017) ("The Commission . . . considers that there is an urgent need to set out key principles that foster a balanced, smooth and predictable framework for SEPs."); Andrei Iancu, Under Sec'y of Com. for Intell. Prop. & Dir. of the United States Patent and Trademark Office, Remarks Delivered at the Standard-Essential Patents Strategy Conference, Solvay Business School, Université Libre de Bruxelles (ULB) (Sept. 10, 2019), www.uspto.gov/about-us/news-updates/remarks-director-iancu-standard-essential-patents-strategy-conference ("Government policy should make clear that good faith negotiations are expected on both sides, and that the presence or absence of good faith during negotiations can be a factor in the setting of remedies for infringement of FRAND-encumbered SEPs.").

[2] US Dep't of Just., Public Comments Welcome on Draft Policy Statement on Licensing Negotiations and Remedies for Standards-Essential Patents Subject to F/RAND Commitments (Dec. 6, 2021), www.justice.gov/opa/pr/public-comments-welcome-draft-policy-statement-licensing-negotiations-and-remedies-standards?utm_campaign = subscriptioncenter&utm_content = &utm_medium = email&utm_name = &utm_source = govdelivery&utm_term=; *Standard Essential Patents and Innovation: Call for Views*, Intellectual Property Office (UK) (Dec. 7, 2021), www.gov.uk/government/consultations/standard-essential-patents-and-innovation-call-for-views/standard-essential-patents-and-innovation-call-for-views; Jacob Schindler, *Japanese Ministry Seeks Industry Input on SEP Negotiation Guidelines*, IAM (Feb. 7, 2022), www.iam-media.com/frand/japanese-ministry-seeks-industry-input-sep-

common theme that emerges is the proposal that each counterparty negotiate a FRAND license in good faith.

Judicial opinions in SEP cases also refer to the duty to negotiate a FRAND license in good faith, but judges so far have failed to explain that duty's precise origin or its metes and bounds. For example, in two European decisions, *Sisvel v. Haier* and *Unwired Planet International Ltd. v. Huawei Technologies Co.*, the German Federal Court of Justice and the High Court of England and Wales, respectively, have emphasized that patent implementers must be willing participants in a FRAND negotiation.[3] In the United States in *TCL v. Ericsson*, Judge James Selna of the US District Court for the Central District of California quoted the relevant part of the FRAND commitment established by the European Telecommunications Standards Institute (ETSI), but he never determined the precise obligations that this contract imposed on the SEP holder, much less any obligation that ETSI imposed on the

negotiation-guidelines (citing Ministry of Economy, Trade, and Industry, *Study Group on the Ideal Trading Environment for Licenses for Standard Essential Patents* (Dec. 15, 2021) (Japan), www.meti.go.jp/shingikai/economy/patent_license/pdf/006_gijiyoshi.pdf?_x_tr_sl = ja&_x_tr_tl = en&_x_tr_hl = ja&_x_tr_pto = wapp); European Commission, *Call for Evidence for an Impact Assessment; Intellectual Property – New Framework for Standard-Essential Patents*, ARES (2022) 1076263 (Feb. 14, 2022) (EU), https://ec.europa.eu/info/law/better-regulation/have-your-say/initiatives/13109-Intellectual-property-new-framework-for-standard-essential-patents_en.

3 Sisvel v. Haier, Bundesgerichtshof [BGH] [Federal Court of Justice] May 5, 2020, KZR 36/17, ¶ 83 (Ger.), www.arnold-ruess.com/fileadmin/user_upload/2020_07_07_FCJ_SisvelvHaier_English ("The obligation of the market dominant patentee to inform the infringing party about the infringement and the possibility of obtaining a licence and to make an offer of a licence to the infringer willing to take a licence is not an end in itself, but is intended to make it easier for the latter to negotiate reasonable conditions with the patentee for his use. For this reason, after the first indication of infringement, it is not sufficient for the establishment of further obligations for the market-dominant patentee if the infringer then merely shows himself willing to consider entering into a licence agreement or to enter into negotiations as to whether and under what conditions the conclusion of a contract is possible for him. Rather, the infringer, for his part, must clearly and unequivocally declare his willingness to conclude a licence agreement with the patent proprietor on reasonable and non-discriminatory terms and must also subsequently participate in the licence agreement negotiations in a target oriented manner.") (citation omitted); Unwired Planet Int'l Ltd. v. Huawei Techs. Co. [2017] EWHC (Pat) 711 [163] (Eng.) ("The implementer must take a FRAND approach to the negotiation and accept a licence on FRAND terms if it wishes to take advantage of the constraint on the patentee's rights imposed by the FRAND undertaking. A FRAND approach to negotiation does not mean that parties cannot negotiate in good faith and a FRAND approach will allow for starting offers which leave room for negotiation. The fact an opening offered rate is higher than the true FRAND rate does not mean of itself that a patentee has failed to take a FRAND approach any more than the converse could be said about an implementer. On the other hand, making extreme offers and taking an intransigent approach which prejudice fair, reasonable and non-discriminatory negotiation is not a FRAND approach."), *aff'd*, [2018] EWCA (Civ) 2344 (Eng.), *aff'd*, [2020] UKSC 37 (Eng.); *id.* [693] ("An alleged infringer who wishes to show they are a willing licensee would do well to make an open offer of the FRAND terms it would be prepared to accept."); *id.* [708] ("[A] willing licensee must be one willing to take a FRAND licence on whatever terms are in fact FRAND.").

implementer as an implicit condition of its being empowered to enforce that FRAND contract as that contract's intended third-party beneficiary.[4]

The agencies issuing guidelines have conspicuously neglected to define good faith negotiation, let alone determine the steps that each party must take (and how quickly each must act) before a court may declare the negotiation to be at an impasse, such that contract formation has failed and the SEP holder may enforce its remedies against the unlicensed implementer as provided in the national law of the jurisdiction that issued the patents in suit. The Antitrust Division of the US Department of Justice (DOJ) has pontificated about the problem of implementers refusing to act in good faith, but it has not proposed any meaningful solution.[5] In a September 2020 supplement to a 2015 business review letter, the Antitrust Division merely requested that "any SDO policy updates should encourage good-faith bilateral licensing negotiation by both patent holders *and* implementers."[6] That supplement was shelved in April 2021, early in the Biden administration, approximately seven months after it had been issued by the Trump administration.[7]

In this section, I seek to make two points regarding a duty to negotiate in good faith. My first point is that mechanism design – a field of study within economics and game theory – can add rigor to the policy prescriptions and nebulous statements of government agencies about good faith negotiation. For the SEP holders and implementers that have experienced protracted litigation over the licensing of SEPs for smartphones, it is possible to draw lessons from what economists managed to fashion from whole cloth two decades ago to create the functioning market for the public auctioning and subsequent transferability of licenses to 3G spectrum. Without the groundwork laid by those economists, the smartphone today, if it

4 TCL Commc'n Tech. Holdings, Ltd. v. Telefonaktiebolaget LM Ericsson, Nos. SACV 14-341 JVS, CV 15-2370 JVS, 2018 WL 4488286, at *6 (C.D. Cal. Sept. 14, 2018), *vacated, reversed in part, and remanded*, 943 F.3d 1360 (Fed. Cir. 2019). For my own extended discussion of the SEP holder's and the implementer's possible duties to negotiate a FRAND license in good faith, see J. Gregory Sidak, *Judge Selna's Errors in TCL v. Ericsson Concerning Apportionment, Nondiscrimination, and Royalties under the FRAND Contract*, 4 Criterion J. Innovation 101, 102–07 (2019) [hereinafter *Judge Selna's Errors*].

5 Makan Delrahim, Assistant Att'y Gen., Antitrust Div., US Dep't of Just., Remarks as Prepared for the Licensing Executives Society (LES) 2019 Annual Meeting: "The Times They Are A'Changin'": The Nine No-No's in 2019, at 6 (Oct. 21, 2019), www.justice.gov/opa/speech/file/1213831/download ("[A]ctual evidence of hold-up remains scant even after a decade has passed since the theory was first introduced. The gulf between the theory and practice is especially troubling as many advocates ignore the real risk of hold-out by potential licensees of the chosen SEP technology. 'Hold-out,' of course, refers to the countervailing problem to hold-up: when an implementer licensee refuses to negotiate in good faith with a patent holder for a license, and instead forces the patent holder either to undertake significant litigation costs or to give up IP enforcement efforts.").

6 Letter from Makan Delrahim, Assistant Att'y Gen., US Dep't of Just., to Sophia A. Muirhead, Gen. Coun. & Chief Compliance Officer, IEEE (Sept. 10, 2020), www.justice.gov/atr/page/file/1315291/download (emphasis in original).

7 DOJ Policy: Department Restores 2015 Business Review Letter Interpreted as Opposing SEP Holders Seeking Injunctive Relief, Capitol Forum (Apr. 14, 2021).

existed at all, would be little more than an expensive pocket camera or portable media player searching for a Wi-Fi signal.[8]

My second point is that courts and policymakers possibly have failed to recognize the inadequate design of the current mechanism for FRAND licensing negotiations because that foundational economic question has become intertwined with, if not obscured by, the question of which jurisdiction's law controls a court's interpretation of the FRAND contract. Simply put, as I explained in my 2018 article "The FRAND Contract," the existing body of American contract law concerning offer, acceptance, and contract formation is concise and relatively unambiguous, and thus it provides a turnkey legal framework for resolving FRAND licensing disputes.[9] In a word, the American jurisprudence on contract formation is efficient. No further guidelines are necessary to apply that jurisprudence productively to interpreting the rights and duties surrounding negotiations for the licensing of SEPs. The wheel needs no reinventing.

### A. *The Phenomenon of Differential Ambiguity in Matters of Contract Formation and Good Faith Negotiation*

Building on the English common law tradition, but insulated from Europe by an ocean, American contract law appears to have evolved in its own distinctly didactic manner, which I attribute to the long shadow cast by Oliver Wendell Holmes[10] and his intellectual heir in American jurisprudence, Richard Posner.[11] With respect to

8 The economists who made major contributions to the design of those spectrum auctions include Nobel laureate Paul Milgrom of Stanford, Paul Klemperer of Oxford, and Ken Binmore of University College London. PAUL MILGROM, PUTTING AUCTION THEORY TO WORK (2004); PAUL KLEMPERER, AUCTIONS: THEORY AND PRACTICE (2004); Ken Binmore & Paul Klemperer, *The Biggest Auction Ever: The Sale of the British 3G Telecom Licenses*, 112 ECON. J. C74 (2002); Paul Milgrom, *Putting Auction Theory to Work: The Simultaneous Ascending Auction*, 108 J. POL. ECON. 245 (2000).

9 J. Gregory Sidak, *The FRAND Contract*, 3 CRITERION J. INNOVATION 1 (2018). In at least one reported case, a RAND commitment was found to be unenforceable, thus mooting the question of contract interpretation. In an investigation before the US International Trade Commission (ITC), Chief Administrative Law Judge Charles Bullock, in the public version of his Initial Determination on Violation of Section 337 and Recommended Determination on Remedy and Bond, found that, on the basis of the specific facts of the case, and pursuant to New York law, the complainant's RAND commitment to the Joint Electron Device Engineering Council (JEDEC) was too ambiguous to constitute an enforceable contract. Inv. No. 337-TA-1023, slip op. at 195 (USITC Nov. 14, 2017) (Initial Determination – Public Version); Sidak, *The FRAND Contract*, *supra* note 9, at 2–6.

10 OLIVER WENDELL HOLMES, JR., THE COMMON LAW (1881).

11 RICHARD A. POSNER, ECONOMIC ANALYSIS OF LAW (9th ed. 2014). As a new federal appellate judge, Posner relished every opportunity to cite an old common law case. Within four months of joining the Seventh Circuit, he found in a diversity case of first impression that Hadley v. Baxendale was the controlling authority for deciding a breach-of-contract claim involving a misdirected electronic funds transfer. EVRA Corp. v. Swiss Bank Corp., 673 F.2d 951, 955–59 (7th Cir. 1982).

the principles of contract formation, that American demand for crisp answers implicitly conduces to what economists call activity rules and closing rules, which simplify the task of definitively confirming whether a meeting of the minds has or has not occurred (much in the spirit, as I shall explain in the following pages, of economists realizing that national governments needed to define unambiguous activity rules and closing rules if they were to succeed in creating a workable market mechanism to auction licenses for 3G spectrum). By default, American law provides a clear closing rule for determining when contract formation has failed. The SEP holder makes an offer that is legitimately FRAND. Either the offer is accepted, or it is rejected explicitly or by counteroffer or by the passage of a commercially reasonable period of time. The licensee is not permitted to initiate rounds of offer and counteroffer. Following the licensee's failure to accept a legitimately FRAND offer, negotiations of course may continue between the parties, but no longer under the FRAND framework. Instead, those negotiations revert to the framework of public patent law.

In contrast to this American veneration of transactional efficiency (and a concomitant abhorrence of ambiguity or euphemism), the bodies of contract law of other jurisdictions (in Europe and the rest of the world) evidently do not offer, and do not aspire to offer, such black-and-white rules on whether and when a contract has been formed.[12] For example, when and for whom does the duty of good faith negotiation commence, and how long does it remain in effect once an offer has

[12] For example, a discussion on contract formation under French law that predates the 2016 revisions of the Civil Code states:

> French law sees a contract as an agreement, and it shares with English law (and indeed all other Western systems) the analysis of that agreement in terms of offer and acceptance. The practical results of that analysis quite often, however, diverge from those found in English law, and where this is so it is usually because French law ... adopts a more subjective approach.

Barry Nicholas, The French Law of Contract 61 (2d ed. 1992). Elsewhere, Nicholas discusses – again, before the 2016 revisions of the French Civil Code – the difference between English and French contract law at a higher level of abstraction:

> It is clear ... that the analysis of contract in terms of a free agreement of wills (or, in English terms, a meeting of minds) is common to both the French and the English classical theories of contract and remains part of the common currency of both systems.
>
> Where the two systems differ ... is partly in the intellectual rigour with which the analysis is carried through to detailed consequences, and partly in the way that agreement is understood: as a subjective meeting of two minds or as the objective appearance of agreement. English law usually favours the latter approach, as being the more practical and the more conducive to the certainty which commercial convenience demands, whereas French law inclines to the former, though sometimes with a corrective which yields much the same practical result as the objective approach.

*Id.* at 35. I would argue, for the reasons that I explain in this chapter, that it is erroneous as an empirical matter to assume that "much the same practical result" will occur when contract formation for the licensing of SEPs is analyzed under a FRAND or RAND obligation

been made? Some scholars of French law impute to the duty to negotiate in good faith an explicitly noneconomic origin.[13] As one treatise on French law has observed, incompatible interpretations among scholars of the purpose and effect of the doctrine of good faith indicate that "the notion of good faith and its use by the courts is likely to remain contested."[14] If SEP holders, implementers, and their attorneys fail to recognize that the monochrome character of American contract law differs from the Technicolor character of contract law in many other nations, they will expose themselves to an unmarked hazard in SEP licensing negotiations and SEP litigation whenever American contract principles do not control.

controlled by French law than when American law (typically New York law) controls the interpretation of contract formation between the SEP holder and the implementer. For further analysis of offer and acceptance under French law before the 2016 revisions, see JOHN BELL, SOPHIE BOYRON, & SIMON WHITTAKER, PRINCIPLES OF FRENCH LAW 302–05 (2d ed. 2008). For analysis of contract formation under French law after the 2016 revisions, see RUTH SEFTON-GREEN, *Formation of Contract: Negotiation and the Process of Agreement*, *in* THE CODE NAPOLÉON REWRITTEN: FRENCH CONTRACT LAW AFTER THE 2016 REFORMS 59 (John Cartwright & Simon Whittaker eds., 2017).

[13] BELL et al., *supra* note 12, at 334 ("[S]ome [French] jurists consider that the principle of good faith is a useful way for French contract law to be or to become more 'social', . . . allowing the Cour de cassation to 'promote a degree of good citizenship in the relationship of parties to a contract, this being preferable to the cynicism which an exclusively economic understanding of contractual relations could bring'. However, other jurists warn against the potentially subjective and uncertain nature of the concept, or deny the vision of contracts as 'a little society where each party works for a common good' on the basis that . . . 'contracts often appear as the result of a tension between antagonistic interests, the striking of a balance between divergent interests'. So, 'the duty of good faith does not oblige a person to protect the interests of another person to the detriment of his own interest, as some of the partisans of the unlikely notion of "contractual solidarity" contend'.") (footnotes omitted) (first quoting DENIS MAZEAUD, *La Politique Contractuelle de la Cour de Cassation*, *in* LIBRE PROPOS SUR LES SOURCES DU DROIT, MÉLANGES EN L'HONNEUR DE PHILIPPE JESTAZ 371, 382 (2006); then quoting FRANÇOIS TERRÉ, PHILIPPE SIMLER, & YVES LEQUETTE, DROIT CIVIL: LES OBLIGATIONS 443 (2005); and then quoting PHILIPPE MALAURIE, LAURENT AYNÈS, & PHILIPPE STOFFEL-MUNCK, LES OBLIGATIONS 373 (2007)); NICHOLAS, *supra* note 12, at 48 ("[W]here the Common law, in the interests of commercial convenience and the security of transactions, looks to the external appearance of consent, French law, influenced no doubt by the doctrine of the autonomy of the will and more concerned for justice in the individual case than for commercial expediency, often takes account of the true state of mind of one of the parties. . . . [Thus,] the requirement of good faith, though explicitly mentioned by the Code [Napoléon] only in connection with the performance of contracts, is introduced into the context of their formation under cover of the requirement of a genuine consent."); SEFTON-GREEN, *supra* note 12, at 60 ("It could be said that good faith is a big empty envelope into which a lot of concrete circumstances can be folded."); Open Sessions Volume IV at 1111:9–18, Certain LTE- and 3G-Compliant Cellular Communications Devices, Inv. No. 337-TA-1138 (USITC Sept. 17, 2019) (Testimony of Bertrand Fages) ("Q [by Counsel for INVT SPE LLC, the SEP holder and complainant]. Under French law, what does good faith require? A. Under French law it's ultimately up to the judge to decide what French law is in the context of each case; but in concrete terms, act in good faith is making serious proposals, which are consistent with the economic value and the purpose of the contract, and generally, to adopt an active attitude to achieve successful negotiations.").

[14] BELL et al., *supra* note 12, at 334.

That hazard exists in very practical business terms because ETSI plays such a dominant role in the setting of wireless standards, and its FRAND commitment is, of course, controlled by French contract law.[15] In contrast, New York law controls the RAND contract of another prominent SSO, the Institute of Electrical and Electronics Engineers (IEEE).[16] It is quite possible that the differences between French contract law and New York contract law – to take only one example – will produce substantively different conclusions about the legal duties owed under the FRAND or RAND contract in question. Indeed, the scope of a SEP holder's obligations and the scope of the rights granted to third-party beneficiaries by virtue of a FRAND or RAND contract depend on both that contract's actual language and the controlling law governing the interpretation (if needed) of that contract. The key point is that, whenever American contract law does not apply, to know when the negotiation has failed to achieve contract formation, we need a closing rule.

This perspective on contract formation causes me to disfavor and avoid using the terminology of "holdup" and "holdout" to describe the presence or absence of good faith during the negotiation to license SEPs pursuant to the FRAND contract. I find it simpler and more germane to ask whether, and when, the offeror and the offeree have discharged whatever duties they bear under the FRAND contract between the SEP holder and the SSO. What is called "holdout" is a manifestation of the failure of the controlling law to declare in a timely manner that the contract negotiations have become futile, such that the SEP holder has discharged its contractual duty to the SSO (and to the implementer as the third-party beneficiary of the SEP holder's FRAND contract with the SSO). Calling the problem "holdout" supplies an epithet, but it does nothing to help answer the legal or economic question; to the contrary, that nomenclature is arguably counterproductive in the sense that it falsely suggests that the SEP holder must make some further evidentiary showing that "holdout" has occurred before it may pursue its legal remedies under the national law of the country that issued the patents in suit.

To begin the task of reducing legal and economic ambiguity concerning the determination of whether a SEP holder and an implementer have conducted a FRAND licensing negotiation in good faith, I propose here the formulation of a specific activity rule and a specific closing rule when American contract jurisprudence does *not* control interpretation of the FRAND contract in question.

[15] HTC Corp. v. Telefonaktiebolaget LM Ericsson, 12 F.4th 476, 484 (5th Cir. 2021) ("[The] FRAND commitment 'is "governed by the laws of France," and is "solely [] contractual [in] nature."'") (alterations in original) (quoting HTC Corp. v. Telefonaktiebolaget LM Ericsson, No. 6:18-CV-00243-JRG, 2019 WL 126980, at *3 (E.D. Tex. Jan. 7, 2019) (quoting European Telecommunications Standards Institute, ETSI Intellectual Property Rights Policy, Annex 6, § 12 (Apr. 3, 2019) [hereinafter ETSI IPR Policy], www.etsi.org/images/files/IPR/etsi-ipr-policy.pdf)).

[16] Institute of Electrical and Electronics Engineers [IEEE], IEEE-SA Standards Board Bylaws § 3, at 3 (Feb. 2022), https://standards.ieee.org/content/dam/ieee-standards/standards/web/documents/other/sb_bylaws.pdf.

My proposed activity rule is that, in each round of offer and counteroffer – and to the extent that the SEP holder has not already discharged its contractual obligation to ETSI (such as by its already having made a legitimately FRAND offer at the very outset of the negotiation) – a party must revise its bid or ask price by the minimum agreed-upon increment for that party to be deemed still to be negotiating in good faith. My proposed closing rule is that a party will be deemed to have made its final offer or counteroffer if it does not, within a commercially reasonable amount of time after receiving an offer or counteroffer, sweeten its price relative to its price in the previous round of offer and counteroffer. These rules of market design are proposals, which will surely benefit from scrutiny and refinement by others, but these proposals should suffice to invite a needed discussion.

Considering how controversial and how consequential these issues have been in the licensing of SEPs for smartphones, I see no reason why they will prove to be simpler to resolve in the licensing of SEPs for connected cars, smart homes, and the multitude of other 5G devices that will constitute the Internet of Things.

Parties in litigation over SEPs in the United States and England routinely solicit the expert opinions of scholars on French contract law to assist the court's interpretation of a SEP holder's FRAND contract with ETSI. Yet, as I have remarked elsewhere, the public trial testimony, expert reports, and judicial decisions describing those expert opinions on French law cast doubt on the determinacy of French contract law.[17] An American observer might conclude that French contract law on its own is incapable of defining good faith, or at least that it is ill equipped to supply a definition. Principles defining good faith in the evidentiary records of these cases are simply not percolating through into the public domain to shed light on how parties should behave or how judges should judge. Economic insights from the field of mechanism design can cure that indeterminacy.

### 1. How Have EU Courts Interpreted the FRAND and RAND Obligations?

Notwithstanding the variation in due care as to the source of and content of a given FRAND or RAND obligation, what is clear is that US courts almost uniformly have been reluctant to conclude that the breach of a FRAND or RAND obligation amounts to a violation of antitrust law.

Contrast the experience in the United States with the experience in the European Union. Having become a favored jurisdiction for SEP enforcement, Germany has established itself as the lodestar in the European Union for the development of jurisprudence on the meaning of FRAND. Courts in Germany conclude that FRAND and RAND are synonymous, and that FRAND is simply a holistic concept ultimately rooted in EU competition law (as opposed to the particular contract between a particular holder of SEPs and a particular SSO acting on behalf of

[17] Sidak, *Judge Selna's Errors*, *supra* note 4, at 104–07.

implementers, who are the intended third-party beneficiaries of that contract). When an implementer in patent litigation in Germany presents its statement of defense, it typically frames its arguments concerning violation of the FRAND obligation as emanating from Article 102 of the Treaty on the Functioning of the European Union (TFEU).[18] The differing interpretations of particular FRAND or RAND obligations under contract law collapse to a uniform interpretation of FRAND as defined by the Court of Justice of the European Union or national courts (or some combination thereof). This source of law or authority for purposes of interpreting the duties of a particular SEP holder pursuant to a specific FRAND or RAND contract is unsatisfying to an American lawyer or jurist, who might be tempted to ask: What treaty or statute or regulation or court decision made these variegated FRAND and RAND obligations synonymous with the jurisprudence of Article 102?

The approach in the European Union is quite different from saying, as one does in the United States, that the FRAND obligation is a contract that is unique to the SEP holder and the SSO; that that contract is not a uniform statute or even a standard-form contract; and that interpretation of that contract is influenced by choice of law. The stark fork in the road between the understanding of the FRAND obligation in Anglo-American law versus EU law has important practical consequences for licensing and for dispute resolution. Rarely (if ever) do courts in the European Union make a finding of economic fact that a particular offer is (or is not) FRAND because it does (or does not) repose within the bargaining range between a SEP holder and a particular implementer. To avoid making such detailed findings of economic fact, courts in the European Union tend instead to analyze whether the antiphonal negotiating conduct of the licensor and the potential licensee exhibits the good faith that *Huawei v. ZTE* requires.[19]

In the United States, advocates cannot properly treat every FRAND or RAND obligation as uniform and fail to specify the precise legal and economic questions that the court must answer. Expert economic testimony on whether a patent holder has discharged its FRAND or RAND obligation is relevant in a strict evidentiary sense only if the expert's instructions are sensitive to the language contained in a given SSO's bylaws. In contrast, in Germany, the "fair" and "reasonable" components of the FRAND obligation merge and require analysis of whether the SEP holder's offer falls within the bargaining range – or whether the SEP holder's conduct is "fair" and "reasonable" within the meaning of *Huawei v. ZTE*. With respect to the "nondiscrimination" component of the FRAND obligation, it seems likely that EU courts will uniformly engraft the meaning that "discrimination" has been given within the jurisprudence applying Article 102(c) of the TFEU onto every

[18] Consolidated Version of the Treaty on the Functioning of the European Union art. 102(c), Oct. 26, 2012, 2012 O.J. (C 326) 47.

[19] Huawei Techs. Co. v. ZTE Corp., Case C-170/13, ECLI:EU:C:2015:477 (July 16, 2015).

FRAND or RAND contract. That is, courts in the European Union will likely find that a SEP holder's offer to license is discriminatory only if the offer (1) treats similarly situated parties in an unjustifiably dissimilar manner, and (2) the dissimilar treatment proximately causes a distortion of competition in a relevant market.

In FRAND litigation and arbitration, an attorney instructing an expert economic witness should tailor his instructions to ensure that the economic evidence is relevant to the mode of interpretation of FRAND that the particular court will apply. Doing so will avoid the presentation of expert testimony that is ultimately unhelpful to the finder of economic fact. Law firms must take care to give rudimentary instructions on the applicable legal principles within which the expert economic witness must frame an economic opinion if that opinion is to be relevant in an evidentiary sense to the question that the finder of fact must decide – as well as being not prejudicial, confusing, or misleading. Once opposing counsel has shown that a party has incorrectly instructed its economic expert on the controlling law (or has failed to correct the expert's own misconception of the controlling law), that expert economic witness cannot blithely ignore, or feign ignorance of, the controlling meaning of FRAND in a particular court.

Courts less accustomed to expert economic testimony (and which do not allow live cross-examination at trial, as is the case in Germany) need to scrutinize such testimony with greater skepticism and rigor than currently appears to be the norm. Does the expert economic witness tacitly hold the condescending view that legal distinctions are insignificant details that do not rise to being relevant considerations that properly inform an intellectually rigorous economic analysis, even when that analysis has been undertaken specifically for the purpose of assisting the court's resolution of a consequential legal dispute? Regardless of whether the expert economic witness can genuinely profess ignorance, plainly the eminent law firm responsible for retaining and instructing the expert cannot profess such ignorance in good faith.

### 2. Is There a Duty to Negotiate in Good Faith?

No American court deciding a dispute over SEPs has publicly explained the origin in French law of the duty to negotiate in good faith. In particular, no American court has explained why the implementer – as *the third-party beneficiary* to the SEP holder's FRAND contract with ETSI, not a party to that contract – has a duty to negotiate in good faith.[20] Indeed, judges and commentators take as given that the duty to negotiate in good faith applies symmetrically to the third-party beneficiary before contract formation between the SEP holder and the third-party beneficiary of

[20] Sidak, *Judge Selna's Errors*, *supra* note 4, at 102–07.

the SEP holder's FRAND contract with ETSI.[21] Every judicial opinion is a public good that can shed light on the law.[22] If the proposition is uncontroversial that a duty of good faith and fair dealing applies to the negotiation between a SEP holder and an implementer of an ETSI standard, then it would be helpful for judges to explain for the benefit of those less perspicacious than they why and whence that duty arises.[23]

Reading ETSI's FRAND contract alongside the Civil Code of France, which specifies in Article 1104 that "[c]ontracts must be negotiated, formed and performed in good faith,"[24] supports at least the following conclusions concerning the duty to negotiate in good faith. For at least four reasons, these conclusions are not to the exclusion of other conclusions that might follow from French law or the law of other nations.

First, to the extent that the FRAND contract between the SEP holder and ETSI is enforceable under French law, the SEP holder must perform the obligations that arise from that FRAND contract in good faith. That is, the SEP holder has an obligation to ETSI to act in good faith in its preparedness to grant a license to an implementer that qualifies as an intended third-party beneficiary. The same conclusion would likely apply under American contract law, because the SEP holder that is offering to license its SEPs to an implementer is performing its contractual obligations under the FRAND contract; and American contract law, which of course is state law and therefore might vary across the United States, generally provides that parties must perform their contractual duties in good faith.[25]

Second, to the extent that the implementer is itself a SEP holder that has entered into a FRAND contract with ETSI, the implementer also has a duty under French

[21] In 2015, I observed:

> With respect to contract performance and enforcement, the Restatement [of Contracts] says that "[e]very contract imposes *upon each party* a duty of good faith and fair dealing . . ." It is not clear why this symmetry of obligations should give way to asymmetry of obligations at the stage of contract formation, assuming that a court is inferring that the common law duty of good faith and fair dealing encompasses contractual negotiations.

J. Gregory Sidak, *The Meaning of FRAND, Part II: Injunctions*, 11 J. COMPETITION L. & ECON. 201, 217 n.67 (2015) (emphasis added) (citation omitted) (quoting RESTATEMENT (SECOND) OF CONTRACTS § 205 (AM. LAW INST. 1981)).

[22] POSNER, *supra* note 11, at 760–62.

[23] Sidak, *Judge Selna's Errors*, *supra* note 4, at 107.

[24] Code Civ. [Civ. Code] art. 1104 (Fr.), *translated in* John Cartwright, Bénédicte Fauvarque-Cosson, & Simon Whittaker, *The Law of Contract, The General Regime of Obligations, and Proof of Obligations* art. 1104 (2016). "This provision is a matter of public policy." *Id.* This translation of the "provisions of the *Code civil* created by *Ordonnance* n° 2016-131 of 10 February 2016 . . . was commissioned by the *Direction des affaires civiles et du sceau, Ministère de la Justice, République française*." *Id.* at 1.

[25] Northwest, Inc. v. Ginsberg, 572 U.S. 273, 295 (2014) ("[M]ost States recognize some form of the good faith and fair dealing doctrine."); Sidak, *Judge Selna's Errors*, *supra* note 4, at 105–06 & nn. 23–27.

contract law to negotiate in good faith. Such a duty would typically arise when counterparties negotiate a cross-license, wherein the implementer offers to license to the SEP holder the implementer's own SEPs that are subject to a FRAND commitment to ETSI. For example, in *HTC v. Ericsson*, Chief Judge Rodney Gilstrap of the Eastern District of Texas found that HTC, the implementer, had a duty to negotiate a cross-license in good faith because it was also a SEP holder that had entered into its own FRAND contract with ETSI.[26] Again, American contract law would support a similar conclusion. An implementer that itself had executed a FRAND contract (because it is also a SEP holder) has a contractual duty to perform in good faith the obligations that it accepted pursuant to that FRAND contract.[27]

Third, to the extent that the negotiation over SEPs between the SEP holder and the implementer culminates in the execution of a license agreement that is subject to the Civil Code of France (or to some equivalent law of a different nation that imposes a duty comparable to Article 1104 of the Civil Code of France), both the SEP holder and the implementer are obligated to negotiate their contract in good faith, because the Code explicitly directs parties to act in good faith when *negotiating* a contract.

Fourth, to the extent that the FRAND contract between the SEP holder and ETSI is properly characterized as *un accord de principe* (an agreement in principle), as a matter of French law, the *accord de principe* might impose the duty on both the SEP holder and the implementer to negotiate in good faith.[28]

[26] HTC Corp. v. Telefonaktiebolaget LM Ericsson, No. 6:18-CV-00243-JRG, 2018 WL 6617795, at *5–6 (E.D. Tex. Dec. 17, 2018).

[27] It is worth questioning whether the negotiation of a license to SEPs declared essential to ETSI is beyond the scope of the *performance* of the SEP holder's FRAND contract with ETSI to be "prepared to grant" a license, and thus outside the scope of ETSI's choice-of-law provision. I leave the task of answering this question to others more familiar with French law.

[28] Certain Wireless Devices with 3G Capabilities and Components Thereof, Inv. No. 337-TA-800, Initial Determination at 422 (USITC June 28, 2013) ("The parties agree that the ETSI IPR Policy is governed by French law. Under French law, the type of obligation set forth in the ETSI undertaking is best described as *un accord de principe* (agreement in principle). This imposes on both negotiating parties a duty to negotiate in good faith. It does not, however, impose an obligation actually to conclude a contract . . . In this regard, French law is consistent with U.S. contract law, under which a generalized 'agreement to agree' is unenforceable, but parties may enter into binding agreements to negotiate.") (citations omitted); Certain LTE- and 3G-Compliant Cellular Communications Devices, Inv. No. 337-TA-1138, Open Sessions Volume IV at 1106:8–21 (USITC Sept. 17, 2019) (McNamara, ALJ) (Testimony of Bertrand Fages) ("Q [by Counsel for INVT SPE LLC, the SEP holder and complainant]. What is an *accord de principe*, or agreement in principle under French law? A. An agreement in principle is characterized by the fact that it entails no obligation to contract, but only an obligation to negotiate in good faith; and this obligation to negotiate in good faith is only an obligation of means that permits behavior, consists of having an attitude in order to achieve successful negotiations by conducting them fairly, and we have a decision from the Paris Court of Appeal that puts it very clearly by saying that there is no obligation to conclude but only a commitment to negotiate."); BELL et al., *supra* note 12, at 305 ("Sometimes parties to contractual negotiations make preliminary agreements before concluding any ultimate contract. French law's attitude to these is generally more favourable than is English law's owing in particular to the absence of

However, if the SEP holder and an implementer are not negotiating in the shadow of the Civil Code of France, and if that implementer has not entered into its own binding FRAND contract with ETSI, it is far from apparent what source of law or what equitable principle would force the implementer to negotiate in good faith. In particular, I am aware of no state in the United States whose contract law imposes a general duty to negotiate in good faith, as does Article 1104 of the Civil Code of France. As Judge Posner explained in 1991, the general contractual duty to negotiate before contract formation in good faith in the United States is so vanishingly small as to be virtually nonexistent.[29] In other words, under American contract law principles, an implementer has no duty to negotiate in good faith a license agreement for FRAND-committed SEPs. If the FRAND agreement does not create a contractual duty to negotiate a

the requirement of consideration ... A very important example of pre-contractual agreements which are enforced as contracts are 'unilateral promises to contract' (*promesses unilatérales de contrat*). Here, a person promises to contract on particular terms with another at the latter's option, this promise being binding once accepted. Of more uncertain status are 'agreements in principle' (*accords de principe*) which usually involve an agreement by the parties on certain matters and that they will continue to negotiate towards final contract ... Even in the absence of any preliminary agreement as to the course or conduct of negotiations, French law holds the parties to a standard of proper conduct, [which is] referred to either positively in terms of the requirements of good faith or negatively in terms of the parties having 'abused their right' to break off negotiations before a contract is concluded.") (footnotes omitted).

29 Market St. Assocs. Ltd. v. Frey, 941 F.2d 588, 593–94 (7th Cir. 1991) ("In fact the law contemplates that people frequently will take advantage of the ignorance of those with whom they contract, without thereby incurring liability. The duty of honesty, of good faith even expansively conceived, is not a duty of candor. You can make a binding contract to purchase something you know your seller undervalues.") (citations omitted); *id.* at 594 ("But it is one thing to say that you can exploit your superior knowledge of the market – for if you cannot, you will not be able to recoup the investment you made in obtaining that knowledge – or that you are not required to spend money bailing out a contract partner who has gotten into trouble. It is another thing to say that you can take deliberate advantage of an oversight by your contract partner concerning his rights under the contract. Such taking advantage is not the exploitation of superior knowledge or the avoidance of unbargained-for expense; it is sharp dealing. Like theft, it has no social product, and also like theft it induces costly defensive expenditures, in the form of overelaborate disclaimers or investigations into the trustworthiness of a prospective contract partner, just as the prospect of theft induces expenditures on locks."); *id.* at 595–96 ("The emphasis we are placing on postcontractual versus precontractual conduct helps explain the pattern that is observed when the duty of contractual good faith is considered in all its variety, encompassing not only good faith in the *performance* of a contract but also good faith in its *formation* and in its *enforcement*. The formation or negotiation stage is precontractual, and here the duty is minimized. It is greater not only at the performance but also at the enforcement stage, which is also postcontractual ... At the formation of the contract the parties are dealing in present realities; performance still lies in the future. As performance unfolds, circumstances change, often unforeseeably; the explicit terms of the contract become progressively less apt to the governance of the parties' relationship; and the role of implied conditions – and with it the scope and bite of the good-faith doctrine – grows.") (emphasis in original) (citations omitted).

license in good faith, then it appears American contract law will not itself create such a duty.[30]

Despite not having a legal obligation to negotiate in good faith under American contract law principles, an implementer that fails to negotiate in good faith might nevertheless face legal consequences for such conduct. For example, a US court might order the implementer to pay enhanced damages for willful infringement of SEPs.[31] Similarly, an implementer that fails to negotiate in good faith might forfeit its right, as an intended third-party beneficiary of a FRAND contract, to receive a FRAND offer. As a separate matter, whether or not the implementer has a preexisting duty to negotiate in good faith with the SEP holder,[32] the implementer could also incur liability in tort law for fraud or deceit.

### 3. Why Does It Matter Whether a Duty Exists to Negotiate in Good Faith?

My purpose in this chapter is not to attempt to answer the preceding questions concerning choice-of-law principles and the source of the possible duty, borne by a third-party beneficiary of the SEP holder's FRAND contract with ETSI, to negotiate a license to the committed SEPs in good faith. Instead, I simply and briefly expose the ambiguity of French law concerning the certainty of whether and when contract formation has occurred in FRAND cases, as well as the substantive implications of that ambiguity.[33] And I now submit that the solution to overcoming that ambiguity lies in reframing the problem as one of efficient market design.

30 I do not attempt to answer the question of whether the duty to negotiate in good faith under contract law (relevant to the license negotiation) is different (narrower or broader) than the contractual duty of good faith negotiation that is presumed to exist under the FRAND contract.

31 J. Gregory Sidak, *Enhanced Damages for Infringement of Standard-Essential Patents*, 1 Criterion J. Innovation 1101 (2016).

32 Furthermore, to the extent that a court were to find ETSI's FRAND commitment *not* to be contractual, there is reason to doubt that the SEP holder's promise could have binding effect under French contract law, which does not recognize promissory estoppel as an available claim. In 2019, Yves-Marie Laithier, a professor of French contract law at the Université Paris 1 Panthéon-Sorbonne, testified in a pretrial hearing in u-blox v. InterDigital that "[t]he doctrine of promissory estoppel is unknown in French contract law. It is indeed untranslatable in French." Declaration of Prof. Yves-Marie Laithier in Support of Motion to Dismiss at 2, u-blox AG v. InterDigital Inc., No. 19-cv-00001-CAB-BLM (S.D. Cal. Feb. 25, 2019), ECF No. 50-5. That conclusion comports with the understanding that promissory estoppel "is peculiar to common law systems." *Id.* (quoting John Cartwright, Formation and Variation of Contracts § 10-01 (2d ed. 2018)).

33 That ambiguity is exemplified by the opposing expert reports of two French law scholars filed in late 2018 and early 2019 in one such case. *Compare* Expert Report of Dr. Philippe Stoffel-Munck ¶ 121, at 29, Koninklijke KPN N.V. v. Sierra Wireless, Inc., No. 17-090-LPS-CJB (D. Del. Nov. 20, 2018) ("[T]he Declarant must grant a license on FRAND terms and must negotiate in good faith irrespective of the outcome of their negotiations and, a fortiori, irrespective of the binding force that their future agreement will be given or not in retrospect. In any event, they must answer for any loss caused to the other party by any breach of their duty to grant a license on FRAND terms and to negotiate in good faith."), ECF No. 205-1, *and*

That such questions would arise in SEP disputes tells us that a rift exists between American contract law and French contract law on matters of contract formation, and that the existence of that divide, much less the depth of its economic significance, again, has eluded courts and scholars. American law, it would seem, differs from the law of much of the rest of the world, evidently even common law jurisdictions, with respect to the negotiation and formation of a contract. American law presents a regimented view of when offer, acceptance, and contract formation each occurs. A more European approach evidently envisions an elongated process permitting multiple rounds of offer and counteroffer. Yet that European process lacks any explicit rule for determining whether a given offer or counteroffer is sufficiently sweetened, relative to the prior offer or counteroffer, to contribute materially to closing the bid–ask spread separating the parties. And that European process also lacks any explicit rule declaring when the negotiation must end because the parties have reached an impasse and therefore deserve to have it recognized as a matter of law that they have failed in their efforts to form a contract.

The quiddity of this characteristic, which materially differs in degree between American contract law and European contract law, I will call *expedition*. It neatly illustrates how Americans and citizens of other advanced nations sometimes understand quite differently a concept so foundational to legal or economic reasoning that it is commonly presumed to admit no dispute. Expedition is the impatient foot tapping of the marketplace. The enemy of indecision, dithering, sloth, torpor, and indolence, expedition despises dilatory guile and circumlocution. In the arena of commerce and all its works and days of hands, expedition is how one acts upon Seneca's admonition: "It is not that we have a short time to live, but that we waste a lot of it."[34]

Being expeditious in the licensing of SEPs increases economic welfare in both the short run and the long run. In the short run, expedition reduces ambiguity, facilitates contract formation, and reduces the need to resort to litigation (as well as the opportunity to use litigation for strategic reasons). In the long run, in the context of licensing SEPs, expedition hastens the creation of consumer surplus and producer surplus from the commercialization of products practicing a new voluntary standard.

The presence or absence of efficient activity rules and closing rules could spell the difference between an SSO's success or its failure and withering away. There has

Reply Expert Report of Dr. Philippe Stoffel-Munck ¶ 77, at 16, Sierra Wireless, No. 17-090-LPS-CJB (D. Del. Feb. 3, 2019) ("Where the negotiations form part of a binding agreement, the duty of good faith becomes more demanding, as comes into play the duty to *perform* in good faith.") (emphasis in original), ECF No. 205-1, *with* Rebuttal Expert Report of Dr. Vernon Valentine Palmer ¶ 133, at 39–40, Sierra Wireless, No. 17-090-LPS-CJB (D. Del. Jan. 8, 2019) ("[T]o the extent any good faith obligation exists, it is simply one to avoid committing a clear abuse of the liberty to negotiate freely – an obligation that would not be understood to require that a debtor put aside its own interests in favor of another party.").

34 SENECA, ON THE SHORTNESS OF LIFE 1 (C.D.N. Costa trans., 1997) (49 A.D.).

been an evident poverty of foresight in regard to designing the *end* of the process of licensing SEPs. In the absence of some other scapegoat coming forward, I will blame that deficiency of market design on the engineers for having failed to recognize the economic and legal significance of setting in place the mechanism necessary to ensure the expeditious completion of negotiations between a SEP holder and third-party beneficiaries of the SEP holder's FRAND contract. Although we can infer that the vast majority of bilateral negotiations for the licensing of SEPs produce successful commercial agreements, those that do not have cost billions of dollars in litigation over the past decade.

### B. *Activity Rules, Closing Rules, and "Best Practices" in Negotiations over FRAND-Committed SEPs*

I have previously argued that time is of the essence in the implementation of a standard – in particular because to waste time in the introduction of an entirely new generation of products featuring standard-dependent technological innovations is to harm the public interest by sacrificing consumer surplus irreparably.[35] To the extent that one can properly impute a duty (or covenant) to negotiate in good faith to an intended third-party beneficiary of ETSI's FRAND contract with a particular SEP holder, that duty reflects the understanding that a public interest inheres in the expeditious negotiation of SEP licenses. Whether the implementer's behavior after receiving a legitimately FRAND offer adheres to the standard of good faith will depend ultimately on how quickly the implementer seeks to close the bid–ask spread and converge on an agreement – which is to say, contract formation.[36] Following that interpretation to its logical conclusion, the point at which the implementer ceases to sweeten its counteroffer from one round of the negotiation to the next defines the point of impasse.

Implicit in this rule is the understanding that the parties also must define how long a given round lasts during their negotiation. How long may a party take to sweeten its bid or ask? If the parties provide no answer of their own to this question, the default answer becomes "a commercially reasonable amount of time." But, rather than have a court rule what amount of time is commercially reasonable, the parties can create considerable value by agreeing on a framework that is both more precise and more expeditious than what is merely commercially reasonable.

[35] J. Gregory Sidak, *What Makes FRAND Fair? The Just Price, Contract Formation, and the Division of Surplus from Voluntary Exchange*, 4 Criterion J. Innovation 701, 725–27 (2019) [hereinafter *What Makes FRAND Fair?*]; Sidak, *The FRAND Contract*, supra note 9, at 13–14 & n.47 (citing J. Gregory Sidak, *Irreparable Harm from Patent Infringement*, 2 Criterion J. Innovation 1 (2017); J. Gregory Sidak, *Is Harm Ever Irreparable?*, 2 Criterion J. Innovation 7, 10 (2017) (Inaugural Address for the Ronald Coase Professorship in Law and Economics, Tilburg University, Tilburg, The Netherlands (Sept. 16, 2011)).

[36] Sidak, *The Meaning of FRAND, Part II: Injunctions*, *supra* note 21, at 218.

The obvious analogy here is to the activity charge required of a bidder to maintain its right to keep bidding in the simultaneous multi-round ascending auction for 3G spectrum licenses in the United Kingdom. Economist Paul Klemperer of Oxford, who advised the UK government, explains:

> Our design entailed multiple rounds of simultaneous bids. In the first round, each bidder makes a bid on one license of its own choice. To remain in the auction, a bidder must be "active" in every subsequent round. An active bidder either currently holds the top bid on a particular license, or else raises the bid on a license of the bidder's choice by at least the minimum bid increment. A bidder who is inactive in any round is eliminated from the rest of the auction.[37]

As I observed at the opening of this part, governmental agencies around the world have promoted "best practices" in negotiations over FRAND-committed SEPs. The most useful thing left undone in such statements of best practices is to endorse the concept of an activity charge for good faith FRAND negotiations, and then to identify an unambiguous economic methodology for determining the minimum bid increment by which an implementer must sweeten its counteroffer to the SEP holder's legitimately FRAND offer for the implementer to be deemed still to be negotiating in good faith.

The next most useful thing left undone in statements of best practices for good faith FRAND negotiation is to identify an unambiguous economic methodology for determining when the negotiation has ended in failure. Again, it bears emphasis that this question is legally relevant only for FRAND obligations not controlled by American-style contract principles of offer and acceptance – which, I have explained earlier, inherently have the admirable (but evidently underappreciated) quality of unambiguously defining a closing rule for a bilateral negotiation.[38] Evan Kwerel, a highly respected economist who spent a career at the Federal Communications Commission and made important contributions to the design and execution of American spectrum auctions there, explained that "[t]he closing rule was one of the major [market] design issues for a simultaneous auction" for spectrum in the United States.[39] Stanford professors Robert Milgrom and Robert Wilson, who shared the Nobel Prize in economics in 2020 for their work on auction theory, "proposed a simultaneous closing rule whereby the auction closes on all licenses only after a round has passed with no bidding on any license."[40] In contrast, conspicuously absent from the current conception of the FRAND negotiation is any guidance on when it reaches its end in terms of rounds of offer and counteroffer. Instead, the

37 Klemperer, *supra* note 8, at 181–82; Milgrom 2004, *supra* note 8, at 5–6, 14 (discussing the activity rule used in US spectrum auctions in 1994).

38 Sidak, *The FRAND Contract*, *supra* note 9, at 15–19.

39 Evan Kwerel, *Foreword*, *in* Milgrom 2004, *supra* note 8, at xvii.

40 *Id.*; Milgrom 2004, *supra* note 8, at 267 ("The auction *closing rule* is especially important: the [simultaneous ascending] auction ends only after a round in which there are no new bids on any license.") (emphasis in original).

negotiation is like a baseball game with an infinite number of innings, or a poker game in which a player may remain in the hand without calling his opponent's bet.

Because of the failure of SSOs or courts or other institutions in positions of authority to impose both an activity rule and a closing rule, a contentious negotiation for a FRAND license, if not controlled by the law of a jurisdiction having American-style principles concerning contract formation, will regrettably resemble Zeno's Dichotomy paradox. The journey's end becomes ever closer in incrementally smaller half steps, but it is never reached.

## II. WHAT MAKES FRAND FAIR?

SSOs generally permit each SEP holder to set a FRAND or RAND royalty for its SEPs through private bilateral negotiations with each implementer, rather than require the SEP holder to post tariffed rates for all customers. Such voluntary exchange benefits both parties, who divide their aggregate gains from trade, which economists call surplus.[41] This economic principle – that voluntary exchange is mutually beneficial – is as profound as it is simple, and for that reason, economists call it "The Fundamental Theorem of Exchange."[42]

In any negotiation, the total surplus from a successful transaction is equivalent to the bargaining range – the distance between the buyer's maximum willingness to pay and the seller's minimum willingness to accept. Put differently, the gains from trade (that is, the gains from voluntary exchange) consist of the sum of consumer surplus and producer surplus.[43] As Jack Hirshleifer, Amihai Glazer, and David Hirshleifer emphasize in their pellucid undergraduate textbook on price theory, this terminology about consumption and production should not detract from the essential characteristic of voluntary exchange: "The names of these measures are somewhat misleading. The benefits stem *from trading*, not from consuming or producing. Instead of Consumer Surplus and Producer Surplus one should, properly speaking, refer to Buyer Surplus and Seller Surplus."[44] Elsewhere within economics, auction theory uses still other terminology – the reserve price or

[41] J. Gregory Sidak, *Fair and Unfair Discrimination in Royalties for Standard-Essential Patents Encumbered by a FRAND or RAND Commitment*, 2 Criterion J. Innovation 301, 333–34 (2017); J. Gregory Sidak, *Bargaining Power and Patent Damages*, 19 Stan. Tech. L. Rev. 1, 20–22 (2015).

[42] Jack Hirshleifer, Amihai Glazer, & David Hirshleifer, Price Theory and Applications: Decisions, Markets, and Information 203 (7th ed. 2005).

[43] *Id.* at 203–04.

[44] *Id.* at 204 n.4 (emphasis in original); Armen A. Alchian & William R. Allen, Exchange and Production: Competition, Coordination, and Control 48–49 (3d ed. 1983) (demonstrating that the total surplus in a negotiation is the sum of the *seller's* gain from trade and the *buyer's* gain from trade).

reservation price – to identify the same concepts, respectively, of the seller's minimum willingness to accept and the buyer's maximum willingness to pay.[45]

One question regarding the bilateral negotiation of SEPs on FRAND terms has received surprisingly little attention in either court decisions or scholarly writings: What is a *fair* division of the surplus generated by a voluntary negotiation successfully concluded between the SEP holder and the implementer?[46]

## A. *Defining the Fair Division of Surplus*

John Rawls famously argued that "fair" means "just."[47] "Justice as fairness," he asserted, "is an example of ... a contract theory."[48] Rawls argued that "[t]he word 'contract' suggests," among other things, "the condition that the appropriate division of advantages must be in accordance with principles acceptable to *all*

45 IAN STEEDMAN, *Reservation Price and Reservation Demand, in* 4 THE NEW PALGRAVE: A DICTIONARY OF ECONOMICS 158 (John Eatwell, Murray Milgate, & Peter Newman eds., 1987); KLEMPERER, AUCTIONS: THEORY AND PRACTICE, *supra* note 8, at 18, 109, 112; MILGROM 2004, *supra* note 8, at 9–11; ROBERT L. PHILLIPS, PRICING AND REVENUE OPTIMIZATION 46 (Stanford University Press 2005); ROBERT S. PINDYCK & DANIEL L. RUBINFELD, MICROECONOMICS 393, 510–11 (9th ed. 2018); JOHN G. RILEY, ESSENTIAL MICROECONOMICS 451–56 (2012); GARRETT J. VAN RYZIN, *Models of Demand, in* THE OXFORD HANDBOOK OF PRICING MANAGEMENT 340, 342–43 (Özalp Özer & Robert Phillips eds., 2012).

46 Two handbooks published by the Cambridge University Press concerning SEPs do not explain how fairness constrains the calculation of a FRAND royalty. PATENT REMEDIES AND COMPLEX PRODUCTS: TOWARD A GLOBAL CONSENSUS (C. Bradford Biddle et al. eds., 2019); THE CAMBRIDGE HANDBOOK OF TECHNICAL STANDARDIZATION LAW: COMPETITION, ANTITRUST, AND PATENTS (Jorge L. Contreras ed., 2017).

47 JOHN RAWLS, A THEORY OF JUSTICE 11 (rev. ed. 1999) (1971); JOHN RAWLS, JUSTICE AS FAIRNESS: A RESTATEMENT (Erin Kelly ed. 2001); John Rawls, *Justice as Fairness: Political Not Metaphysical*, 14 PHIL. & PUB. AFF. 223 (1985). Rawls' biographer, Thomas Pogge, reminds us that Rawls began A THEORY OF JUSTICE with this proposition: "Justice is the first virtue of social institutions." THOMAS POGGE, JOHN RAWLS: HIS LIFE AND THEORY OF JUSTICE 28 (Michelle Kosch trans., 2007) (quoting RAWLS, A THEORY OF JUSTICE, *supra* note 47, at 3). Pogge then explains how this proposition relates to the making and keeping of promises. By referring to "social institutions,"

> Rawls means to refer to the practices and rules that structure relationships and interactions among agents. This sense [of Rawls' use of "social institutions"] is exemplified by a social institution of promising. Its rules lay down what interactions between two agents count as creating a promise, what promisee conduct (if any) counts as releasing the promisor from the promise, what circumstances (if any) can be invoked as justification or excuse for nonperformance, and so on.

*Id.* By Pogge's account, Rawls' *Theory of Justice* has greater relevance to contract interpretation than might immediately appear to be the case for a book on political philosophy.

48 RAWLS, A THEORY OF JUSTICE, *supra* note 47, at 14; *id.* at 14–15 ("The merit of the contract terminology is that it conveys the idea that principles of justice may be conceived as principles that would be chosen by rational persons, and that in this way conceptions of justice may be explained and justified.").

parties."[49] One would expect the same of the FRAND contract. How can we go about imputing to the fairness component of the FRAND contract – a meaning that is intellectually rigorous in both legal and economic respects?

This question of the meaning of a fair price turns out to have very real legal ramifications in the present day. Rarely do I disagree with Judge Posner, but I do with respect to his view that "fair" is surplusage in the FRAND contract. Judge Posner, sitting by designation as the trial judge in *Apple, Inc. v. Motorola, Inc.* in 2012 in the Northern District of Illinois said that, in the context of FRAND, "the word 'fair' adds nothing to 'reasonable' and 'nondiscriminatory.'"[50] My previous writings have followed this convention of making no legal or economic distinction between FRAND and RAND royalties, though I have never excluded the possibility that someone might eventually make a compelling argument for why "fair" is not a throwaway word for parties to insert into a contract.[51] And so, for example, I have previously analyzed at length the differences between actual FRAND contracts and actual RAND contracts with respect to how fairness creeps into the constraint to license SEPs on nondiscriminatory terms.[52] This part of this chapter will show why courts should take the distinction between FRAND contracts and RAND contracts more seriously.

More than 30 years ago, Robert Frank of Cornell University proposed a precise economic definition that is directly relevant to the question of what makes a FRAND royalty fair:

> Using the notions of reservation price and surplus, we can construct the following operational definition of a fair transaction: A *fair transaction is one in which the surplus is divided (approximately) equally. The transaction becomes increasingly unfair as the division increasingly deviates from equality.*[53]

Frank then explained the problem that unfairness presents: "People will sometimes reject transactions in which the other party gets the lion's share of the surplus, even though the price at which the product sells may compare favorably with their own reservation price."[54]

[49] *Id.* at 15 (emphasis added). I have elsewhere traced my theory of fairness to the writings on the just price by Saint Thomas Aquinas. Sidak, *What Makes FRAND Fair?*, *supra* note 35, at 701–02, 710–21.

[50] 869 F. Supp. 2d 901, 912 (N.D. Ill. 2012), *rev'd in part on other grounds*, 757 F.3d 1286 (Fed. Cir. 2014). Judge Marsha Berzon previously wrote for the Ninth Circuit that a FRAND obligation was "legally equivalent" to a RAND obligation. Microsoft Corp. v. Motorola, Inc., 696 F.3d 872, 877 n.2 (9th Cir. 2012).

[51] Sidak, *Fair and Unfair Discrimination in Royalties for Standard-Essential Patents Encumbered by a FRAND or RAND Commitment*, *supra* note 41, at 308.

[52] *Id.* at 308–11.

[53] Robert H. Frank, Passions within Reason: The Strategic Role of the Emotions 165 (1988) (emphasis in original).

[54] *Id.* at 167 (emphasis suppressed).

This reasoning is very close to the conclusion I had reached before benefiting, late in the process of revising an earlier incarnation of this text over the course of several years, from reading Frank's 1988 book. Frank and I each find ourselves using Judge Posner as our foil, though for different reasons. Frank criticized Judge Posner's writings through the mid-1980s, as denying what Frank argued was the considerable explanatory power of fairness considerations in law and economics.[55] In contrast, I gently chide Judge Posner for overlooking roughly 25 years later that, by the private ordering of contract law, some SSOs had chosen to impose an obligation of fairness so that (according to my economic interpretation) those SSOs could nudge parties into exercising the degree of moderation in their negotiation demands that is necessary to achieve contract formation reliably and expeditiously.

The irony is that my interpretation of why the word "fair" must have an independent meaning within the FRAND contract is quintessentially Posnerian: A division of surplus that is perceived by both parties to be fair maximizes the probability of contract formation over some defined time horizon, which in turn immediately benefits the parties to the contract. Thus, fairness clearly promotes static allocative efficiency. Moreover, across time the fairness constraint on the division of surplus also benefits countless consumers, whom the grand edifice of the FRAND contract is surely intended to benefit (though not necessarily by the formal machinery of conferring on those consumers legally enforceable rights of a third-party beneficiary, as the FRAND contract does confer on implementers). As Joseph Schumpeter taught us, it is the consumption of innovative products in the future that delivers radical – not marginal – gains in consumer surplus.[56] Thus, the fairness constraint promotes dynamic efficiency as well.

[55] Frank argues that in "the self-interest model," which he identifies with Judge Posner,

> the division of the surplus simply plays no role in determining whether a transaction will take place. It will occur provided each party gets some positive share of the surplus, no matter how small. When Posner says fairness "has no content," this feature of the traditional model must be at least in part what he has in mind. Yet, as we will presently see, concerns about fairness repeatedly cause people to reject transactions with positive surplus.

*Id.* (attributing Judge Posner's quotation to Paul Barrett, *Influential Ideas: A Movement Called "Law and Economics" Sways Legal Circles*, WALL ST. J. 1, 16 (Aug. 4, 1986). Frank argues that Judge "Posner and other rationalists would hardly deny that people *say* they care about fairness [,] [b]ut hardheaded economists treat such statements as mere verbiage, devoid of any power to predict behavior." *Id.* (emphasis in original).

[56] J. Gregory Sidak & David J. Teece, *Dynamic Competition in Antitrust Law*, 5 J. COMPETITION L. & ECON. 581, 602 (2009) ("Dynamic competition is powered by the creation and commercialization of new products, new processes, and new business models. As [Joseph] Schumpeter said, competition fueled by the introduction of new products and processes is the more powerful form of competition: 'competition from the new commodity, the new technology, the new source of supply, the new type of organization – competition which commands a decisive cost or quality advantage and which strikes not at the margins of the profits and the output of existing firms, but at their foundations and their very lives.' Advocates of strong competition policy must surely favor dynamic competition, for static competition is anemic in

In this respect, Posner's emphasis on efficiency and Frank's emphasis on fairness are reconcilable. A lopsided division of surplus is a cost imposed on efficient transactions to the extent that it prevents some otherwise promising negotiations from achieving successful contract formation; if that cost can be eliminated or mitigated, a larger number of efficient transactions will occur. Therefore, regardless of whether one prefers to call it a quest for fairness or a quest for efficiency, an SSO's constraint on the SEP holder that a royalty for its SEPs be fair is a privately ordered feature of contract – a self-imposed cattle prod – that contributes to a result that proponents of fairness and proponents of efficiency can both applaud.

## B. *Fairness and Contract Formation*

One can formalize a simple theory of fairness and contract formation. Imagine a decision tree depicting the expected surplus of a contract negotiation as the sum of the expected values of two mutually alternative outcomes: $EV = pS + (1 - p)0$, where $p$ is the probability of contract formation and S is the surplus created by a successful transaction. The size of the surplus S is separately identified by the bargaining range, which is bounded by the reservation prices of the parties to the negotiation. But the expected value of the surplus is necessarily smaller than S because the division of the surplus might cause one of the parties to walk away. A simple and intuitive formulation of the relationship comes from defining as $R$ the ratio of the seller's share of the surplus (X) to the buyer's share of the surplus (Y): $R = X / Y = X / (1 - X)$. $R$ is bounded below by zero and above by infinity. As $R$ approaches zero, $p$ approaches zero. As $R$ approaches infinity, $p$ again approaches zero. In either case, it becomes more likely that contract formation will fail, and consequently the parties will forfeit the surplus from the transaction.

At this point, it is instructive to consider the Ultimatum Game, a bargaining game in which a player makes a single take-it-or-leave-it offer, rather than multiple offers and counteroffers.[57] The game ends in either an agreement to the unaltered terms of the first offer or no agreement at all. If the second party rejects the offer, neither party benefits – the first party does not keep any portion of the asset but rather forfeits it all. Thus, both parties have an incentive to agree, and the division of surplus (which in this particular game is assumed to be a windfall, not a return on either party's investment) will depend on a fair offer having been made. As I previously explained in 2013, the Ultimatum Game is interesting in analyzing the FRAND contract not because a FRAND negotiation represents an Ultimatum Game.[58] After all, in

comparison.") (quoting JOSEPH A. SCHUMPETER, CAPITALISM, SOCIALISM AND DEMOCRACY 8 (1942)).

57 Richard H. Thaler, *Anomalies: The Ultimatum Game*, 2 J. ECON. PERSP. 195 (1988).

58 J. Gregory Sidak, *The Meaning of FRAND, Part I: Royalties*, 9 J. COMPETITION L. & ECON. 931, 1047 (2013). In 2013, I said that the Ultimatum Game sheds light on the reasonableness of an offer as well. *Id.* I no longer believe that proposition to be correct. I have concluded instead

FRAND licensing, there are typically repeated rounds of offer and counteroffer, the identities of the parties are known (perhaps because the parties have previously negotiated a licensing contract), and the reputation of the players matters because they will face the prospect of repeated play in subsequent licensing over future standards. Instead, the Ultimatum Game is interesting for FRAND licensing because the results of economic experiments based on the Ultimatum Game shed light on which divisions of surplus the parties to a stylized negotiation would consider fair. Surveying the experimental economics literature as it existed in 2000, Ernst Fehr and Simon Gächter reported that "[a] robust result in [the Ultimatum Game] experiment, across hundreds of trials, is that Proposers who offer the Responder less than thirty percent of the available sum are rejected with a very high probability."[59]

If there are any positive spillovers for society as a whole from successful contract formation, as there of course would be if the contract is one for the licensing of patents essential to practice an industry standard, then those externalities are forfeited as well when the negotiation collapses. In contrast to the scenarios of negotiation impasse that I previously described, as $R$ goes to one, $p$ approaches one, and thus contract formation becomes increasingly certain. An impartial spectator nudging the parties to maximize the expected value of the surplus of their contemplated transaction would prescribe "maximize $p$ with respect to $R$," since $S$ is already exogenously determined.

Although any possible division of the surplus created by voluntary exchange is mutually beneficial, that fact does not imply that every price along the bargaining range (which defines the locus of "reasonable" royalties) is equally likely to yield an agreement. How does a given split of the surplus between the SEP holder and the implementer influence the probability of their successful contract formation within a specified period of time? One interpretation of a fair royalty is that it leads more expeditiously to contract formation than some other division of the gains from trade. That is, the fairness component of the FRAND contract between the SEP holder and the SSO takes on independent meaning by giving teeth to the proposition that time is of the essence in achieving contract formation between the SEP holder and the implementer. Fairness promotes economic efficiency in the sense of hastening

that the reasonableness component of the FRAND (or RAND) obligation concerns the identification of the size of the surplus, not the question of how the parties will find it mutually acceptable to divide the surplus. Sidak, *What Makes FRAND Fair?*, *supra* note 35, at 708 n.30.

59 Ernst Fehr & Simon Gächter, *Fairness and Retaliation: The Economics of Reciprocity Source*, 14 J. ECON. PERSP. 159, 161 (2000) (citing Werner Güth, Rolf Schmittberger, & Bernd Schwarze, *An Experimental Analysis of Ultimatum Bargaining*, 3 J. ECON. BEHAVIOR & ORG. 367 (1982); Colin Camerer & Richard Thaler, *Anomalies: Ultimatums, Dictators and Manners*, 9 J. ECON. PERSP. 209 (1995); Alvin Roth, *Bargaining Experiments*, in 1 THE HANDBOOK OF EXPERIMENTAL ECONOMICS 253 (Alvin E. Roth & John H. Kagel eds., 1995)). A 29–71 split of the surplus would correspond to a value of $R$ equal to 0.41.

voluntary exchange, which is the prerequisite to the expeditious exploitation of the standard.

## C. *Dividing Surplus Fairly*

The probability of a successful voluntary exchange increases as each party signals its willingness to accept a lesser share of the surplus that the transaction will create. Thus emerges a simple understanding of fairness, which can be expressed in comparative terms: The price corresponding to a given bilateral division of the surplus from a voluntary exchange is fairer than the price corresponding to some alternative bilateral division of that surplus if the first division is more likely than the second to lead the parties to agree to a successful transaction within some specified period of time.

My proposed definition of a fair price echoes, but is not identical to, certain themes found in the economic literature examining justice and fairness. Most notably, my definition resembles the proposition that fairness requires the approximately equal division of surplus, which Robert Frank proposed in 1988 in *Passions within Reason*.[60] However, my rationale for that definition differs from what I understand to be Frank's reasoning.

### 1. The Established Royalty and the Bid–Ask Spread

If SEPs were bushels of wheat, one could observe a multitude of market transactions in the aggregate that would obviate the forensic attempt by expert witnesses and judges to divide the surplus between buyer and seller. The bid–ask spread would be a sliver, and that fact would be considered a virtuous indicator of market efficiency. The liquidity needs, patience, and bargaining skill of any given seller would be irrelevant to the market's price formation. The idiosyncratic valuations of both the buyer and the seller also would be irrelevant. The conditions permitting an informed market price would obviate any inquiry into the outcome of a hypothetical bilateral negotiation between any two given parties at a given moment. Instead, we

[60] Frank, Passions within Reason, *supra* note 53, at 165. For a (now somewhat dated) survey of that literature, see James Konow, *Which Is the Fairest One of All? A Positive Analysis of Justice Theories*, 41 J. Econ. Literature 1188 (2003). "Justice arguments are now widely invoked to improve theoretical and empirical analysis in nearly every field of economics," wrote Konow in 2003, a development that "contrasts with the traditional belief of many economists that justice is chimerical or amorphous." *Id.* at 1188. "Despite the emerging consensus in economics over the relevance of fairness, though, no … agreement yet exists among economists or, for that matter, among psychologists, political scientists, sociologists, or philosophers, about the proper theory of justice." *Id.* at 1189. Ernst Fehr & Klaus M. Schmidt, *A Theory of Fairness, Competition and Cooperation*, 114 Q.J. Econ. 817 (1999). Earlier surveys on the economics of fairness appear in Edward Zajac's two books. Edward E. Zajac, Political Economy of Fairness (1995); Edward E. Zajac, Fairness or Efficiency: An Introduction to Public Utility Pricing (1978).

would simply consult the observed market price for the given asset on the day specified. In patent law, courts call this kind of market-determined price the "established royalty" for a patent.[61]

### 2. Fairness in Executing Licenses to SEPs

When no established royalty is apparent, the court must determine how to divide fairly the surplus from licensing the SEPs. As I have argued earlier, a given interpretation of "fairness" for purposes of SEP royalties might actually be an efficiency rationale in disguise that nudges the parties toward a successful voluntary transaction when some emotion threatens to interfere with the maximization of economic surplus. By analogy, research on the Ultimatum Game suggests that emotions such as envy, anger, or spite might upset a negotiation and thus cause the parties to forgo the benefits of dividing the positive surplus from a successful contract formation.[62] (Recall, for example, how Hal Varian, building on work by Duncan Foley,[63] defined a given allocation of resources as fair if it is both Pareto efficient and free from envy,[64] and William Baumol defined a "superfair" allocation of resources as one free from envy.[65])

[61] Georgia-Pacific Corp. v. U.S. Plywood Corp., 318 F. Supp. 1116, 1120 (S.D.N.Y. 1970), *modified and aff'd*, 446 F.2d 295 (2d Cir. 1971).

[62] One large-scale experiment of the Ultimatum Game found support for the prediction that "informed, knowledgeable respondents may react to small ultimatum offers by perceiving them as unfair, feeling anger, and acting spitefully." Madan M. Pillutla & J. Keith Murnighan, *Unfairness, Anger, and Spite: Emotional Rejections of Ultimatum Offers*, 68 Org. Behav. & Human Decision Processes 208, 208 (1996). Silicon Valley has had its titans who have expressed anger and spite toward their commercial adversaries. *See, for example,* Walter Isaacson, Steve Jobs (2011). Relatedly, in the business context, negotiators might be adverse to an "unfair" (that is, extremely skewed) split of the surplus based on the purely unemotional reason that it would set an unfavorable precedent for future negotiations or (in the event of information leakage) concurrent negotiations with other parties. The lopsided division of surplus might be used as comparable royalty evidence in future litigations.

[63] Duncan K. Foley, *Resource Allocation and the Public Sector*, 7 Yale Econ. Essays 45 (1967).

[64] Hal Varian wrote: "Consider the problem of dividing a fixed amount of goods among a fixed number of agents. If, in a given allocation, agent *i* prefers the bundle of agent *j* to his own, we will say *i envies j*. If there are no envious agents at allocation *x*, we will say *x* is *equitable*. If *x* is both pareto efficient and equitable, we will say *x* is *fair*." Hal R. Varian, *Equity, Envy, and Efficiency*, 9 J. Econ. Theory 63, 63 (1974) (emphasis in original); Konow, *supra* note 60, at 1204 ("The theory of fairness with the purest economic pedigree, and the usual definition of equity in welfare economics, is the absence of envy criterion."). *But see id.* at 1205 ("Absence of envy is questionable not only as a description of justice but also of what is meant by envy in common parlance: it seems quite possible that I would like to have another person's allocation, but that I do not experience the resentful feeling about his advantage that the word envy typically connotes.").

[65] William J. Baumol, Superfairness: Applications and Theory 15 (1986).

Matthew Rabin has studied how explicit consideration of fairness influences monopoly pricing.[66] An important caveat that he makes is that the players "make only mutually beneficial offers."[67] This restriction is fortuitous for purposes of analogizing his analysis to a FRAND or RAND royalty negotiation because (in my opinion) the succinct economic meaning of the reasonableness constraint is to demand that the SEP holder's offered price would produce a positive surplus for the offeror and for the offeree. Rabin then asks: "What is the highest price consistent with a fairness equilibrium at which this product could be sold?"[68] He finds that "the highest equilibrium price is lower than the conventional monopoly price when fairness is added to the equation."[69] This result – which is consistent with the earlier experimental findings of Daniel Kahneman, Jack Knetsch, and Richard Thaler – implies, in Rabin's words, that "a monopolist interested in maximizing profits ought not to set price at 'the monopoly price,' because it should take consumers' attitudes toward fairness as a given."[70]

One might conjecture that the purpose of a contractual obligation to make a "fair" division of surplus is to keep on a short leash the human emotions that might upset a mutually beneficial transaction. This interpretation of "fair" seems to be an acknowledgment that some principle ostensibly rooted in fairness is in actuality a lubricant to facilitate efficient voluntary exchange. So viewed, the constraint that a price embody fairness is in truth a means to an end. Just as a reduction in transaction costs can facilitate the expeditious conclusion of a voluntary exchange, so too can the absence of certain kinds of provocative (or strategic) behavior reduce the likelihood that one party will walk away in anger or spite from a voluntary negotiation that, if completed, would create surplus in which both parties would share.[71] This possibility is consistent with the observation of Kahneman, Knetsch,

[66] Matthew Rabin, *Incorporating Fairness into Game Theory and Economics*, 83 Am. Econ. Rev. 1281, 1292 (1993).

[67] *Id.*

[68] *Id.*

[69] *Id.* at 1293.

[70] *Id.* (citing Daniel Kahneman, Jack L. Knetsch, & Richard H. Thaler, *Fairness as a Constraint on Profit Seeking: Entitlements in the Market*, 76 Am. Econ. Rev. 728 (1986); Daniel Kahneman, Jack L. Knetsch, & Richard H. Thaler, *Fairness and the Assumptions of Economics*, 59 J. Bus. S285 (1986)).

[71] A negotiation might reach an impasse because of "the tendency for parties to arrive at judgments that reflect a self-serving bias – to conflate what is fair with what benefits oneself." Linda Babcock & George Loewenstein, *Explaining Bargaining Impasse: The Role of Self-Serving Biases*, 11 J. Econ. Persp. 109, 110 (1997). "Such self-serving assessments of fairness," Babcock and Loewenstein warn, "can impede negotiations and promote impasse in at least three ways." *Id.* The first way that they describe actually results from the incorrect identification of the bargaining range owing to a false understanding of one's true outside option: "if negotiators estimate the value of the alternatives to negotiated settlements in self-serving ways, this could rule out any chance of settlement by eliminating the contract zone (the set of agreements that both sides prefer to their reservation values)." *Id.* In contrast, the next two ways concern biases affecting the division of a positive surplus whose size is commonly understood:

and Thaler in 1986 that earlier arguments by Arthur Okun, George Akerlof, and Kenneth Arrow "to account for apparent deviations from the simple model of a profit-maximizing firm is that fair behavior is instrumental to maximization of long-run profits."[72] "In these approaches," they write, "the rules of fairness define the terms of an enforceable implicit contract: Firms that behave unfairly are punished in the long run."[73]

Fairness in executing licenses to today's SEPs can serve as a credible commitment to one's doing so with respect to tomorrow's SEPs, which are currently unknown. Fairness in the execution of licenses can produce increased acceptance of the SEP holder's technologies in future standards, increased participation by future implementers and holders of complementary SEPs, and an increased probability that the standard will achieve the scale necessary to be commercially successful. Fairness might be a commitment not to pull back the veil of ignorance, and thus fairness might discourage actions such as the IEEE's 2015 bylaw revisions, which large implementers favored and large SEP holders opposed.[74]

### D. *Licensing SEPs on Terms Consistent with the Fairness Constraint of the FRAND Contract between the SEP Holder and the SSO*

By definition, any price within the bargaining range is mutually beneficial. But that fact does not imply that every such price is equally likely to yield an agreement. How does a given split of the surplus between the SEP holder and the implementer influence the probability of their successful contract formation within a specified period of time? Is the distinguishing characteristic of a fair royalty that it leads more expeditiously to contract formation than some other division of the same gains from trade?

> Second, if disputants believe that their notion of fairness is impartial and shared by both sides, then they will interpret the other party's aggressive bargaining not as an attempt to get what they perceive of as fair, but as a cynical and exploitative attempt to gain an unfair strategic advantage. Research in psychology and economics has shown that bargainers care not only about what the other party offers, but also about the other party's motives. Third, negotiators are strongly averse to settling even slightly below the point they view as fair. If disputants are willing to make economic sacrifices to avoid a settlement perceived as unfair and their ideas of fairness are biased in directions that favor themselves, then bargainers who are "only trying to get what is fair" may not be able to settle their dispute.

*Id.* (citation omitted).

72 Kahneman et al., *Fairness as a Constraint on Profit Seeking: Entitlements in the Market*, *supra* note 70, at 728.

73 *Id.* at 728–29.

74 J. Gregory Sidak, *The Antitrust Division's Devaluation of Standard-Essential Patents*, 104 Geo. L. J. Online 48 (2015); J. Gregory Sidak, *Testing for Bias to Suppress Royalties for Standard-Essential Patents*, 1 Criterion J. Innovation 301 (2016).

To analyze this question, let us normalize the bargaining range so that it runs from 0 to 100. Normalizing the bargaining range simplifies the application of this analysis to different prospective implementers of the SEPs belonging to a given SEP holder. A license agreement struck at a normalized price of 0 gives the implementer 100% of the surplus. That is, an agreement at a normalized price of 0 is equivalent to a license bearing a royalty rate equal to the SEP holder's minimum willingness to accept and not a penny more. In contrast, an agreement at a normalized price of 100 gives the SEP holder 100% of the surplus. That is, an agreement struck at a normalized price of 100 is equivalent to a license bearing a royalty rate equal to the implementer's maximum willingness to pay and not a penny less.

For any license agreement struck at a normalized price between 0 and 100, each party will gain some of the surplus generated. For any possible agreement at a single given normalized price between 0 and 100, some probability exists that, within a specified period of time, the implementer will accept that price and enter into an agreement, and some different probability exists that the SEP holder will accept that same price and enter into an agreement. If both parties accept the same price, then an agreement is reached, and contract formation occurs.

The probability that the implementer will agree to terms decreases as the negotiated price moves farther from the SEP holder's minimum willingness to accept and closer to the implementer's maximum willingness to pay. Conversely, the probability that the SEP holder will agree to terms increases as the negotiated price moves farther from the SEP holder's minimum willingness to accept and closer to the implementer's maximum willingness to pay. The implementer has a "bid function" that determines the implementer's probability of agreeing to terms (within a specified period of time) at any given price over the bargaining range. Similarly, the SEP holder has an "ask function" that determines the SEP holder's probability of agreeing to terms (within the same specified period of time) at any given price over the bargaining range.

If the bid function and the ask function are symmetric, then the most common agreement will occur where the parties divide the gains from trade evenly. This 50–50 outcome is merely the arithmetic implication of the bid function's being the mirror image of the ask function. It is important to emphasize that this result does not rely on the Nash bargaining solution, which predicts a 50–50 split of the surplus in a bilateral negotiation using cooperative game theory.[75] Nor does this result rely

[75] John F. Nash, Jr., *The Bargaining Problem*, 18 ECONOMETRICA 155 (1950). In his 1950 article, John Nash proposed a solution to what he called the "bargaining situation" – an economic game in which two parties "have the opportunity to collaborate for mutual benefit in more than one way." *Id.* at 155. A solution to that game maximizes "the amount of satisfaction each [party] should expect to get from the situation." *Id.* According to Nash's model, an increase in the value of a party's position absent an agreement improves the party's bargaining position and therefore results in an improvement in that party's value of the bargain.

Before deriving his solution, Nash made certain assumptions about the game's participants: that each bargaining party is "highly rational," "can accurately compare [its] desires for various

on the familiar cake-cutting principle described by Rawls and others – "You cut, I choose"[76] – which Peyton Young notes "is fair because the outcome is envy-free."[77]

However, there is no economic reason to expect that the bid function and the ask function will be symmetric. In practice, risk aversion, discount rates, or other economic factors will influence the specific shapes for the bid function and the ask function.

The point royalty within the range of reasonable royalties upon which the SEP holder and the implementer will agree – that is, how they will divide the surplus from voluntary exchange – will be determined by the parties' relative bargaining power. The party that suffers least from delaying the agreement – that is, the party that is most patient – will typically have more bargaining power. Parties can have different levels of "patience" during a FRAND licensing negotiation while still negotiating in good faith, and it is common for SEP negotiations to take multiple years. For example, a SEP holder that lacks liquidity might need an immediate resolution of the negotiations. Or the implementer might be on the verge of releasing a standard-compliant product and therefore quickly needs a license to the SEPs before releasing a noninfringing product. Conversely, the SEP holder might not need an immediate license to the SEPs, which would increase its bargaining power. The near-absence of injunctive relief in SEP infringement litigation and the limited likelihood of enhanced damages might lead one to conclude, all other factors remaining the same, that implementers are more

things," is "equal [to the other] in bargaining skill," "has full knowledge of the tastes and preferences of the other," and "wishes to maximize the utility to [itself] of the ultimate bargain." *Id.* at 155, 159. Nash further assumed the independence of irrelevant alternatives – that is, if a bargainer faces a choice between A and B and prefers A to B, then that bargainer must also prefer A to B if faced with a choice between A, B, and C. *Id.* at 156. In 1953, Nash extended his 1950 article in a manner that "tells the players what threats they should use in negotiating." John Nash, *Two-Person Cooperative Games*, 21 Econometrica 128, 130 (1953). He summarized: "Supposing A and B to be rational beings, it is essential for the success of the threat that A be *compelled* to carry out his threat *T* if *B* fails to comply. Otherwise it will have little meaning." *Id.* (emphasis in original).

American courts have been skeptical of the real-world applicability of the Nash bargaining solution as expert economic testimony and thus bristle at its invocation as a basis for predicting a 50–50 division of surplus in a bilateral negotiation. As the Federal Circuit explained in the context of measuring reasonable royalty damages for patent infringement, "[t]he Nash [bargaining] theorem arrives at a result that follows from a certain set of premises" but "itself asserts nothing" about the real-world reliability of those premises. VirnetX, Inc. v. Cisco Sys., Inc., 767 F.3d 1308, 1332 (Fed. Cir. 2014) (analyzing Nash, *The Bargaining Problem*, *supra* note 75).

76 Rawls, A Theory of Justice, *supra* note 47, at 74; Ken Binmore, Game Theory and the Social Contract: Just Playing 382–83 (1998); H. Peyton Young, Equity in Theory and Practice 135 (1994); H. Peyton Young, *Dividing the Indivisible*, 38 Am. Behavioral Scientist 904, 911–12 (1995); Hugo Steinhaus, *The Problem of Fair Division*, 16 Econometrica 101 (1948).

77 Young, Equity in Theory and Practice, *supra* note 76, at 135.

"patient." It is well established in the economic literature that the cost that each party bears from a delay is measured by its respective discount rate.[78]

A more precise model could use different assumptions concerning the probability of contract formation. For example, are the probabilities for the two parties independent of one another, or is each probability conditional on the expected reaction of the counterparty (and, if so, for how many future rounds of the negotiation)? These questions are appropriate to ask if an economist wants to model the probability of contract formation in precise mathematical terms – for example, along the lines of the Rubinstein bargaining model, which is based on noncooperative game theory.[79] But my goal here is more modest and more heuristic. So those particulars about the precise nature of the probabilities are unnecessary to resolve to make the larger point (which I believe a judge or jury could readily understand intuitively) – namely, that it is reasonable to expect that the speed of contract formation will depend on the relative parity or disparity of the shares by which each party to a negotiation proposes to divide the surplus from a successful licensing transaction.

One proposed division of surplus might be substantially more likely than another to yield successful contract formation within a specified period of time spent negotiating. For example, it seems intuitively clear that a 60–40 split of the surplus would more readily be accepted by both parties than would a 99–1 split. If so, we would say that the 60–40 split is fairer than the 99–1 split.

What would be the threshold for a judge or jury to make the qualitative determination that a particular division of surplus would be *unfair*? Perhaps the experimental results of the Ultimatum Game, which I discussed earlier, will suggest a useful line of analysis.

78 PINDYCK & RUBINFELD, *supra* note 45, at 562; ROBERT GIBBONS, GAME THEORY FOR APPLIED ECONOMISTS 68–71 (1992).

79 Ariel Rubinstein, *Perfect Equilibrium in a Bargaining Model*, 50 ECONOMETRICA 97 (1982).

# PART III

# Patent Holdout and the Rise of "Efficient" Infringement

# 5

# Efficient Infringement in the SEP Space

*Kristen Osenga*

## I. INTRODUCTION

Many of the most innovative companies in the 5G and Internet of Things (IoT) spaces rely on patent rights. Not only do patents provide property rights over the technology invented and submitted to standards development organizations (SDOs), but the ability to license their patents allows these companies to recoup their research expenditures and invest in further development of innovative technologies. Unfortunately, due to a variety of legal and policy decisions, these companies' ability to effectively license their patents is being hindered; instead, manufacturers of standards-compliant goods are opting to engage in efficient infringement before, or even in lieu of, negotiating and accepting a license to use these patents.

To understand efficient infringement, it is helpful to think first about patent infringement more generally. A patent grants to the owner an exclusive right, or the ability to exclude others from making, using, selling, or offering for sale the subject matter of the patent without permission.[1] If a patent owner believes a party is infringing the patent, the patent owner can bring a lawsuit for patent infringement in a US District Court. The accused infringer will often defend by asserting that the patent claims are invalid. The court will then determine infringement and invalidity, as well as reviewing other raised defenses. If the patent is determined to be infringed and not invalid, the court will generally award damages, either actual damages or a reasonable royalty, for past infringement and issue a permanent injunction to prohibit infringement going forward. Through the damages award,

[1] US Const., art. I, § 8, cl. 8; 35 U.S.C. § 154(a)(1) ("Every patent shall contain . . . a grant to the patentee, his heirs or assigns, the right to exclude others from making, using, offering for sale, or selling the invention throughout the United States or importing the invention into the United States . . . .").

the patent owner is made whole for previous trespasses upon their exclusive right, and through the injunction, the patent owner's exclusive right is restored.

Not only is injunctive relief entirely consistent with the patent grant of an exclusive right, but it also plays a role in maintaining an effective private ordering system surrounding patent rights. Specifically, the knowledge that an injunction would result in the case of patent infringement serves as a good deterrent to those considering whether to infringe. Anticipating that their actions could lead to injunctive relief, most potential infringers would instead seek a license to use the patent or, alternatively, endeavor to design around the patent, rather than face having to revamp their entire manufacturing and distribution system ex post to abide by the injunction.[2] In addition to encouraging pre-infringement negotiations between potential infringers and patent owners, the possibility of injunction also incentivizes pre-litigation negotiations and settlements of lawsuits.[3] In short, the availability of injunctive relief creates a viable ecosystem for nonjudicial transactions surrounding patent rights. This ecosystem of transactions surrounding patent rights then provides a viable business model for innovative companies wishing to participate in SDOs.

On the other hand, if there is no credible threat of injunction for patent infringers, this balanced ecosystem is upset. Instead, a more nefarious course of action is allowed to flourish that forces patent transactions into the judicial system at great cost to patent owners. Efficient infringement is the idea that, if an injunction is unlikely to issue in the future, it may be more rational for a party to "infringe now, pay later." Because the worst expected outcome for the infringer will be to pay damages for past infringement and an ongoing royalty for any future infringement, the infringer is no worse off than it would have been had it negotiated a license before commencing infringement. In fact, because damages are not awarded until after the trial, which could be years after the infringement began, the infringer ends up with what is essentially an interest-free loan for the period of time preceding the lawsuit. Perversely, the infringer may even end up in a better position, if the royalty rate set by the court is lower than the parties would have arrived at via negotiation.[4]

Of course, the calculus behind efficient infringement is only attractive if the infringer is relatively certain that an injunction will not be issued should a court make a finding of infringement. Relevant to this book, in fields where technology standards are prevalent – including 5G and IoT – the availability of injunctive relief is not certain and, in fact, injunctions are rarely granted. While injunctive relief used to be essentially guaranteed in cases of patent infringement, in part due to the

[2] Karen Sandrik, *Reframing Patent Remedies*, 67 U. Miami L. Rev. 95, 98 (2012).

[3] John M. Golden, *Litigation in the Middle: The Context of Patent Infringement Injunctions*, 92 Tex. L. Rev. 2075, 2081 (2014) (explaining how injunctive relief affects the expected gains and losses from litigation and may alter settlement decisions).

[4] Adam Mossoff & Bhamati Viswanathan, *Explaining Efficient Infringement*, C-IP2 (May 11, 2017), https://cip2.gmu.edu/2017/05/11/explaining-efficient-infringement/.

acknowledgment of a patent's grant of exclusive rights, the Supreme Court's 2006 decision in *eBay, Inc. v. MercExchange, LLC*[5] altered this landscape. Stating that injunctive relief should not be categorically granted, as the US Court of Appeals for the Federal Circuit (Federal Circuit) had long done, nor categorically denied, as the district court in the *eBay* case had presumed, the Supreme Court instead imposed a four-factor test to be used when deciding to issue injunctive relief. The four factors to be considered include: (i) irreparable harm to the patent owner; (ii) remedies at law (damages) are inadequate; (iii) balance of hardship; and (iv) public interest.[6] In most cases, these factors weigh in favor of granting an injunction because the patent owner's exclusive right is irreparably harmed by infringement and money damages cannot restore that exclusivity after the fact.[7] Further, the public has a strong interest in a reliable and effective patent system, which respects the patent owner's exclusive rights.

Despite the exhortation that categorical grants or denials of injunctions are inappropriate, Justice Kennedy concurred to explain that injunctions perhaps should not issue in cases where the patent owner is a non-practicing entity, where the patent covers only a small component of a much larger product, and where the patented technology is a business method.[8] Kennedy warned against injunctions being used in cases of "patent holdup," although he did not reference it as such. Specifically, he noted that "an injunction, and the potentially serious sanctions arising from its violation, can be employed as a bargaining tool to charge exorbitant fees to companies that seek to buy licenses to practice the patent," and that "[w]hen the patented invention is but a small component of the product the companies seek to produce," injunctive relief is "employed simply for undue leverage in negotiations."[9] Because of these issues highlighted in Justice Kennedy's concurrence and because of how courts have applied the four-factor *eBay* test, there are two particular groups of patent owners for whom injunctive relief is uncertain or possibly even unlikely – non-practicing entities and owners of standard-essential patents (SEPs).

Efficient infringement is of particular concern in the SEP space, including 5G and IoT, because injunctive relief is generally unavailable to SEP owners. This chapter explains that courts are unlikely to grant requests for permanent injunctions and SEP owners are unlikely to even seek injunctive relief, setting up the necessary condition for efficient infringement to flourish. Unsurprisingly, there is evidence that infringers are selecting to "infringe now, pay later" when it comes to SEPs, and

5 547 U.S. 388 (2006).

6 *Id.* at 391.

7 *Id.* at 395 (Roberts, C.J., concurring) (noting that injunctions should continue to issue in the "vast majority" of patent cases, as the right to exclude is difficult to capture via monetary damages).

8 *Id.* at 395–97 (Kennedy, J., concurring).

9 *Id.* at 396 (Kennedy, J., concurring).

this chapter sets out the case why this is problematic for a well-functioning innovation system.

## II. COURTS ARE UNLIKELY TO GRANT INJUNCTIVE RELIEF IN SEP CASES

It is difficult to accurately quantify whether efficient infringement is occurring because it presumes a mental calculus by the infringer that infringing first, paying later would be a viable option. However, it is possible to determine if the conditions are favorable for efficient infringement by assessing the relative threat of being enjoined in the face of infringement. This section explains why the *eBay* factors present difficulties for SEP owners and then provides some exemplary cases where this analysis was performed by a court. This section concludes with some data about the current state of injunctive relief in patent infringement cases.

### A. *SEPs and the eBay Factors*

Based on the four-factor test set forth in the Supreme Court's *eBay* opinion, SEP owners are unlikely to obtain injunctive relief in most cases due to the FRAND commitment the SEP owner made to the SDO. This section explains how the FRAND commitment interacts with the *eBay* factors.

SDOs generally have intellectual property (IP) rights policies that address issues surrounding the IP of the companies who participate in the SDO. One typical provision of IP rights policies requires SDO participants to disclose patents and patent applications covering technology that the participant has submitted to the SDO for potential incorporation into the standard. If that technology becomes part of the standard and the participant deems the patent or patent application to be essential for the practicing of a standard, it becomes known as an SEP. Another common provision of an SDO's IP rights policy is a commitment by SDO participants to license any SEPs related to the standard on fair, reasonable, and nondiscriminatory (FRAND) terms to any company who wants to make or offer standards-compliant goods or services.[10] The purpose of the FRAND commitment is to facilitate widespread adoption of the standard while still providing SEP owners an opportunity to recoup their investments in the research and development of the technology, as well as to promote the standardization process.

Courts are likely to deny injunctive relief, based on the *eBay* factors, to an SEP owner that has made a FRAND commitment. Specifically, courts will generally find that SEP owners fail to meet the first two *eBay* factors – irreparable harm and money

[10] Kristen Osenga, *Ignorance over Innovation: Why Misunderstanding Standard Setting Organizations Will Hinder Technological Progress*, 56 U. Louisville L. Rev. 159, 183–87 (2018).

damages are inadequate. The argument is that, by accepting a FRAND commitment, an SEP owner has voluntarily relinquished its exclusive rights in exchange for having its patented technology incorporated into the standard. Because the SEP owner has committed to license its patent to all-comers, it cannot claim irreparable harm in the form of loss of exclusive rights, and similarly, because the SEP owner has accepted the FRAND commitment, it has signaled to all that money is sufficient to make it whole. As one commentator summed it up, an SEP owner that has obligated itself to FRAND "is, by definition, willing to license rather than exclude, and benefits from the widespread adoption of its technology resulting from standardization."[11]

Some courts have also found that the FRAND commitment also causes the third and fourth *eBay* factors to support a denial of injunctive relief. The balance of the hardships argument weighs in favor of the infringer because the FRAND commitment means that the SEP owner has given up its rights, while the infringer would be greatly harmed if they were unable to make standards-compliant products like everyone else. The public interest argument also weighs in favor of the infringer being allowed to continue infringing, because the public has an interest in having access to standards-compliant products from multiple implementers. This interest of the public in innovative devices takes precedent over both the public's interest in a robust standardization ecosystem and a strong and reliable patent system.

While superficially appealing, the aforementioned application of the *eBay* factors to SEP owners is too facile and is exactly the type of categorical denial that the Supreme Court warned against in that case. Although the FRAND commitment obligates the patent owner to offer and accept a license to use its SEP on fair, reasonable, and nondiscriminatory terms, there is no concomitant obligation on the potential infringer to accept, or even negotiate in good faith for, these terms. While this one-sidedness is more accurately a problem with the overall construct of the FRAND commitment, it does play a role in both the *eBay* analysis as well as the infringer's decision to infringe. As to the *eBay* factors, FRAND affects the acceptability of monetary damages as sufficient and, because the infringer has no obligation to negotiate, implicates the irreparability of harm to the patent owner's exclusive rights. Finally, while the public does have an interest in standards-compliant products, the fact that SEPs are licensed on FRAND terms to any desiring manufacturer means that the infringer is unlikely to be the only source of any standards-compliant product. An injunction issued against any particular manufacturer will likely not deprive the public of its much-loved and useful devices.

Looming behind the flawed, or at least thoughtless, application of the *eBay* factors to SEPs is Kennedy's callout to patent holdup in his concurrence in that case.

[11] Thomas F. Cotter, *Reflections on Holdup and Royalty Stacking, Part 1*, Compar. Pat. Remedies (June 11, 2014, 4:20 AM), http://comparativepatentremedies.blogspot.com/2014/06/reflections-on-holdup-and-royalty.html.

Patent holdup is the idea that a patent owner could force a firm wishing to make or offer standards-compliant goods or services to pay an excessively high royalty rate, relying on the fear of injunctive relief if the infringer fails to pay the royalty. This is particularly acute where the patented technology may be just a small portion of a standard encompassing hundreds or thousands of patents.[12] Although there has been much research and debate over whether patent holdup is merely theoretical or actually occurs, courts have openly reflected a concern about holdup when deciding not to grant injunctive relief for infringement of SEPs.[13]

Despite objections to how courts view SEP owners under the *eBay* factors and a largely misplaced concern about patent holdup, the examples in the next section demonstrate that courts will generally decline to issue an injunction for infringement of an SEP.

### B. *Courts, eBay, and SEPs*

Very few cases exist that illustrate how courts assess the *eBay* factors after a finding of patent infringement of a SEP. The dearth of cases may be, in part, because SEP owners do not seek injunctive relief – even as they avail themselves of other remedies. This failure to seek an injunction could be due to an SEP owner's determination that they are unlikely to obtain this remedy or because another consideration prevents it. Both of these situations are described in the following sections. The cases that exist, however, provide interesting insight into how courts think about SEPs and injunctions. The following section provides a few examples of these insights.

#### 1. Hynix Semiconductor Inc. v. Rambus Inc.

While this case is more commonly known for a number of other issues, how the court assessed injunctive relief is instructive.[14] The SEP in question covered DRAM technology, patented by Rambus and found to be infringed by Hynix. The bulk of Hynix's DRAM fell under the JEDEC standard. Rambus developed and licensed

[12] J. Gregory Sidak, *Holdup, Royalty Stacking, and the Presumption of Injunctive Relief for Patent Infringement: A Reply to Lemley & Shapiro*, 92 Minn. L. Rev. 714, 714 (2008).

[13] For just a few examples, see Jonathan M. Barnett, *Has the Academy Led Patent Law Astray?*, 32 Berkeley Tech. L. J. 1313 (2017); Damien Geradin & Miguel Rato, *Can Standard-Setting Lead to Exploitative Abuse? A Dissonant View on Patent Hold-Up, Royalty-Stacking, and the Meaning of FRAND*, 3 Eur. Comp. J. 101, 101–02 (2007); Kirti Gupta, *The Patent Policy Debate in the High-Tech World*, 9 J. Competition L. & Econ. 827 (2013); F. Scott Kieff & Anne Layne-Farrar, *Incentive Effects from Different Approaches to Holdup Mitigation Surrounding Patent Remedies and Standard-Setting Organizations*, 9 J. Competition L. & Econ. 1091 (2013); Maureen K. Ohlhausen, *The Federal Trade Commission's Path Ahead*, 2 Criterion J. Innovation 31, 33 (2017).

[14] 609 F. Supp. 2d 951 (N.D. Cal. 2009).

chipset technology; although Rambus did not make a product that competed with Hynix, there was competition because Rambus' design competed with the JEDEC standard.

In determining whether to grant Rambus injunctive relief, the court begins by recounting the *eBay* factors, but then instantly proceeds to a section entitled "Injunctions Should Not Encourage Holdup."[15] The court cites the holdup language from Justice Kennedy's concurrence in *eBay*, but then proceeds to examine some very old cases (from 1882 to 1900) that address holdup and arrives at the propositions that injunctive relief poses a disproportionate cost on the infringer and that injunctions are not meant to be punitive. The court then returns to analyze the *eBay* factors, specifically the factors of irreparable harm and inadequacy of monetary damages. The court determined that, although Rambus does not lose sales to Hynix because Rambus does not sell competing products, Rambus did stand to "lose" due to Hynix's continued infringement because the JEDEC standard would "win" over Rambus' proprietary design. The court, however, determines this to be a minimal harm, because Rambus has multiple other licensees.[16] Rambus also argues irreparable harm and inadequacy of money damages due to diminished royalty rates and harm to its innovation-based model, after the *CSIRO* case (described in Section 4). The court was unpersuaded by these arguments, stating that Rambus was in the business of seeking money because it has a licensing program and thus money damages were indeed sufficient.[17] As to the remaining *eBay* factors of balance of the hardships and public interest, the court found that Hynix would go out of business in the face of injunctive relief, which would be disproportionately more harmful than the "slight" harm faced by Rambus.[18]

In deciding the *Hynix* case, the judge drew comparisons between that case and *Broadcom v. Qualcomm*, calling it the most significant case since *eBay*.[19] The *Broadcom* case also involved telecom chipsets falling under different standards. The court there found irreparable harm, even where the chipsets were under different standards, because the competition between the companies was not sale for sale, but for domination of design.[20] The judge then issued injunctive relief on a sunset basis, giving the infringer 20 months to stop infringing, while paying an ongoing royalty, as a means to protect the public.[21] For the *Broadcom* court, the bottom line was that a company who builds their business on infringement cannot complain that their business will fail if an injunction is granted. However, the *Hynix* court noted that if the infringement was clearly not willful, or where there exist

[15] *Id.* at 966.
[16] *Id.* at 981.
[17] *Id.* at 983–85.
[18] *Id.* at 984–85.
[19] *Id.* at 969.
[20] Broadcom Corp. v. Qualcomm Inc., 543 F.3d 683, 686–87 (Fed. Cir. 2008).
[21] *Id.* at 687–88.

serious questions about the patent's validity, the balance may tip back in favor of not enjoining the infringer.[22] Ultimately, in the *Hynix* case, the court denied injunctive relief.

### 2. Apple Inc. v. Motorola Inc.

In this case involving cross-infringement suits,[23] Judge Posner, sitting by designation at the district court, determined that neither side would be entitled to an injunction or damages even upon a finding of infringement, and so dismissed the case.[24] Although the case is a bit messy because Judge Posner found that neither side had presented an adequate case with respect to damages, the case does present a commonly held viewpoint with respect to injunctions. As Judge Posner noted: "I do not see how, given FRAND, I would be justified in enjoining Apple from infringing the '898 unless Apple refuses to pay a royalty that meets the FRAND requirement."[25] Essentially, in the face of outright refusal, courts are unable to see how an SEP owner can meet the *eBay* factors for injunctive relief. On appeal to the Federal Circuit, that court took note of Judge Posner's statement presented earlier.[26] The Federal Circuit then acknowledged that while it is difficult to establish irreparable harm in cases involving SEPs, there is no categorical prohibition on awarding injunctive relief.[27] Specifically, injunctions may be justified where the infringer unilaterally delays FRAND negotiations or rejects FRAND offers.[28] Although this interpretation is more generous than that of the *Hynix* court that limited injunctive relief to willful infringement, it still creates a quite narrow set of difficult facts that an SEP owner must prove to obtain injunctive relief.

### 3. Microsoft Corp. v. Motorola, Inc.

While the *Hynix* case provided a fairly standard perspective on the *eBay* factors and SEPs, it did allow for a small ray of hope in a case where the patent owner could show willfulness and the *Apple v. Motorola* case gave two additional pathways to injunctions. On the other hand, the *Microsoft v. Motorola* case takes the idea of injunctive relief for infringement of SEPs in an entirely different direction.[29] Mid-case, Microsoft moved to dismiss Motorola's request for injunctive relief based on the argument that the *eBay* factors could not be met in the case.[30] After discussing

[22] *Hynix*, 609 F. Supp. 2d at 970.
[23] Apple Inc. v. Motorola, Inc., 869 F. Supp. 2d 901 (N.D. Ill. 2012).
[24] *Id.* at 924.
[25] *Id.* at 913–14.
[26] Apple Inc. v. Motorola, Inc., 757 F.3d 1286, 1331 (Fed. Cir. 2014).
[27] *Id.* at 1332.
[28] *Id.*
[29] No. C10-1823JLR, 2012 U.S. Dist. LEXIS 170587 (W.D. Wash. Nov. 29, 2012).
[30] *Id.* at *30.

SEPs and Motorola's FRAND commitments under the H264 and 802.11 standards, the court determined without much analysis that there is no irreparable harm given the FRAND commitment and further that monetary damages are perfectly sufficient to make Motorola whole in this case.[31]

The situation became much more interesting when, a year later, Microsoft argued to a jury that Motorola breached its FRAND commitment in even seeking injunctive relief.[32] The jury found in favor of Microsoft on this point, and the judge denied Motorola's motion for judgment as a matter of law, leaving intact the finding that a mere request for injunctive relief by a SEP owner was incompatible with a FRAND commitment.[33] The US Court of Appeals for the Ninth Circuit (Ninth Circuit) went on to affirm this finding.[34] This notion that an SEP owner may not even seek an injunction for fear of breaching a duty of good faith and fair dealing with respect to the FRAND obligation creates a difficult choice for an SEP owner.[35]

### 4. *CSIRO v. Buffalo Tech Inc.*

There is, of course, an exception that proves the rule. Although the *CSIRO* case is better known as a rare decision where an injunction is granted to a non-practicing entity, it is also a case involving an SEP.[36] Commonwealth Scientific and Industrial Research Organisation (CSIRO) is a research arm of the Australian government, not unlike the National Science Foundation or National Institutes of Health in the United States. CSIRO received a patent on WLAN technology, and IEEE reached out to CSIRO to inquire whether the organization would commit to FRAND licensing of this patented technology. When CSIRO agreed, IEEE incorporated the technology into the 802.11a and 802.11 g standards, although the more popular 802.11b standard does not include CSIRO's technology.[37] CSIRO then sued Buffalo Tech for infringement based on their making and selling of WLAN products compliant with 802.11a and 802.11 g. In the infringement case, the patent was found to be not invalid and infringed on summary judgment.[38] CSIRO then sought a permanent injunction.

The court applied the *eBay* factors. As to irreparable harm, CSIRO argued that it is difficult to incentivize a manufacturer to take a license when other manufacturers

[31] *Id.* at *26–30.

[32] Microsoft Corp. v. Motorola Inc., 795 F.3d 1024, 1034 (9th Cir. 2015) (recounting the district court proceedings).

[33] *Id.*

[34] *Id.* at 1047–48.

[35] The same concept, that seeking injunctive relief is a breach of FRAND, was also approved of by Realtek Semiconductor Corp. v. LSI Corp., 964 F. Supp. 2d 998, 1008 (N.D. Cal. 2013).

[36] 492 F. Supp. 2d 600 (E.D. Tex. 2007).

[37] *Id.* at 601–02.

[38] *Id.* at 602.

are infringing. As expected in these cases, Buffalo Tech argued that there is no irreparable harm due to the FRAND commitment and that CSIRO is in the business of licensing patents. The court recognized that the harm to CSIRO was not merely financial, because the organization used its licensing program to fund its research and development activities, compete for ideas and scientists, and more. Because funding makes these activities possible, infringement represents a lost opportunity and an irreparable harm.[39] For similar reasons, money damages were insufficient to make CSIRO whole.[40] The balance of the hardships was found to favor CSIRO, as the infringing products were only a small amount of Buffalo Tech's business and Buffalo Tech opted to infringe rather than license.[41] Finally, the court recognized that the public has a strong interest in an effective patent system, particularly where there are no health or safety concerns at issue.[42]

While this case could represent a path forward for SEP owners to obtain injunctive relief, as was tried in the *Hynix* case described earlier, the *CSIRO* court drew two distinctions that make it difficult to extend that case to more typical SEP cases. First, the court noted that CSIRO's patented technology was not a small component of a larger product, but was instead the whole of the invention.[43] This is generally not true of today's standardized technologies, such as 5G and IoT. Second, the court found CSIRO's mission as a research organization particularly compelling, noting that "the work of research institutions has produced enormous benefits to society" and is "fundamental to scientific advancement."[44] It is unlikely courts would extend the same sort of reasoning to innovative for-profit firms, despite any societal benefits or scientific advancements.

### C. *Injunctions Are Being Denied*

Efficient infringement is only an attractive option when the infringing firm believes it is unlikely to be enjoined, even if adjudged to be infringing a valid patent. Numerous studies have been conducted post-*eBay* to understand whether and when injunctive relief is being denied. Although most of these studies are focused on non-practicing entities, rather than SEP owners, the reasons provided for denying injunctive relief to non-practicing entities overlaps with the reasons given to SEP owners. That courts are willing to deny injunctive relief in a significant portion of patent infringement cases provides the necessary foundation for efficient infringement to occur.

39 *Id.* at 603–04.
40 *Id.* at 605–06.
41 CSIRO v. Buffalo Tech Inc., 492 F. Supp. 2d 600, 606 (E.D. Tex. 2007).
42 *Id.* at 607.
43 *Id.* at 606.
44 *Id.* at 607.

Studies of injunction grants prior to *eBay* found that injunctive relief was granted in 95–100% of cases where patent infringement was found.[45] Studies done in the first decade after *eBay* found that requests for permanent injunction were granted in approximately three-quarters of the cases where patent infringement was found (and an injunction was requested). In a widely cited study covering the time period between *eBay* in 2006 and 2013, Professor Christopher Seaman found that requests for permanent injunction were granted in 72.5% of cases.[46] Similar numbers were shown in studies by Professors Colleen Chien and Mark Lemley (70% covering June 2006 through August 2011) and Professor Jay Kesan and Kirti Gupta (80% between June 2006 and December 2012).[47] In the study conducted by Professor Seaman, discussed previously, requests for injunction were granted 16% of the time when the patent holder was not practicing the patent.[48] Similar low grant rates for non-practicing entities were seen in the studies by Chien and Lemley, as well as Kesan and Gupta.

Current research shows that these numbers have remained steady since the *eBay* case. This chapter's author has looked at decided patent infringement cases since the *eBay* decision where permanent injunctive relief was disputed. Cases decided by default (due to defendant's failure to answer) or cases where an injunction was issued by consent, stipulation, or settlement were not considered. This resulted in 342 cases with written opinions. Of those 342 cases, 249 resulted in the issuance of an injunction and 93 cases where injunctive relief was denied. This is approximately a 73% grant rate for permanent injunction requests and is generally consistent with the earlier noted studies ( Figure 5.1).

From this data, it is fair to assert that courts are willing to deny injunctive relief in cases that do not, in the court's estimation, satisfy the *eBay* factors. However, it is difficult to say much else, particularly about SEPs. Of the cases represented in the data, only seven implicate SEPs and only one resulted in the grant of a permanent injunction (the *CSIRO* case). There are significant selection effects in the data overall, in that very few cases of patent infringement are actually litigated and even fewer reach the stage where a court must decide a dispute about injunctive relief. Beyond that, there are concerns that, especially in SEP cases, the SEP owner does not seek injunctive relief, as will be discussed later. Further, some SEP cases address the injunction question even before tackling infringement and validity (such as the *Apple v. Motorola* case) and so were not captured in the data set. Given the data and

45 Ryan T. Holte, *The Misinterpretation of eBay v. MercExchange and Why: An Analysis of the Case History, Precedent, and Parties*, 18 Chap. L. Rev. 677, 719 (2015).

46 Christopher B. Seaman, *Permanent Injunctions in Patent Litigation after eBay: An Empirical Study*, 101 Iowa L. Rev. 1949, 1983, fig. 1, fig. 3 (2016).

47 Colleen V. Chien & Mark Lemley, *Patent Holdup, the ITC, and the Public Interest*, 98 Cornell L. Rev. 1, 11 (2012); Kirti Gupta & Jay P. Kesan, *Studying the Impact of* eBay *on Injunctive Relief in Patent Cases* (July 10, 2015) (unpublished manuscript), http://papers.ssrn.com/sol3/papers.cfm?abstract_id=2629399.

48 Seaman, *supra* note 46.

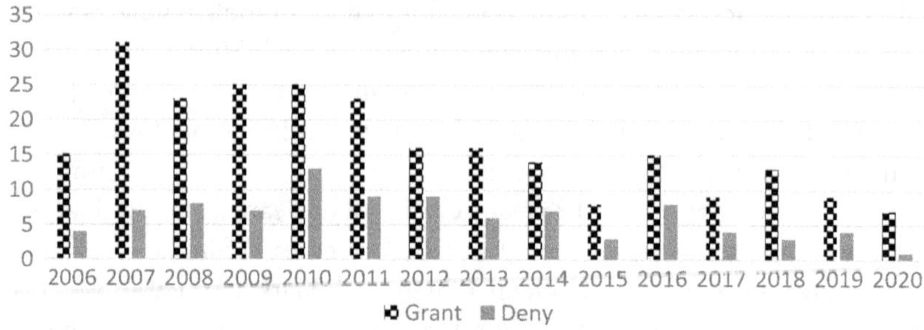


FIGURE 5.1. Graph of injunction grants & denials June 2006–December 2020

the discussion previously, parties considering efficient infringement may find that the likelihood of being enjoined upon infringing an SEP is sufficiently low enough that it is a prudent decision.

## III. SEP OWNERS ARE UNLIKELY TO SEEK INJUNCTIVE RELIEF

One of the difficulties in gathering data about efficient infringement in the SEP space, or at least understanding the favorability of the conditions surrounding it, is that SEP owners do not always (or even often) seek injunctive relief when bringing a patent infringement lawsuit. Of course, if an SEP owner does not seek injunctive relief, none will be granted – again setting up conditions where efficient infringement is likely to flourish. This section will discuss why SEP owners are unlikely to seek an injunction, as well as consider.

### A. *SEP Owners Are Dissuaded by eBay and SEP Case Law from Seeking Injunctive Relief*

As described previously, case law precedent and general conceptions surrounding the *eBay* factors are likely to dissuade SEP owners from seeking injunctive relief. It is quite difficult for an SEP owner to successfully argue irreparable harm or insufficiency of monetary damages, unless that SEP owner happens to be a governmental research organization. Except in very narrow circumstances, the presence of a FRAND commitment is likely to prove fatal to any sort of injunction being issued. These narrow circumstances – including willful infringement or failure to either negotiate or pay – are difficult to prove and, as will be illustrated later, do not seem to encourage SEP owners to seek injunctive relief in any case.

### B. *Additional Considerations Affecting Seeking Injunctive Relief for SEP Infringement*

SEP owners are not just discouraged from seeking injunctions by *eBay* and the SEP cases that have followed. There are also additional considerations that have made injunctive relief nearly impossible for SEP owners to obtain and, in some cases, even prohibit SEP owners from seeking injunctive relief. Some of these considerations are driven by the government, from the White House to administrative agencies like the Federal Trade Commission (FTC) and the Department of Justice's Antitrust Division (DOJ). Other considerations come straight from SDOs as part of their IP rights policies. In addition to directly impacting an SEP owner's decision to seek injunctive relief, these considerations also have influenced courts' decisions about whether to grant injunctions in favor of SEP owners.[49]

### 1. Government Interventions against SEPs

Government entities are increasingly calling for injunctive relief to be unavailable to SEP owners, although these appeals date back nearly a decade or longer. In 2013, the DOJ and the United States Patent and Trademark Office (PTO) issued a policy statement strongly suggesting that injunctive relief was generally inappropriate in patent infringement cases involving SEPs.[50] Specifically, the statement stated that, absent extraordinary circumstances, an injunction for infringement of an SEP "may be inconsistent with the public interest."[51] This notion, coupled with the idea that an SEP owner who has committed to FRAND licensing cannot suffer irreparable harm and is made whole by money damages, makes application of the *eBay* factors a near certainty for the infringer. The basis for this position, as described by the statement, is a concern over patent holdup, although no evidence about the existence and extent of patent holdup was presented in the statement.[52]

Under the previous administration, there was a shift away from a near-categorical bar to injunctive relief for infringement of SEPs. The DOJ and PTO withdrew their support for the 2013 policy statement and jointly issued a new policy statement on remedies for SEPs subject to FRAND obligations in December 2019, together with the National Institute of Standards and Technology (NIST).[53] This statement noted that "All remedies available under national law, including injunctive relief and adequate damages, should be available for infringement of standards-essential patents subject to a F/RAND commitment."[54] One stated purpose of this position is to encourage good faith licensing negotiations between SEP owners and

[49] Apple Inc. v. Motorola, Inc., 757 F.3d 1286, 1332, 1333 (Fed. Cir. 2014) (both the majority and dissent cite the 2013 Policy Statement, *infra* note 50, for their position).

[50] US Dep't of Just. & US Pat. & Trademark Office, Policy Statement on Remedies for Standards-Essential Patents Subject to Voluntary F/RAND Commitments (Jan. 8, 2013), www.justice.gov/atr/page/file/1118381/download.

[51] *Id.* at 6.

[52] *Id.* at 4–5.

[53] U.S. Dep't of Just., U.S. Pat. & Trademark Office, & Nat'l Inst. for Standards and Tech., Policy Statement on Remedies for Standards-Essential Patents Subject to Voluntary F/RAND Commitments (Dec. 19, 2019), www.justice.gov/atr/page/file/1228016/download.

[54] *Id.* at 3–4.

companies wishing to make or offer standards-compliant goods or services.[55] Although unstated, the policy was also intended both to make efficient infringement less attractive and to reinvigorate a viable ecosystem for nonjudicial transactions surrounding patent rights.

With the Biden Administration, however, government policy has shifted back to disfavoring injunctive relief for infringement of SEPs. In July 2021, President Joseph Biden issued the Executive Order on Promoting Competition in the American Economy, where, among other things, he "encouraged [the Attorney General and the Secretary of Commerce] to consider whether to revise their position on the intersection of intellectual property and antitrust laws, including by considering whether to revise" the 2019 DOJ-PTO-NIST joint policy statement.[56] In response, the three agencies issued a Draft Policy Statement on Licensing Negotiations and Remedies for Standards-Essential Patents Subject to Voluntary F/RAND Commitments in December 2021.[57] Under this new guidance, injunctions are again unlikely to issue for SEP infringement: "Where a SEP holder has made a voluntary F/RAND commitment, the *eBay* factors, including the irreparable harm analysis, balance of harms, and the public interest generally militate against an injunction."[58]

This latest policy statement does recognize the importance of a viable ecosystem for patent licensing, acknowledging the value of "widespread and efficient licensing of SEPs" and recognizing the "efficient negotiation of F/RAND licenses is likely to improve standardization efforts and support competition and innovation."[59] The agencies, however, fail to understand that efficient infringement is an attractive option where injunctive relief is unavailable and that efficient infringement is the antithesis of efficient licensing. In any case, where the government has come out repeatedly and strongly against the availability of injunctive relief, the likelihood of a court granting an injunction or an SEP owner seeking injunctive relief is quite low.

As of June 8, 2022, the three agencies withdrew the 2019 DOJ-PTO-NIST joint policy statement and, at this point, have not adopted the 2021 Draft Policy Statement.[60] In announcing the withdrawal, the agencies noted that the issues of injunctive relief will be addressed on a case-by-case basis, but no additional guidance has been provided, and the direction of the path forward is unclear.

[55] *Id.* at 3.

[56] Exec. Order No. 14036, 86 Fed. Reg. 36987, 36991 (July 9, 2021).

[57] US Dep't of Just., US Pat. & Trademark Office, & Nat'l Inst. for Standards and Tech., Draft Policy Statement on Licensing Negotiations and Remedies for Standards-Essential Patents Subject to Voluntary F/RAND Commitments (Dec. 6, 2021), www.justice.gov/atr/page/file/1453471/download.

[58] *Id.* at 9.

[59] *Id.* at 10.

[60] US Dep't of Just. Office of Public Affairs, Justice Department, the United States Patent and Trademark Office and National Institute of Standards and Technology Withdraw 2019 Standards-Essential Patents (SEP) Policy Statement (June 8, 2022), www.justice.gov/opa/pr/justice-department-us-patent-and-trademark-office-and-national-institute-standards-and.

### 2. SDO's Prohibition on Injunctive Relief

SDOs have also weighed in on whether injunctive relief should be permitted for infringement of SEPs. While some SDOs have remained neutral, at least one major SDO has restricted the ability of its participants to seek injunctive relief. This certainly has a great impact in making efficient infringement a reasonable choice for makers of standards-compliant products.

In 2015, under the purported basis of addressing patent holdup, the Institute for Electronics and Electrical Engineers (IEEE) amended its IP rights policy to, among other things, address injunctive relief.[61] Unfortunately, the process by which these amendments were made was not the usual open, consensus-driven activity that is typical for SDO activity, but was driven largely by implementers, or those that make standards-compliant products, over the objection of SEP owners.[62] Specifically, the amended IEEE policy stated that a FRAND commitment to the IEEE "precludes seeking or seeking to enforce" an injunction unless the infringer "fails to participate in, or to comply with the outcome of, an adjudication including an affirming first-level appellate review" or "where the failure to request a Prohibitive Order in a pleading waives the right to seek a Prohibitive Order at a later time."[63] This prohibition is even more restrictive than what had been seen in the courts, where at least there was a possibility of injunctive relief where the infringer had refused to negotiate, as described earlier. In September 2022, the IEEE withdrew this language from its policy, effective January 1, 2023.

## C. *Showings Indicative of Efficient Infringement*

Although it is difficult to fully grasp how often efficient infringement is occurring, there are anecdotes and circumstantial data that support its existence. Anecdotes, while difficult to find, often provide very compelling illustrations of infringers making conscious choices to not negotiate licenses with SEP owners. Circumstantial data, on the other hand, is easier to obtain but complicated to interpret. It is possible to gauge how often SEP owners seek injunctive relief by considering their filed complaints, but this does not often address why they made this choice. Similarly, SEP owners often allege willful infringement when companies refuse to negotiate a license or in the wake of failed negotiations, but this too does not fully capture whether and how much efficient infringement is occurring in

[61] J. Gregory Sidak, *The Antitrust Division's Devaluation of Standard Essential Patents*, 104 Geo. L.J. Online 48, 50 (2015).

[62] J. Gregory Sidak, *Testing for Bias to Suppress Royalties for Standard-Essential Patents*, 1 Criterion J. Innovation 301, 302 (2016).

[63] *IEEE-SA Standards Board Bylaws*, IEEE Standards Ass'n § 6.2 (Dec. 2015), http://standards.ieee.org/develop/policies/bylaws/sb_bylaws.pdf.

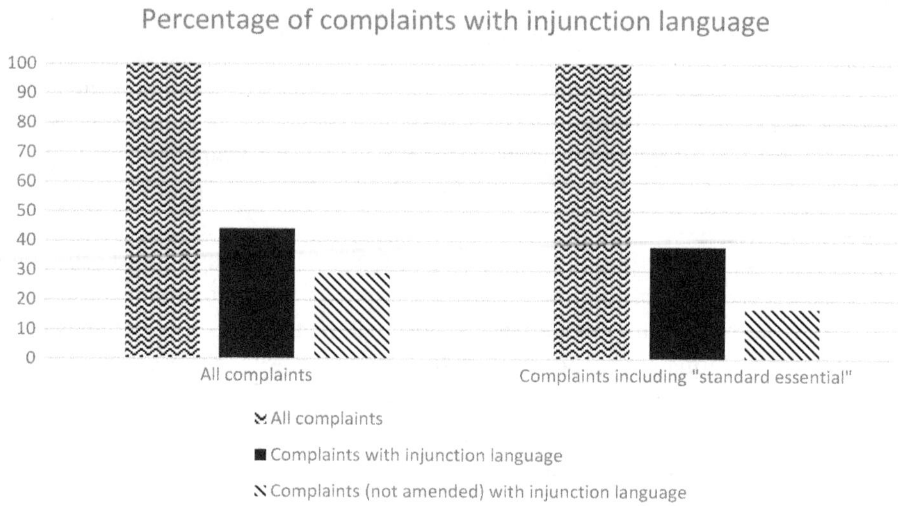


FIGURE 5.2. Graph of complaints seeking injunctive relief

the SEP space. In lieu of better information, however, this section briefly considers these data points.

In another chapter in this book – Restoring *Deterrence: The Case for Enhanced Damages in a No-Injunction Patent System* by Jonathan M. Barnett and David J. Kappos – the authors provide a number of anecdotes illustrating efficient infringement. One of the more striking is worth recounting here – Core *Wireless v. LG Electronics.*[64] In that case, Core Wireless, a joint endeavor of Microsoft and Nokia, assigned its portfolio to Conversant Intellectual Property Management.[65] Conversant initiated licensing talks with LG Electronics, which ultimately responded with a "terse one-page presentation stating that a lawsuit was ... 'preferable' to a license and that LG would prefer to wait until another major cell phone manufacturer licensed the portfolio" before taking a license.[66]

Following the precedent and other considerations described previously, it would be expected that SEP owners seek injunctive relief less often than other patent owners. A quick text search of complaints filed in patent cases between *eBay* and March 1, 2022, seems to bear this out,[67] although more in-depth analysis is warranted and beyond the scope of this chapter (Figure 5.2).

[64] Core Wireless Licensing S.A.R.L. v. LG Elecs., Inc., No. 2:14-cv-912-JRG, 2016 WL 10749825 (E.D. Tex. Nov. 2, 2016).

[65] *Id.* at *1.

[66] *Id.*

[67] Out of all complaints filed in patent cases for that date range (85,531 documents), approximately 44% of complaints (38,207 documents) include the phrases, "permanent injunction" or "injunctive relief." If the complaints are narrowed to exclude amended complaints (removing largely duplicative filings), approximately 29% (25,379 documents) of the complaints appear to

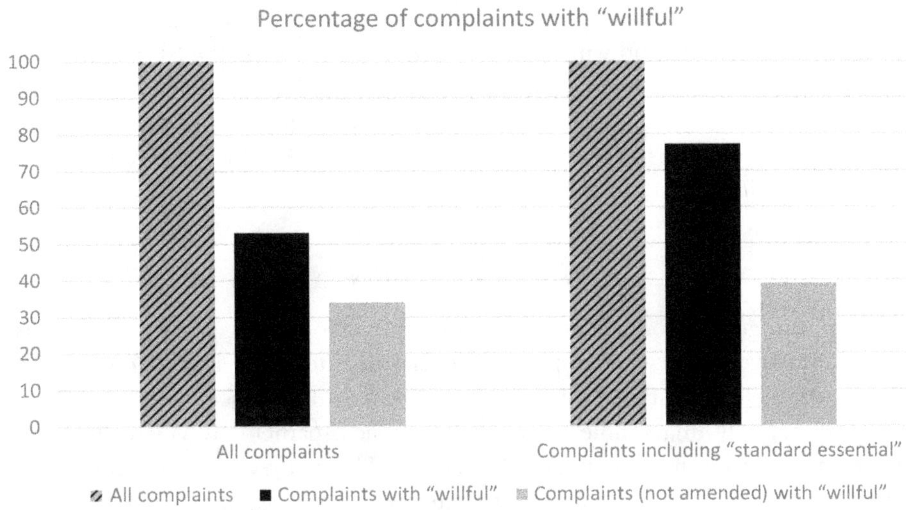


FIGURE 5.3. Graph of complaints indicating "willful"

It is difficult to draw many conclusions from this data, other than if a complaint includes the term "standard essential," it is slightly less likely to also include the terms "permanent injunction" or "injunctive relief." The data is, of course, subject to significant selection effects, as described previously, and also is simply a text-based search for relevant terms. However, this could signal that SEP owners are less likely to seek injunctive relief in some circumstances, particularly when coupled with the case law precedent and additional considerations detailed in the previous sections.

Performing a similar textual search based on the presence of "willful" in patent infringement complaints also may provide some insight about efficient infringement, subject to the same caveats and need for further research analysis (Figure 5.3).[68]

Again, it is difficult to draw any certain conclusions from this data, but in reading from the complaints for patent infringement of an SEP that include the term "willful," there are many cases where licensing negotiations are hindered by the

seek injunctive relief. Looking again at all complaints filed in patent cases, if the set is narrowed to those that include the phrase "standard essential," there are 234 documents total. Running the same text searches for "permanent injunction" or "injunctive relief" yields 38% (89 documents) and excluding amended complaints yields 17% (40 documents), including the terms for injunctive relief.

68 Out of all complaints filed in patent cases for that date range (85,531 documents), approximately 53% of complaints (45,831 documents) include the phrase "willful." If the complaints are narrowed to exclude amended complaints (removing largely duplicative filings), approximately 34% (29,463 documents) of the complaints appear to seek injunctive relief. Looking again at all complaints filed in patent cases, if the set is narrowed to those that include the phrase "standard essential," there are 234 documents in total. Running the same text searches for "willful" yields 77% (182 documents) and excluding amended complaints yields 39% (93 documents), including willfulness.

infringer, where negotiations continue for multiple years often attributed to delay tactics by the infringer, and where multiple offers are made by the SEP owner but are declined by the infringer.[69] What is interesting is that, as these negotiations continue and are, in essence extended by the infringer, that same party is continuing to use, without paying, the patented technology of the SEP. This is the very essence of efficient infringement.

## IV. CONCLUSION

Efficient infringement, or the decision by an infringer to "infringe now, pay later," is only an attractive option where injunctive relief is unavailable to a patent owner who has successfully demonstrated infringement of its patent. One sector where injunctions are routinely unavailable is in the SEP space, particularly where the SEP owner has obligated itself to FRAND licensing. Under these circumstances, it is difficult for the SEP owner to make a satisfactory showing against the *eBay* factors, used by courts to grant injunctive relief. Policymakers and SDOs have also weighed in, in favor of denying injunctions to SEP owners, concerned about the unproven phenomenon of patent holdup. SEP owners are often discouraged, or even prohibited, from seeking injunctive relief.

In these conditions, where injunctive relief is routinely unavailable (or even not sought), efficient infringement is able to flourish. Although it is difficult to measure efficient infringement, there are indications that it is occurring in the SEP space. From anecdotes to alleged facts in complaints to court determinations, it is clear that at least some makers of standards-compliant goods and services are willing to delay or even decline licensing of SEPs and to take their chances in court instead. Because the worst that could happen for these infringers is a damages award reflecting past infringement and providing a royalty for future infringement, the option of efficient infringement seems rational. What is missing from these anecdotes and the circumstantial data is how efficient infringement is harming SEP owners. It is time to stop acting as though efficient infringement does not exist; there are plenty of theoretical reasons and actual indicia that it does. Instead, it is time to study the negative effects it is having on standardization and innovation.

[69] Complaint, Ericsson Inc. v. TCL Commc'n Tech. Holdings, 2:14-CV-667 (E.D. Tex. June 3, 2014), ECF No. 1; Complaint, Nippon Tele. & Tele. Corp. v. Acer Inc., 6:20-CV-769 (W.D. Tex. Mar. 25, 2020), ECF No. 1.

# 6

# Restoring Deterrence

## The Case for Enhanced Damages in a No-Injunction Patent System

*Jonathan M. Barnett and David J. Kappos*

### I. INTRODUCTION

In May 2006, the US Supreme Court issued a decision that dramatically changed the landscape of patent enforcement, and with it the respect paid to patent rights. In *eBay, Inc. v. MercExchange, LLC,*[1] the Court departed from the long-standing principle that a patent owner is presumptively entitled to an injunction once it defends validity and demonstrates infringement. While the decision was unanimous, it produced two concurring opinions, one of which (authored by Chief Justice Roberts) emphasized the historical practice of usually granting injunctions to prevailing patentees and the other of which (authored by Justice Kennedy) emphasized that non-practicing entities (NPEs) abuse patent litigation to "hold up" users[2] for windfall payouts. In post-*eBay* case law, the latter opinion has not only prevailed but also been applied expansively. As a result, proliferating categories of patent owners, extending significantly beyond NPEs, no longer have any reasonable expectation of securing an injunction against infringers.

In this chapter, we ask a simple question. If a patent owner has no or a low expectation of securing injunctive relief against infringers, is it necessary to enhance damages for infringement to fully compensate the patent owner and deter infringement?

The Kennedy concurrence, and the bulk of the post-*eBay* case law, has implicitly answered this question in the negative. Courts have generally adopted the view that a patent licensing entity (or even an operational entity in certain circumstances) is made whole by monetary damages appropriately calculated based on the "reasonable royalty" standard. We contest this reasoning. For patent owners that have no

1 547 U.S. 388.

2 Throughout this chapter, "users" refers to intermediate users of patent rights, such as manufacturers, distributors, telecom carriers, or other entities that occupy intermediate positions in a technology supply chain.

realistic expectation of securing injunctive relief, reasonable royalty damages must be enhanced to achieve full compensation and deter infringement. This is true even if a court could calculate without error reasonable royalty damages in a hypothetical willing buyer-willing seller negotiation. So long as the infringer is sufficiently well-resourced (and can therefore fund a prolonged litigation), and there is a sufficiently low likelihood that the court will shift attorneys' fees or award treble damages, the infringer will usually find that agreeing to pay a license fee for use of a patented technology is economically irrational. Absent concerns about preserving goodwill with business partners, the infringer is better off using the technology and effectively negotiating the royalty rate through the litigation and settlement process, with some possibility that the patent will be invalidated altogether.

This reasoning is not merely theoretical. As we show through case studies of selected litigations involving owners of standard-essential patents (SEPs), for whom the likelihood of securing injunctive relief approaches zero, device makers and other intermediate users adopt the "use, then litigate" strategy, rather than paying a license fee up front. This behavior has an important implication. Rather than reducing patent litigation, a judicial standard that eliminates or significantly limits the availability of injunctive relief encourages infringement, promotes stalling tactics by users, and therefore *increases* litigation. These counterproductive effects are exacerbated in cases where infringers have greater litigation funding and lower opportunity costs than patent owners, who may settle for an amount that undervalues the patented technology or, by anticipation, never bother to enter the market at all. Contrary to the Kennedy concurrence and much of the post-*eBay* case law, a no-injunction regime is unlikely to leave patent owners – even a patent owner that relies principally on licensing revenues – in an economically equivalent position. So long as litigation is costly and uncertain and the risks of fee-shifting and treble damages are low, infringers have weak incentives to agree to a license and, depending on litigation resources and opportunity costs, patent owners are likely to agree to undercompensatory settlements. Given that all licensing takes place "in the shadow" of potential infringement remedies, a no-injunction regime with a low likelihood of enhanced damages is prone to distort even negotiated royalty rates to the advantage of licensees.

To translate our analysis into an immediately actionable policy proposal, we assume that neither the Supreme Court nor Congress is likely to take action to restore the historical presumption favoring injunctive relief for prevailing patent owners. Given this background assumption, we propose that, in any infringement litigation in which the injunction remedy is unlikely to be granted as a matter of law or practice, courts should apply a multiplier to enhance the monetary damages owing to the patentee under the reasonable royalty standard. The multiplier is designed to mitigate or correct the underdeterrence and undercompensation effects of a no-injunction regime by making even a well-resourced infringer worse off by electing to infringe (and invite litigation) rather than negotiating a license fee up front. Additionally, we discuss how courts can set the multiplier at an appropriate level to minimize any overcompensation and overdeterrence effects under an enhanced damages regime. While this proposal departs from current doctrine by using a damages multiplier in the conventional rather than exceptional case (in infringement litigation without an injunction option), it is consistent (as we explain in Section IV) with the original purpose of treble damages in US patent law – namely, to compensate patentees who were unable to petition for injunctive relief (due to procedural reasons that are now obsolete). Post-*eBay* case law has placed increasingly large portions of the patentee population in the same position.

Our discussion is organized as follows. In Section II, we discuss in theory how a no-injunction regime induces "efficient infringement" by users, resulting in undercompensation and underdeterrence effects in a wide range of circumstances. In Section III, we provide brief case studies of selected litigations, showing how well-resourced infringers act strategically to defer negotiation and payment of license fees to patent owners that cannot credibly threaten to secure an injunction remedy. In Section IV, we present our proposed adjustment to patent remedies to mitigate undercompensation and underdeterrence effects in a no-injunction regime.

## II. *EBAY* AND "EFFICIENT" INFRINGEMENT

### A. *The Disappearance of the Patent Injunction*

To fully appreciate the *eBay* decision, it is helpful to recall the context in which it was rendered.

A few months prior to the *eBay* decision, Research in Motion (more commonly known as "RIM"), the maker of the then-ubiquitous Blackberry device, had agreed to a $612 million settlement of a patent infringement litigation brought by NTP, Inc., a NPE cofounded by an individual inventor. The litigation and settlement had been widely reported as a windfall for the patent owner who reportedly "held up" RIM by threatening to shut down the Blackberry network given the lack of an immediately available design-around for the infringing component. Given this characterization, it might naturally follow that, at least in certain circumstances, injunctions should be limited in order to deter patent owners from engaging in opportunistic litigation and settlement strategies. This was precisely the motivating principle behind the *eBay* decision, issued later that same year.

Following the law of unintended consequences (at least for the four Justices who supported the decision but did not sign on to the Kennedy concurrence), the *eBay* decision has been applied expansively by the lower courts to encompass proliferating categories of patent owners that are perceived to pose an elevated holdup risk. With some qualification in specific cases, there is now generally a low likelihood that NPEs, SEP owners, and the owners of patents on components of a larger complex product can secure injunctive relief. It should be noted that the latter two categories encompass significant portions of the information technology markets. Even in a patent infringement litigation involving Apple and Samsung, two operating companies in head-to-head competition, the district court initially denied Apple even a phased-in "sunset" injunction on the grounds that, under the *eBay* test, "the principles of equity do not support the issuance of an injunction."[3] While the district court's decision was ultimately overturned by the Federal Circuit, it nonetheless illustrates the extent to which *eBay* has been applied well beyond the limited circumstances in which at least some members of the Court most likely envisioned their decision would alter patent remedies.

### B. *The Normalization of Patent Infringement*

Any license can be understood as a preemptive settlement of a potential infringement litigation. Hence the impact of *eBay* (or more precisely, post-*eBay* case law) extends well beyond the remedies issued in infringement litigation and permeates everyday licensing and other transactions involving patent-protected assets. Even if litigation is infrequent, sophisticated business parties must take into account whether there is any credible threat that the patent owner could ultimately secure an injunction against an infringing user. Generally speaking, the greater the threat of an injunction, the greater the bargaining leverage enjoyed by the patent owner in negotiating license terms, and vice versa. Given the limited availability of injunctive relief under post-*eBay* case law (and assuming that treble damages are a low-probability outcome), firms that are primarily users of patented technology may often, if not typically, conclude that the expected net payoff from infringement exceeds the expected net payoff from negotiating and paying an up-front license fee. While eliminating injunctive relief may deter patent owners from adopting holdup strategies that can yield settlement payouts that overvalue the relevant technology, it encourages well-resourced users to adopt holdout strategies that lead to license fees that undervalue the relevant technology. In effect, *eBay* has rendered patent infringement a rational business strategy for significant categories of intermediate users.

We can identify more precisely the conditions under which this "use, then maybe litigate" strategy will be preferred by users versus a "license, then use" strategy in a no-injunction regime. These conditions are as follows:

1. It is costly for the user to acquire or develop a design-around substitute for the patented technology.
2. The user has sufficient resources to fund a prolonged infringement litigation.
3. The user will retain access to, and therefore derive revenues or other economic benefits from use of, the patented technology during the litigation.

[3] Apple, Inc. v. Samsung Elecs. Co. Ltd., 909 F. Supp. 2d 1147 (N.D. Cal. 2012).

4. There is a sufficiently low likelihood that a court would award attorneys' fees or treble damages.
5. Reasonable royalty damages are likely to approximate, or at least will not exceed significantly, the royalty that the user and the patent owner would likely negotiate prior to entering into litigation.

If at least some of these conditions are satisfied, then the user is likely to conclude that infringement delivers a higher expected net payoff than agreeing to a license and avoiding the risks and costs of an extended litigation with the patent owner. The rationale is as follows. By electing to invite litigation, rather than settle it preemptively through a negotiated license, the user incurs the direct and indirect costs of litigation but, in exchange, "purchases" the opportunity to either invalidate the patent entirely (in which case the royalty falls to zero) or negotiate a reduced or equivalent royalty rate. Even in the scenario in which the patent owner ultimately prevails on liability, the previously stated conditions (in particular, the low likelihood of treble damages and fee-shifting) imply that the user can expect to pay a damages award equal to the royalty fee it would have incurred initially, plus prejudgment interest (if included).[4] The only incremental cost incurred by the infringer are attorneys' fees, which constitute a fee paid by the user for the opportunity to reduce or eliminate the royalty rate through the litigation and settlement process.

The likelihood that infringement outperforms licensing increases if the user is a more "patient" litigator than the patent owner. This will be the case if the user has substantially greater resources to fund litigation as compared to the patent owner and the patent owner relies substantially on licensing revenues from the patented technology. Moreover, given the absence of any risk of preliminary injunctive relief, the infringer extracts revenues and other economic benefits from use of the patented technology while the patent owner, who incurred the costs of developing the technology (or acquiring the patent covering the technology), receives nothing. Assuming this asymmetry in both litigation resources and opportunity costs – up to and including insolvency on the part of the patent owner – bargaining leverage shifts to the infringer in any settlement discussions that take place concurrently with litigation, likely resulting in a negotiated royalty that is lower than the amount that would be awarded in a fully adjudicated litigation or would have been awarded in a pre-litigation negotiation.

These observations have an important implication for the ultimate consequences of the limitations on injunctive relief that have arisen in post-*eBay* case law.

[4] Prejudgment interest is contemplated by the Patent Act; see 35 U.S.C. § 284 (stating that a successful patentee is entitled to "damages adequate to compensate for the infringement, but in no event less than a reasonable royalty . . . together with interest and costs as fixed by the court"). In interpreting the statute, the Supreme Court has held that "prejudgment interest should ordinarily be awarded" but also stated that courts had discretion not to do so when appropriate; see General Motors Corp. v. Devex Corp., 461 U.S. 648, 657 (1983).

While *eBay* may have identified a situation in which the availability of injunctive relief can enable a patent owner to secure an "overvalued" royalty from a well-resourced user, the *un*availability of injunctive relief sometimes enables a well-resourced user to negotiate an "undervalued" royalty with the patent owner. Unless there is reason to believe that patent owners are systematically securing royalty rates that are excessive relative to a socially optimal royalty rate,[5] this is clearly an inefficient state of affairs that runs counter to the public interest in providing a level playing field for open-market negotiation of royalty rates for IP assets among willing licensors and licensees.

### C. *Existing Damages Enhancements*

The likelihood that infringement will be privately efficient for the user, as compared to negotiating and paying a license fee up front, depends on whether a user has a sufficiently low expectation of being held liable for damages enhancements, such as a court shifting attorney fees or applying a multiplier to a reasonable royalty damages award. As discussed subsequently, this result is consistent with applicable case law, which emphasizes that these tools are reserved for exceptional cases, and available data, which shows that an informed infringer should anticipate a low likelihood of financial exposure to damages enhancements.

#### 1. Attorney Fees

Civil litigation in US courts has a long-standing commitment against shifting attorney fees to the losing party. Patent law shares this commitment. The patent statute provides that "the court in *exceptional* cases may award reasonable attorney fees to the prevailing party."[6] In 2005, the Federal Circuit set a high bar for shifting fees. Specifically, the court held that a court may only award attorney fees "if both (1) the litigation is brought in subjective bad faith, and (2) the litigation is objectively baseless."[7] Additionally, the court clarified that litigation is objectively baseless only if it is "so unreasonable that no reasonable litigant could believe it would succeed," and that litigation is made in subjective bad faith if the plaintiff "actually know[s] that it is objectively baseless."[8] In *Octane Fitness, LLC v. Icon Health & Fitness, Inc.*, decided in 2014, the Supreme Court rejected this standard, holding that a case

[5] This would arise if either (i) patentees were acting collusively in setting licensing rates, or (ii) patents were being consistently issued that were excessively large in scope, duration, or other parameters relative to the hypothetical patents that would be issued by an omniscient and benevolent social planner. There is no evidence to support either scenario on a systematic basis.

[6] 35 U.S.C. §285 (emphasis added).

[7] Brooks Furniture Mfg., Inc. v. Dutailier Int'l, Inc., 393 F.3d 1378, 1381 (Fed. Cir. 2005).

[8] *Id.* at 1377.

is "exceptional" and therefore merits fee-shifting when the case "stands out from others with respect to the substantive strength of a party's litigating position . . . or the unreasonable manner in which the case was litigated."[9] While the Court's decision provides courts with some additional latitude to award attorney fees, a prevailing litigant seeking to shift its fees to the other litigant still faces significant hurdles and fee-shifting remains an atypical outcome in patent infringement litigation. Moreover, fee-shifting is most often applied in favor of prevailing defendants, so if anything, in the aggregate it would tend to encourage users to litigate rather than take licenses.[10]

### 2. Treble Damages

The patent statute provides that a "court *may* increase the damages up to three times the amount found or assessed."[11] While not stated in the statute, courts have widely required a finding of willfulness as a predicate condition for even considering whether to exercise the statutory option to award enhanced damages.[12] Critically, a willfulness finding is a predicate condition for *considering*, not *awarding*, enhanced damages. Hence, it is entirely possible (and common) that a court may decline to award enhanced damages even if willfulness has been found. Moreover, as discussed subsequently, it is entirely possible (and again common) that a court may select a damages multiplier that is *less than* three times the damages amount.

The evidentiary threshold for showing willfulness has varied. In a 1983 opinion, *Underwater Devices Inc. v. Morrison-Knudsen Co.*, the Federal Circuit lowered the threshold by holding that the willfulness standard implied that a potential infringer had "an affirmative duty to exercise due care to determine whether or not he is infringing."[13] In a 2007 opinion, *In re Seagate Tech LLC*, the Federal Circuit rejected this standard and raised the bar for finding willfulness, adopting a two-part test that required patentees to show that (1) "the infringer acted despite an objectively high likelihood that its actions constituted infringement of a valid patent," and (2) the risk of infringement "was either known or so obvious that it should have been known to the accused infringer."[14] In 2016, the Supreme Court rejected the *Seagate* test in *Halo Electronics, Inc. v. Pulse Electronics, Inc.*, emphasizing that the statute "gives district courts the discretion to award enhanced

9 Octane Fitness, LLC v. Icon Health & Fitness, Inc., 572 U.S. 545, 554 (2014).

10 Scott M. Flanz, *Octane Fitness: The Shifting of Patent Attorneys' Fees Moves into High Gear*, 19 Stan. Tech. L. Rev. 329, 353 (2016).

11 35 U.S.C. § 284 (emphasis added).

12 Eko Brands, LLC v. Adrian Rivera Maynez Enters., Inc., 946 F.3d 1367, 1378 (Fed. Cir. 2020) (stating that a district court may consider awarding enhanced damages "once an affirmative finding of willfulness has been made").

13 Underwater Devices Inc. v. Morrison-Knudsen Co., 717 F.2d 1380, 1389 (Fed. Cir. 1983).

14 497 F.3d 1360, 1371 (Fed. Cir. 2007).

damages against those guilty of patent infringement."[15] Yet, the shift put in place by the *Halo* decision should not be exaggerated. Addressing arguments that the *Seagate* standard had protected inadvertent infringers from treble damages, the Court emphasized that the lower courts' exercise of discretion under §284 should be guided by the "sound legal principles developed over nearly two centuries of application and interpretation of the Patent Act."[16] Additionally, the Court emphasized that treble damages "are not to be meted out in a typical infringement case, but are instead designed as a 'punitive' or 'vindictive' sanction for egregious infringement behavior."[17] As we discuss subsequently, a 2021 Federal Circuit decision, *SRI International Inc. v. Cisco Systems*,[18] appears to heighten the standard for finding willfulness consistent with the guidance set forth in *Halo*.

Empirical studies of judicial determinations of willfulness and enhanced damages in infringement litigation show that these determinations are sensitive to the governing threshold for finding willfulness.[19] During 1983–1999, when willful infringement was determined under the plaintiff-friendly "affirmative duty" standard, juries found willfulness in 71% of litigations in which they considered the issue and judges found willfulness in 53% of litigations in which they considered the issue.[20] Among litigations that found willfulness, enhanced damages were awarded 63% of the time if a jury found willfulness and 95% of the time if a judge found willfulness.[21] In the aggregate, this data implies that, out of all fully adjudicated infringement trials during this period, 18% reached a positive willfulness determination and 12% then resulted in enhanced damages.[22] In the three years prior to the 2007 *Seagate* decision (September 2004–August 2007), which adopted the more demanding "objective recklessness" standard, courts had found willfulness in 48.2% of litigations in which they considered the issue, compared to 37.2% in the three

[15] 136 S. Ct. 1923, 1935 (2016).
[16] *Id.* at 1935.
[17] *Id.* at 1932.
[18] No. 20-1685 (Fed. Cir. Sept. 28, 2021).
[19] For purposes of this discussion, we do not take into account "selection effects" that may qualify the interpretation of changes in willfulness findings following changes in the governing legal standard. However, we note that, due to the time lag inherent to the litigation process, any such selection effects cannot plausibly impact the interpretation of changes in willfulness findings shortly following any such change in the governing legal standard. For further discussion of this point, see Christopher B. Seaman, *Willful Patent Infringement and Enhanced Damages after* In re Seagate: *An Empirical Study*, 97 Iowa L. Rev. 417, 442–43 (2012).
[20] Kimberly A. Moore, *Judges, Juries, and Patent Cases – An Empirical Peek Inside the Black Box*, 99 Mich. L. Rev. 365, 391 (2000).
[21] *Id.* at 394.
[22] These figures were derived by the authors based on data in the 2000 Moore study. *See id.* at 383 n.76 (providing number of fully adjudicated patent trials) and at 394, Tbl. 5 (providing percentages of such trials in which willfulness was found and enhanced damages were awarded).

years after *Seagate* (August 2007–July 2010).[23] Following the 2016 *Halo* decision, which rejected *Seagate* and lowered the threshold for finding willfulness (but without reverting to the *Underwater Devices* "affirmative duty" standard), there was again an increase in willfulness findings and enhanced damages awards. Out of all district-court litigations that determined willfulness and enhanced damages during December 2013–December 2018, the percentage of such decisions that found willfulness increased after *Halo* (decided in June 2016) from 22.8% to 55.7%, and the percentage of such decisions that awarded enhanced damages increased from 10.1% to 29%.[24]

It is important to emphasize that the likelihood that a *fully adjudicated infringement suit* will result in an enhanced damages award is inherently greater, and almost certainly significantly greater, than the likelihood that a *filed infringement suit* or a *particular act of infringement* will ultimately result in enhanced damages. There are several reasons why this would be the case: (1) the patent owner may not bring suit due to lack of knowledge, resources, or economic interest; (2) the patent owner brings suit, but the parties settle (probably the most common outcome); or (3) the parties do not settle, but the defendant prevails on invalidity, noninfringement , or other grounds so damages are a moot issue.[25] Even among cases that do not settle, the likelihood of enhanced damages is limited since adjudicated patent litigations only result in a finding of infringement liability about one-third of the time on average. During 1998–2017, only 34% of all patent litigations that proceeded to a final decision resulted in a finding of infringement,[26] of which (as shown by the data discussed earlier) a minority then resulted in both a willfulness finding and enhanced damages. Since the vast majority of filed infringement suits never reach adjudication (approximately 94% based on one estimate),[27] an infringer faces an insignificant likelihood of incurring an enhanced damages award at the time an infringement suit is filed (even without discounting for the less-than-certain likelihood that a patent owner will detect infringement and elect to bring suit in response).

While these estimates are necessarily imprecise to some extent and may vary on a case-specific basis depending on the strength of a particular infringement claim and

[23] Seaman, *supra* note 19, at 444, Tbl. 2. Other commentators observed that the *Seagate* decision enabled infringers to defeat willfulness claims by finding an "objectively reasonable" legal theory that the contested patent was invalid or had not been infringed. *See* Dmitry Karshtedt, *Enhancing Patent Damages*, 51 U.C. Davis L. Rev. 1427, 1459 (2018).

[24] Veena Tripathi, *Halo from the Other Side: An Empirical Study of District Court Findings of Willful Infringement and Enhanced Damages Post-Halo*, 103 Minn. L. Rev. 2617, 2637–40 (2019).

[25] For similar views, *see id.* at 2636.

[26] PriceWaterhouseCoopers, 2018 Patent Litigation Study 13, 18 (2018).

[27] Kimberly A. Moore, *Empirical Statistics on Willful Patent Infringement*, 14 Fed. Cir. Bar. J. 227, 234 (2004) (based on patent infringement litigations during 1999–2000). For the period 1983–1999, the same author found that 6.9% of all patent infringement suits went to trial. *See* Moore, *supra* note 20, at 383.

the litigation resources available to a patent owner, it is nonetheless clear that informed infringers can in general anticipate a low likelihood of enhanced damages liability when electing whether to infringe upon, or take a license to, a patented technology, or to settle a patent infringement suit. Those expectations are consistent with our theoretical analysis that a prospective licensee in a no-injunction environment will expect that an infringement litigation would most likely result in either (1) zero liability (due to a finding of invalidity or noninfringement), or (2) a damages award approximately equivalent to the royalty rate that would have been paid in a negotiated transaction (plus prejudgment interest if awarded): in each case, excluding the infringer's legal fees. These anticipated outcomes of a fully adjudicated litigation in turn impact settlement outcomes, which can even result in a settlement amount that leaves the infringer better off than if it had agreed initially to pay a royalty, given the fact that settlements do not typically include prejudgment interest to reflect the time value of money.[28] So long as the infringer is willing to bear the expected legal fees (which deliver a potential gain in the form of a zero royalty rate due to a finding of invalidity or noninfringement or a reduced royalty rate by settlement), the incentive to infringe is self-evident.

## III. EFFICIENT INFRINGEMENT IN ACTION

SEPs cover critical technologies behind 3G, 4G/LTE, and now 5G mobile communication standards. It is widely asserted that SEP owners have incentives to "hold up" potential licensees by demanding exorbitant royalty rates, which would in turn inflate retail prices for consumers at the end of the technology supply chain. To remedy this purported risk (which has yet to be empirically demonstrated), courts and regulators in the United States, European Union, and other jurisdictions have construed an SEP owner's commitment to "fair, reasonable, and nondiscriminatory" (FRAND) licensing as implying a waiver by the SEP owner of its right to seek injunctive relief against infringers in most circumstances. As a result, SEP owners can only credibly threaten infringers with the prospect of monetary damages determined by a court based on a "reasonable royalty" standard (for which there are in turn a variety of calculation methodologies). Hence, SEP owners operate under the equivalent of the no-injunction regime imposed by *eBay* and post-*eBay* case law.

This nearly complete ban on injunctive relief for SEP owners logically incents well-resourced users to "hold out" by stalling licensing negotiations and compelling patentees to bear the costs and delay involved in pursuing litigation as a means of securing remuneration for use of their technology. The business case for

[28] Based on one of the authors' personal experience in practice. On this point, see also Sanofi-Aventis, et al. v. Apotex Inc. and Apotex Corp., Case No. 2011-1048 (Fed. Cir. 2011) (upholding a patent litigation settlement agreement that barred the patent owner from seeking prejudgment interest).

infringement in a no-injunction environment is compelling. The former head of patent licensing at Apple has explained the logic, stating that "efficient infringement, where the benefits outweigh the legal costs of defending against a suit, could almost be viewed as a 'fiduciary responsibility,' at least for cash-rich firms that can afford to litigate without end."[29]

To explore these strategies in more detail (within the scope of this chapter), we describe four selected litigations between SEP owners and alleged infringers in US and UK courts. In each case, we indicate in parentheses the principal court and the period during which the litigation took place, which provides a sense of the delay involved when pursuing a royalty through infringement proceedings (which in turn often follows a substantial period of licensing negotiations). Each litigation provides qualitative evidence illustrating how the absence of injunctive relief leads device manufacturers to engage in stalling tactics that require patent owners to undertake costly and protracted litigation in an effort to secure a royalty through settlement or adjudication.

### A. *Core Wireless v. LG (US District Court for the Eastern District of Texas, 2014–2016)*

In 2014, Core Wireless (a subsidiary of Conversant Intellectual Property Management), the holder of patents relating to user interfaces, battery life, and voice recognition in smartphone devices, brought two infringement suits against LG, a leading manufacturer of tablets, handsets, and other electronic devices.[30] When each of the cases went to trial in 2016, Core Wireless prevailed on patent validity and infringement in both litigations, and juries awarded Core Wireless $3.5 and $2.28 million in damages, respectively.[31] In the second of the two trials, the judge awarded enhanced damages of $456,000 in light of evidence that LG had "undisputed" knowledge of Core Wireless's patents and "abruptly terminated" licensing discussions.[32] Specifically, the judge cited weak invalidity and

[29] *The Trouble with Patent-Troll Hunting*, ECONOMIST (Dec. 14, 2019) (citing Boris Teksler, described as "Apple's former patent chief").

[30] Complaint for Patent Infringement, Core Wireless Licensing S.A.R.L v. LG Elec., Inc., No. 2:14-cv-911 (E.D. Tex. Sept. 26, 2014); Complaint for Breach of Contract, Declaratory Relief, and Patent Infringement, Core Wireless Licensing S.A.R.L v. LG Elec., Inc., No. 2:14-cv-912-RSP (E.D. Tex. Sept. 26, 2014).

[31] Verdict Form, Core Wireless Licensing S.A.R.L v. LG Elec., Inc., No. 2:14-cv-911-RSP (E.D. Tex. Mar. 24, 2016) (finding infringement and awarding $3.5 million in damages); Verdict Form, Core Wireless Licensing S.A.R.L v. LG Elec., Inc., No. 2:14-cv-911-JRG-RSP (E.D. Tex. Sept. 16, 2016) (finding infringement and awarding $2.28 million in damages). The $3.5 million damages award was relitigated in a second trial, which upheld the award; see Verdict Form, Core Wireless Licensing S.A.R.L v. LG Elec., Inc., No. 2:14-cv-912-JRG (E.D. Tex. Feb. 27, 2019).

[32] Final Judgment, Core Wireless Licensing S.A.R.L v. LG Elec., Inc., No. 2:14-cv-912-JRG (E.D. Tex. Nov. 01, 2016).

noninfringement defenses and evidence that, after an extended negotiation period, LG had invited Core Wireless representatives to its offices in South Korea purportedly to resolve the matter, but "[r]ather than make an offer or engage in serious, good faith negotiations, LG delivered a terse one-page document stating that a lawsuit at that time between the parties was 'preferable' to a license."[33] Among recent SEP litigations, this is perhaps the only case in which a prospective licensee explicitly adopted a policy of ignoring the patent and inviting litigation, even arguing in court that "infringement is an expected part of the standard-setting model."[34] This may explain why, to our knowledge, it is the only SEP litigation in which the court awarded enhanced damages. Few implementers are likely to repeat this mistake.

### B. *Qualcomm v. Apple (US District Court for the Southern District of California, 2017–2019)*

This complex sequence of litigations involved Qualcomm, the leading chip supplier and innovator in the wireless communications market, and Apple, one of the world's leading handset device manufacturers and the most valuable brand in this market. In the context of a dispute over rebate payments allegedly owed by Qualcomm, Apple sued Qualcomm in January 2017 for engaging in licensing practices that allegedly violated the antitrust laws.[35] In response, Qualcomm filed in May 2017 a breach of contract suit against Apple's contract manufacturers, which had stopped paying fees under their licensing agreements with Qualcomm, purportedly at Apple's direction.[36] In July 2017, Qualcomm filed a patent infringement suit against Apple for use of its patents in certain Apple devices.[37] These litigations took place while the Federal Trade Commission (FTC) concurrently pursued an antitrust suit against Qualcomm, seeking a dramatic remedy that purported to operate on a worldwide basis and would have required Qualcomm to renegotiate virtually all of its licensing agreements. The result would have been a reengineering of a critical element of the contractual infrastructure behind the global wireless communications industry.

To appreciate the dynamics in the Apple/Qualcomm litigation, it is important to note three key facts. First, the size of Apple, which is regularly ranked as the world's largest or second-largest company by market capitalization, means that it faced

33 *Id.*
34 Core Wireless Licensing S.A.R.L. v. LG Elecs., Inc., No. 2:14-cv-911, 2016 WL 4596118, at *2 (E.D. Tex. Sept. 3, 2016) (Gilstrap, C.J.).
35 Redacted Complaint, Apple Inc. v. Qualcomm Inc., No. 17CV0108GPCNLS (S.D. Cal. Jan. 20, 2017).
36 Redacted Complaint, Qualcomm Inc. v. Compal Electronics, Inc. et al. (S.D. Cal. May 17, 2017).
37 Complaint for Patent Infringement, Qualcomm Inc. v. Apple Inc., No. 17CV1375JAHAGS (S.D. Cal. July 06, 2017).

virtually no constraints on litigation funding. Second, while Qualcomm derived no revenue from Apple's use of its patent-protected assets during the two-year litigation, Apple enjoyed billions of dollars in revenue through the sale of devices that used and relied upon Qualcomm's technology. Third, given the low likelihood of an injunction, Apple never faced any material prospect that it would be forced to withdraw devices for which it had already incurred the costs of manufacture and distribution.

These factors naturally tilted bargaining leverage in favor of Apple, which effectively made a litigation "investment" in an effort to reduce its input costs (and increase its profit margins) from a technology supplier through the vehicle of an antitrust litigation.[38] For Apple, the costs of the litigation were nominal relative to the potential benefits in the form of reduced royalty fees during the lifetime of the current wireless technology generation. This was not true for Qualcomm, which was deprived during the litigation of a principal revenue stream and, together with the antitrust suit brought against it concurrently by the FTC, faced legal challenges that, if successful, would have threatened the viability of its business model. This asymmetry can be observed in the fact that, following settlement of the Apple/Qualcomm litigation on April 16, 2019, Qualcomm's stock rose 23%, while Apple's stock only rose 1%.[39]

To be clear, Apple's actions are neither nefarious nor surprising; rather, they represent a rational business response to a truncated property rights environment that favors users over originators of IP assets. As observed by Apple's former head of licensing,[40] a well-resourced user that faces no real threat of injunctive relief is best off infringing and then litigating the royalty rate in court (or settling the rate in the context of litigation), rather than entering initially into a license. In a post-*eBay* environment, Apple (and any other sufficiently resourced user) would be foolish to do otherwise.

### C. *Unwired Planet v. Huawei (UK, 2014–2020)*

In 2013, Unwired Planet, the owner of patents relating to wireless communications, approached Huawei, a leading handset manufacturer, to enter into discussions

[38] This strategy is not a hypothetical construct. In an internal Apple slide deck that Qualcomm presented in oral arguments at trial (immediately prior to settlement), one slide stated, "Goal: Reduce Apple's Net Royalty to Qualcomm"; other slides listed goals such as "Devalue SEPs," "Limit Injunction," "leverage our purchasing power," and "captur[e] IP value with purchase price." Other documents evidenced a strategy of negotiating agreements in which Qualcomm's obligation to supply Apple with modems applies irrespective of whether Apple's contract manufacturers were still making royalty payments to Qualcomm for licensing its patents. On this evidence, see J. Gregory Sidak, *Monopoly, Innovation, and Due Process:* FTC v. Qualcomm – and *the Imperative to Destroy*, 6 Criterion J. Innovation 1, 125–26 (2020).

[39] Kif Leswing, *Why Apple Was So Upset with Qualcomm – and Why It Finally Had to Give in*, CNBC (Apr. 17, 2019).

[40] *The Trouble with Patent-Troll Hunting*, *supra* note 29.

concerning a license agreement for use of Unwired Planet's SEP-protected technology. After several inquiries, Huawei's IP department entered into protracted discussions with Unwired Planet over the terms of a nondisclosure agreement (a straightforward document that is typically uncontroversial). Unable to conduct constructive negotiations with Huawei, Unwired Planet elected to file an infringement suit in March 2014 against Huawei, Samsung, and Google for the infringement of six UK-issued patents, including five claimed SEPs.[41] During the litigation, Unwired Planet made several license offers to the defendants. Google and Samsung settled while Huawei made counteroffers, but the parties were unable to resolve the dispute.

The court ultimately held that Huawei had infringed two valid patents held by Unwired Planet and, for purposes of damages, calculated the FRAND royalty to which the patent owner was entitled. The court also rejected the view that Unwired Planet had breached its FRAND commitment by initiating infringement litigation and seeking injunctive relief. Most importantly, the court held that, while the FRAND commitment in general precludes SEP owners from seeking injunctive relief, this bar is lifted once the alleged infringer is deemed to be an "unwilling licensee" who acts opportunistically to prolong licensing negotiations.[42] On the basis of this principle (which refined a holding by the European Court of Justice in a 2015 decision[43]), the court issued an injunction against further use by Huawei of the infringed patents, unless Huawei elected to enter into a license based on the FRAND royalty as determined by the court. Upon appeal, the UK Supreme Court upheld the lower court's royalty determination and agreed that Unwired Planet had not acted "abusively" given its stated willingness to license upon FRAND-compliant terms.[44] This decision represents one of the few cases in which a court has appreciated seriously the "holdout" risk faced by innovators that are practically precluded from seeking a legal order to block infringement.

### D. *Optis Wireless v. Apple (UK, 2017–Present; US District Court for the Eastern District of Texas, 2019–2021)*

Optis Wireless holds patents that have been declared essential to the 4G LTE standard. In 2017, Optis had approached Apple offering a license for use of the patented technology. In February 2019, after two years of unresolved negotiations, Optis Wireless filed suit against Apple for infringement based on alleged use of the

[41] This description is largely based on Unwired Planet International Ltd. v. Huawei Technologies Co. Ltd. et al. [2017] EWHC 711 (Pat).

[42] *Id.*

[43] Huawei Technologies Co. Ltd. v. ZTE Corp. and ZTE Deutschland GmbH, Case C-170/12 (Court of Justice of the European Union 2015).

[44] Unwired Planet International Ltd v. Huawei Technologies (UK) Co. Ltd. [2020] UKSC 37, at 53.

patented technology in Apple's iPhone, iPad, and Apple Watch devices.[45] In August 2020, a jury reached a finding of willful infringement and awarded Optis $506 million in reasonable royalty damages, which did not include a damages enhancement.[46] In light of jury instructions concerning the FRAND obligation that were deemed to be defective, the court subsequently ordered a new damages proceeding.[47] In August 2021, that proceeding resulted in a reduced award of $300 million.[48] The judge declined to award enhanced damages.[49]

Optis concurrently filed infringement suits against Apple in the United Kingdom. In one of these suits, the court found in October 2020 that Optis' patent claims were valid and infringed.[50] Additionally, the court granted Optis' motion for a separate proceeding to determine whether Apple is an "unwilling licensee," on the ground that it has purportedly declined to commit to pay a FRAND-compliant royalty once that royalty is determined through litigation.[51] Under the *Unwired Planet* decision described previously, an "unwilling licensee" finding is the only circumstance in which SEP owners under British law may be entitled to injunctive relief. In September 2021, the High Court stated that Apple may be enjoined from selling the infringing products in the UK market unless it commits up front to taking a FRAND-compliant license from Optis, the terms of which will be determined at trial.[52] (Apple had previously threatened to withdraw its products out of the British market if the trial resulted in a "commercially unacceptable" royalty award.[53]) In June 2023, the High Court determined a global FRAND rate for the relevant portion of Optis' SEP portfolio, which Apple must accept or it may face an injunction barring sales of some iPhone and iPad models in the UK market.[54]

45 Original Complaint, Optis Wireless Technology, LLC et al. v. Apple Inc., No. 2:19-cv-66 (E.D. Tex. Feb. 25, 2019).

46 Verdict Form, Optis Wireless Technology, LLC et al. v. Apple Inc., No. 2:19-cv-66 (E.D. Tex. Aug. 11, 2020). The jury's findings were upheld by the court; see Final Judgment, Optis Wireless Technology, LLC v. Apple Inc., No. 2:19-cv-66 (E.D. Tex. Feb. 26, 2021).

47 Order Granting New Trial, Optis Wireless Technology, LLC et al. v. Apple Inc., No. 2:19-cv-66-JRG (E.D. Tex. Apr. 14, 2021), at 9.

48 Verdict Form, Optis Wireless Technology, LLC et al. v. Apple Inc., No. 2:19-cv-00066-JRG (E.D. Tex. Aug. 13, 2021).

49 Caleb Drickey, *Gilstrap Won't Triple $300 M 4G Patent Verdict against Apple*, LAW 360, Sept. 9, 2021.

50 Optis Cellular Technology LLC et al. v. Apple Retail UK Limited, Case No. HP-2019-000006 [2020] EWHC 2746 (Pat).

51 Optis Cellular Technology LLC et al. v. Apple Retail UK Limited, Case No. HP-2019-000006 [2020] EWHC 2425 (Pat).

52 Optis Cellular Technology LLC et al. v. Apple Retail UK Limited, Case No. HP-2019-000006 [2021] EWHC 2464 (Pat).

53 Joff Wild, *Apple's Threat to Leave British Market over FRAND Royalty Is Not Credible, Says Judge*, IAM (Sept. 28, 2021).

54 Pinsent Masons, *Global FRAND rate set by UK court in Optis v Apple patent dispute*, Out-Law, June 9, 2023, www.pinsentmasons.com/out-law/news/global-frand-rate-uk-optis-apple-patent

## IV. HOW TO MAKE EFFICIENT INFRINGEMENT INEFFICIENT

As apparent from both our theoretical analysis and our summary of exemplary cases, a patent regime that eliminates the possibility of injunctive relief but makes no offsetting change in the customary menu of patent remedies – namely, a reasonable royalty award that includes damages enhancement – will inherently result in a combination of undercompensation and underdeterrence effects in a wide range of circumstances. As discussed, qualitative evidence from selected infringement litigations, in which the likelihood of injunctive relief was essentially zero, is consistent with these expectations. In the following discussion, we show that these adverse effects can be significantly mitigated by mandating enhanced damages in cases where patent owners have no realistic expectation of injunctive relief. While the prospect of enhanced damages can give rise to windfall awards that invite opportunistic litigation in specialized circumstances, we show that this adverse effect can be mitigated through appropriate adjustment of the damages multiplier based on existing case law.

### A. *Historical Background*

The notion that infringers should be subject to enhanced damages is not new.[55] In an amendment made in 1793 to the patent statute (only three years after its original enactment), Congress required that infringers pay damages *at least* equal to three times "the amount the patentee usually received for either selling the patented invention or licensing the invention."[56] That is: Treble damages were mandatory, and judges had authority to select an even higher damages multiplier. The amendment is thought to have reflected the fact that injunctions were rarely awarded by federal courts (due in part to the Anti-Injunction Act, which generally precluded federal courts from exercising equitable powers reserved for state courts) and therefore treble damages were deemed necessary to correct for undercompensation.[57] In 1800, the statute was amended to set treble damages as the maximum multiplier. In 1819, Congress allowed federal courts to exercise equitable jurisdiction in all patent cases, and in 1836, the patent statute was amended so that treble damages became a discretionary, rather than mandatory, component of patent damages, as remains the law today. Since that time, changes in the incidence of enhanced damages have arisen as a result of changes in the standards adopted by courts to determine the threshold for finding willfulness and awarding enhanced damages, as discussed in Section I.

[55] This paragraph is based largely on James Ryan, *A Short History of Patent Remedies*, 6 CYBARIS 150 (2015).

[56] *Id.* at 156.

[57] *Id.* at 158–59 (stating that "[t]he purpose of the treble damages provision was to provide adequate remedies to those who do not have access to equity").

## B. *Mandating Enhanced Damages*

Legal scholars have long recognized that supercompensatory damages can be a sound policy tool in cases where the identification and prosecution of individual legal violations is sufficiently costly and would not meet a cost–benefit test. In these circumstances, the "windfall" enjoyed by any individual plaintiff is the price paid to maintain deterrence generally since, without a credible threat of supercompensatory damages, prospective violators would have no rational incentive to comply with the relevant legal obligation. Under a patent regime in which injunctive relief is unavailable and infringers' maximum "downside" is a reasonable royalty damages award (plus interest if awarded), the patent regime is unlikely to deter infringers with sufficient litigation resources. Absent concerns about preserving goodwill with actual or potential business partners (including the necessity of securing complementary know-how from the patentee), a well-resourced party will rationally choose to use the patented technology and invite the patentee to initiate infringement litigation. As discussed previously, this underdeterrence effect is likely to lead to undercompensation where patentees with limited litigation funding or high opportunity costs are prone to agree to royalty rates that undervalue the relevant technology. This is especially likely to be the case with smaller entrants, including disruptive start-ups, universities, and independent inventors. As a broader consequence, a downward distortion in royalty rates would then be expected to arise in *all* licensing negotiations involving patent owners that are effectively ineligible for injunctive relief. This effectively transfers wealth from entities that specialize in generating innovations to well-resourced entities that specialize in using them, a result that runs counter to the policy objective behind the patent system.

Assuming it is not feasible to institute a legal presumption favoring injunctive relief for prevailing patentees (which would require overturning *eBay* by judicial or legislative action), the deterrence and compensatory functions of the patent system can be restored (at least in part) by requiring that courts award enhanced damages in any litigation in which injunctive relief is highly unlikely as a matter of case law or judicial practice. Courts would then select a multiplier based on an adaptation of the "*Read*" factors that courts already use to determine the "egregiousness" of the defendant's conduct, which in turn impacts whether enhanced damages are awarded and, if so, the size of the selected multiplier.[58]

Some of the most relevant *Read* factors in this context include "whether the infringer deliberately copied the ideas or design of another; whether the infringer, when he knew of the other's patent protection, investigated the scope of the patent and formed a good-faith belief that it was invalid or that it was not infringed; . . .

[58] *See* Read Corp. v. Portec, Inc., 970 F.2d 816, 826–27 (Fed. Cir. 1992) (internal citations omitted).

[d]efendant's size and financial condition; [and] [c]loseness of the case."[59] In a typical "efficient infringement" scenario, at least two of these factors would favor selecting a high multiplier for purposes of enhanced damages: the infringer is well-resourced financially and deliberately copies the patentee's patented technology. The court could then adjust the multiplier upward or downward depending on the extent to which the facts indicated whether the infringer had investigated the scope of the patent and formed a good faith belief that the patent was invalid. While our proposal goes beyond existing statutory and case law in mandating enhanced damages in all cases in which injunctive relief is not practically available, the factors that would be used to calibrate the multiplier largely track the existing *Read* factors and arguably encompass the "affirmative duty" standard that, as discussed earlier,[60] courts had used to assess willfulness following the *Underwater Devices* decision in 1983 until the *Seagate* decision in 2007.

Enabling patentees to credibly threaten infringers with the prospect of enhanced damages in lieu of injunctive relief would discourage well-resourced users from infringing by increasing the user's anticipated "downside" losses in the event the patentee prevailed in an infringement litigation. This would correct for the under-deterrence and undercompensation effects that arise in a legal environment in which patent owners cannot credibly threaten to deny access to their patented technology. Whether or not a user would determine that negotiating a license outperforms infringing and litigating would then depend on its level of confidence in being able to show that the patent is invalid or not infringed, rather than being dependent on the user's litigation resources and opportunity costs relative to the patent owner. This would constructively lead potential infringers to proactively invest resources in investigating the validity and scope of relevant patents, which may often lead infringers to conclude that negotiating a license, rather than expending funds on litigators, is the preferred business option. Hence, our proposal has the virtuous effect that it renders the user's *license v. infringe* decision dependent on the strength of the patent, rather than the relative litigation resources available to the user and patent owner. This levels the playing field by precluding well-resourced implementers from leveraging the costs and delay of the litigation and settlement process to secure a downward adjustment in the royalty rate from a less well-resourced innovator, irrespective of the value of the underlying patent.

The prospect of enhanced damages (and, in particular, courts' ability to adjust the multiplier upward in response to infringer opportunism) would also unwind the distortionary effects that the current "almost no" injunction patent regime exerts in the patent licensing market. Negotiated royalty rates would more closely track a patent's economic value, since those rates would no longer reflect the artificially depressed levels that can arise under the threat of protracted litigation with well-

[59] *Id.* at 826–27 (internal citations omitted).

[60] *See supra* note 13 and accompanying discussion.

resourced infringers. Over time, improved accuracy in negotiated royalties would likely improve the accuracy of judicially determined royalty awards since courts could more confidently rely on established royalties as an accurate measure of economic value.[61]

## C. *Calibrating Enhanced Damages*

It can nonetheless be objected that awarding enhanced damages would invite a return to opportunistic litigation from certain patent owners for the purpose of securing enhanced damages or favorable settlement payouts from cash-rich users. That is: While substituting enhanced damages for injunctive relief mitigates "holdout" behavior by prospective licensees, it restores the risk of "holdup" behavior by patent owners. To be clear, we do not expect that mandating enhanced damages would give rise to holdup behavior in general. Insisting on exorbitant royalty rates would be a self-defeating strategy for repeat-play innovators that seek to maximize returns by inducing adoption of their existing technology and accruing reputational goodwill to induce adoption of their future technologies.[62] Holdup incentives may arise, however, in the case of certain patent owners that do not have R&D investments at stake, hold a patented technology to which there is no cost-feasible substitute, and are not repeat players in the technology ecosystem. In that specific set of circumstances, long-term incentives to maintain reputational goodwill might not discourage the short-term use of opportunistic litigation strategies and hence, mandating enhanced damages could invite entry by opportunistic litigants, resulting in overcompensation and overdeterrence effects.

These concerns are addressed by design through our proposal, which mandates enhanced damages in "almost no" injunction scenarios but always invests courts with discretion to select the specific multiplier.[63] We anticipate that courts would calibrate the multiplier to achieve a rough trade-off between correcting undercompensation and underdeterrence effects (which favor a higher multiplier) and overcompensation and overdeterrence effects (which favor a lower multiplier), which would in turn depend on the facts demonstrated in any particular litigation.

[61] Currently, courts sometimes adjust reasonable royalties upward to account for the possibility that "comparable" royalties in the market are distorted by the anticipated costs and risks of litigation. *See, for example*, ResQNet.com, Inc. v. Lansa, Inc., 594 F.3d 860, 972 (Fed. Cir. 2010), *citing* Fromson v. W. Litho Plate & Supply Co., 853 F.2d 1568, 1577 (Fed. Cir. 1988) ("[T]his court has long recognized that a reasonable royalty can be different than a given royalty when, for example, widespread infringement artificially depressed past licenses").

[62] On the reputational feedback effects that constrain patent licensing and litigation strategies, see Jonathan M. Barnett, *The "License as Tax" Fallacy*, 28 Mich. Tech. L. Rev. 197 (2022).

[63] While it lies outside the scope of this contribution, we note that courts already have several tools to deter opportunistic use of the civil litigation process, including shifting attorneys' fees to losing plaintiffs in the case of frivolous litigation.

This concept is already reflected in the patent statute, which empowers courts to select a multiplier within a bounded range, and case law outcomes, which exhibit variation in the multipliers used by courts when awarding enhanced damages and show that courts infrequently reach the upper bound.

More specifically, as noted earlier, courts already use the *Read* factors to determine the size of the damages multiplier and our proposal would build upon this existing framework. Specifically, one of the *Read* factors takes into account whether the infringer "investigated the scope of the patent and formed a good-faith belief that it was invalid or that it was not infringed."[64] Courts could apply a lower damages multiplier if an infringer showed that it had satisfied this factor. Additionally, we anticipate that courts would select lower damages multipliers if the infringer could demonstrate that it was unaware of the contested patent after due diligence or was aware of the patent and had made good faith efforts to negotiate a license from the patent owner. This would also preserve incentives for good faith users to *reject* a license when, after due diligence, they are reasonably confident that the patent owner would be unlikely to defend validity and demonstrate infringement (in which case the probability that the patent owner would secure a liability finding, a willfulness finding, and a meaningful enhanced damages award would most likely reach asymptotic levels).[65]

Another *Read* factor is the duration of the defendant's misconduct.[66] The longer the duration of the defendant's misconduct, the higher the enhanced damages. In the context of our proposal for mandatory enhanced damages, courts may take into account the period of time during which negotiations and litigation have played out, enhancing damages to account for the increasing holdout costs borne by the patentee during a protracted negotiation and litigation process. These costs extend beyond mere attorney fees, as the delays incurred by negotiation and litigation can have dramatic adverse effects on a firm's position in the market (for example, it can forfeit the first mover advantage that is often critical in technology markets).

Closely related to the duration of negotiations and litigation and the defendant's good faith belief is whether one of the parties made a credible offer of arbitration as to either the entire dispute or the royalty determination Arbitration is less time-consuming and costly than litigation, so a defendant's refusal to participate in arbitration may be grounds to increase the enhanced damages multiplier, and a

64 970 F.2d at 827.

65 Interestingly, in response to the *Halo* decision (which, as discussed earlier, lowered the threshold for finding willfulness), instances of willfulness and enhanced damages findings increased, but the average damages multiplier declined (from 2.5 before *Halo* to slightly more than 2.0 after *Halo*). *See* PRICEWATERHOUSECOOPERS, 2018 PATENT LITIGATION STUDY 17 (2018) (based on a study of district court decisions in patent infringement litigation from 1998 through 2018). This illustrates how courts can adjust the multiplier to mitigate potential overcompensation and overdeterrence effects that can arise with a change in legal standards that increases the likelihood of enhanced damages.

66 970 F.2d at 827.

plaintiff's refusal may be grounds to decrease the enhanced damages multiplier. Of course, a party may have a legitimate reason for preferring federal court litigation, and showing a legitimate reason (such as a defendant's good faith belief that the patent was invalid or not infringed) can rebut any increase or decrease in enhanced damages.

Appropriately applied, mandating enhanced damages subject to a judicially applied multiplier (in the absence of injunctive relief) would not only deter litigations induced by bad faith users that infringe against likely valid patents but also litigations brought by bad faith *holders* of likely invalid patents. The net result would likely be a reduction in the quantity, and an improvement in the "quality," of patent infringement litigation on the judicial docket, offsetting some of the unintended consequences of the *eBay* decision.

### D. *The Legislative Fix*

There is no perfect solution to the inevitable trade-off between undercompensation and underdeterrence effects, which can arise in certain circumstances when courts infrequently award enhanced damages or typically apply low damage multipliers, and overcompensation and overdeterrence effects, which can arise in certain circumstances when courts regularly award enhanced damages or typically apply high multipliers. Following our proposal, courts can roughly balance these effects by combining mandatory enhanced damages with variable damages multipliers, which courts can adjust downward if there is sufficient evidence of good faith motivations behind the infringer's action or upward if there is sufficient evidence of bad faith motivations behind the infringer's conduct. Courts already have authority to make these adjustments under the enabling language in the patent statute, which provides that courts "*may* increase the damages *up to* three times" (our emphasis). These discretionary powers are illustrated by the district court's decision in *Core Wireless Licensing S.A.R.L. v. LG Electronics, Inc.*,[67] a litigation brought by a SEP owner and therefore without any prospect of injunctive relief. Following the jury's finding of willfulness, the judge elected to impose a 20% damages enhancement based on evidence of bad faith negotiation conduct by the infringer and weak patent invalidity and noninfringement defenses. In *SRI International Inc. v. Cisco* Systems,[68] the Federal Circuit upheld an enhanced damages award on the ground that the defendant not only had infringed willfully but also had engaged at trial in "aggressive

[67] Final Judgment, Core Wireless Licensing, S.A.R.L. v. LG Elecs., Inc., No. 2:14-cv-912 (E.D. Tex. Nov. 4, 2016). The district court also rejected the argument that SEPs should never give rise to enhanced damages. For further discussion of the decision, see J. Gregory Sidak, *Enhanced Damages for Infringement of Standard-Essential Patents*, 1 CRITERION J. INNOVATION 1101 (2016).

[68] No. 20-1685 (Fed. Cir. Sept. 28, 2021).

tactics," such as making dubious arguments that appeared to be contradicted by the infringer's internal documents.

These decisions might suggest that our proposal could be largely implemented through increased judicial willingness to apply the existing damages multiplier for purposes of awarding enhanced damages following a finding of willfulness. *SRI International* shows why this would not be sufficient to restore the missing deterrence effects in the current patent system. In that case, the Federal Circuit clarified that enhanced damages can only be awarded based on a finding that the defendant had engaged in *both* willful infringement and "wanton and malicious" conduct. Given this arguably heightened threshold for awarding enhanced damages, meaningful implementation of our proposal could not rely on courts' discretion under existing case law to award enhanced damages for the simple reason that courts would often not have the opportunity to exercise such discretion. Our proposal therefore requires amending the patent statute to clarify that courts *must* select a damages enhancement, subject to the existing statutory maximum of treble damages, in any infringement litigation in which the patent owner prevails on validity and infringement *and* there is no reasonable likelihood of injunctive relief.[69]

Interestingly, German legislators have recently implemented a version of our proposal. A recent statutory amendment to the German Patent Act precludes injunctive relief "if the claim would lead to disproportionate hardship for the infringer or third parties."[70] This amendment, which codifies German case law,[71] allows courts to deny injunctive relief in special circumstances – injunctive relief is no longer "automatic." While the German Patent Act amendment does not go as far as *eBay*, which (as interpreted by the lower courts) flipped the United States from an "automatic" injunction regime to an "almost no" injunction regime, the German Patent Act amendment does play a similar role by making injunctive relief more difficult to obtain. What is significant, however, is that the shift in the German patent injunction regime is paired – as this chapter proposes – with an increased ability to obtain enhanced damages. In the same amendment making injunctive relief more difficult to obtain, the German Patent Act was also amended to provide "[i]n th[e] case [where injunctive relief is denied], the injured party shall be granted appropriate financial compensation [that] shall not affect the claim for damages pursuant to Paragraph 2 [traditional patent remedies of actual damages, unjust

69 This view is confirmed by a Federal Circuit decision that specifically precludes courts from awarding enhanced damages solely to rectify what is perceived to be inadequate damages, absent a finding of willfulness; *see* Jurgens v. CBK, Ltd., 80 F.3d 1566, 1570 (Fed. Cir. 1996), *citing* Beatrice Foods Co. v. New England Printing & Lithographing Co., 923 F.2d 1576, 1579 (Fed. Cir. 1991).

70 Translated from original Patentgesetz [Patents Act] § 139(1) ("soweit die Inanspruchnahme aufgrund der besonderen Umstände des Ein zelfalls für den Verletzer oder Dritte zu einer unverhältnismäßigen").

71 Federal Court of Justice (decision of 10 May 2016, docket no. X ZR 114/13 – "Heat Exchanger") (Germany).

enrichment and reasonable royalties]."[72] Thus, as the German system transitions away from an "automatic" injunction regime, legislators had the foresight to implement an enhanced damages regime to deter infringers from engaging in holdout tactics. Hopefully, Congress can look to Germany as an example of our proposal in action.

Absent restoration of the historical presumption favoring injunctive relief for prevailing patentees (which would more directly correct the underdeterrence effect), we encourage Congress to consider making this adjustment to the patent statute. In 1793, Congress recognized the necessity for mandating enhanced damages for patent owners when injunctive relief is unavailable; today that same rationale applies once again.

## V. CONCLUSION

The signature element of a property right is the ability to have effective legal recourse to maintain exclusivity of the underlying asset in response to unauthorized users. This principle is as true in intangible goods markets as in tangible goods markets. Contrary to widespread characterizations, the fundamental effect of a robust IP regime is not to entrench "idea monopolists" and enable them to extract maximal rents from intermediate and end users. In most cases, that would be an ill-advised business strategy that would invite some combination of infringing use, underuse, or competitive entry. Rather, the property rights "backstop" supplies a legal platform on which business parties can engineer a myriad of value-creating transactional arrangements that structure licensing, joint venture, and other relationships between parties that hold complementary assets and capacities. While the *eBay* decision may have targeted a specific type of opportunistic litigation, its effects have reverberated across the IP ecosystem, converting patents in many sectors from a property right priced by the market to a quasi-compulsory license priced in court. Absent legislative intervention to correct the "*eBay* effect," we have proposed a simple remedy. If patentees have no realistic expectation of securing injunctive relief, even after having incurred the significant costs and delay involved in defending validity and demonstrating infringement, then the infringer must pay enhanced damages to restore in part the deterrence and compensation effects that have been eroded under the current patent regime.

[72] Translated from original Patentgesetz [Patents Act] § 139(1) ("In diesem Fall kann der Verletzte einen Ausgleich in Geld verlangen, soweit dies angemessen erscheint. Der Schadensersatzanspruch nach Absatz 2 bleibt hiervon unberührt.").

PART IV

# Transactional Solutions

## *Redesigning SEP Licensing Markets*

# 7

# Designing SEP Licensing Negotiation Groups to Reduce Patent Holdout in 5G/IoT Markets

*Ruud Peters, Igor Nikolic, and Bowman Heiden*

## I. INTRODUCTION TO LNGS IN 5G/IOT MARKETS

Standard-Essential Patent (SEP)-enabled cellular standards have experienced a large degree of market success. In 2016, the number of cellular subscriptions exceeded the world population.[1] In 2020, the mobile industry's contribution to world GDP was estimated at $4.4 trillion.[2] By 2035, the impact of 5G is predicted to grow to $13.2 trillion in gross output worldwide.[3] However, while the total estimated revenue from cellular SEP licensing is less than one half percent of the size of the mobile economy,[4] the market for SEP licensing has remained contentious with both SEP licensees and licensors claiming inefficiencies, characterized as patent holdup and holdout, respectively.

In response to concerns that inefficiencies in SEP licensing may have a negative systemic impact on the development of emerging 5G and Internet of Things (IoT) markets, the European Commission (EC) convened an Expert Group on Licensing and Valuation of Standard Essential Patents (SEP Expert Group), resulting in a report including 79 proposals aimed at improving the SEP licensing market.[5] This chapter is focused on Proposal 75 – Licensing Negotiation Groups (LNGs), which was formulated by an individual member of the SEP Expert Group. The goal is not

1 Bowman Heiden, *The Value of Cellular Connectivity – From Mobile Devices to the Internet-of-Things (IoT)* (Aug. 9, 2020), https://papers.ssrn.com/sol3/papers.cfm?abstract_id=3670222.

2 *The Mobile Economy*, GSMA (2021), www.gsma.com/mobileeconomy/wp-content/uploads/2021/07/GSMA_MobileEconomy2021_3.pdf.

3 Campbell et al., *The 5G Economy – How 5G Will Contribute to the Global Economy*, IHS Markit (2019), www.qualcomm.com/content/dam/qcomm-martech/dm-assets/documents/the_ihs_5g_economy_-_2019.pdf.

4 Bowman Heiden, Jorge Padilla, & Ruud Peters, *The Value of Standard Essential Patents and the Level of Licensing*, 49 AIPLA Q.J. 1 (2021).

5 Justus Baron et al., *Group of Experts on Licensing and Valuation of Standard Essential Patents (E03600)*, Eur. Union Comm'n (Jan. 2021), https://ec.europa.eu/transparency/expert-groups-register/screen/expert-groups/consult?lang = en&groupID = 3600.

to critique the specifics of Proposal 75 but to take the general concept of an LNG and formulate a specific institutional and organizational design.

The concept of LNGs or similar collective buying arrangements has been proposed previously in the context of Standard Development Organizations (SDOs) as a means to counteract the perceived market power of SEP holders and reduce the risk of patent holdup and royalty stacking.[6] Ironically, we hypothesize the opposite, that the main social benefit of LNGs is the potential reduction of transaction costs and patent holdout. This is not only because patent holdup and royalty stacking have never been empirically proven to have a systemic impact in SEP-enabled markets,[7] but that the most likely challenge of SEP licensing in IoT is overcoming the collective action problem among a large number of similarly situated SEP implementers.[8]

The deployment of 5G/IoT is expected to result in a large increase in SEP implementers across diverse industries. Similarly situated SEP implementers are market actors competing against one another through prices in the product market, so SEP royalties are seen as an input cost. While these IoT-based SEP implementers are all incentivized to reduce input costs (that is, SEP royalty rates), no implementer is incentivized to take a license at all if they are not assured that all other competing firms will also take a license on comparable terms. Thus, even if an agreement on the standard of a "Fair, Reasonable, and Nondiscriminatory" (FRAND) rate can be achieved,[9] there is a rational, systemic disincentive by SEP implementers (that is, a collective action problem) to take an independent license, which facilitates patent holdout. This challenge is fundamentally different from the simple reduction of transaction costs that accompanies the elimination of redundant bilateral negotiations through collective action, though these savings can also be substantial, as shown in Figure 7.1. Concomitantly, a large reduction in transaction costs could facilitate the licensing of more SEP implementers, especially those in the long tail that have traditionally been able to hold out due to their small size.[10]

6 Jorge L. Contreras, *Aggregated Royalties for Top-Down FRAND Determination: Revisiting "Joint Negotiation"*, 62 Antitrust Bull. 690–709 (2017); Luke McDonagh & Enrico Bonadio, *Standard Essential Patents and the Internet of Things, European Parliament, Policy Department for Citizen's Rights and Constitutional Affairs*, Eur. Union Parliament (Jan. 2019), www.europarl.europa.eu/RegData/etudes/IDAN/2019/608854/IPOL_IDA(2019)608854_EN.pdf.

7 Alexander Galetovic, Stephen Haber, & Ross Levine, *An Empirical Examination of Patent Holdup*, 11 J. Competition L. & Econ. 549–78 (2015); Alexander Galetovic et al., *An Estimate of the Average Cumulative Royalty Yield in the World Mobile Phone Industry: Theory, Measurement and Results*, 42 Telecomm. Pol'y 263–76 (2018);Bowman Heiden, *Valuing Standard Essential Patents in the Knowledge Economy: A Comparison of F/RAND Royalty Methodologies in U.S. Courts*, 13 Int'l J. Standardization Rsch. 19–46 (2015).

8 Bowman Heiden & Nicolas Petit, *Patent "Trespass" and the Royalty Gap: Exploring the Nature and Impact of Patent Holdout*, 34 Santa Clara High Tech. L.J. 179–249 (2018).

9 For the background and nature of FRAND licensing, see the chapters in Parts I and II of this book.

10 *Id*. The "long tail" in this context refers to a distribution of market actors where there exists a small population of actors that hold larger shares (the head) and a large population of actors that hold smaller shares (the tail).

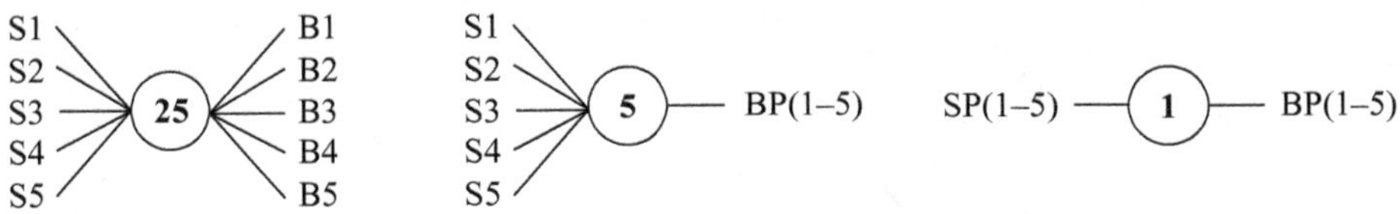


FIGURE 7.1. Theoretical transaction cost reduction through pooling both sellers and buyers.
S, Seller; B, Buyer; SP, Seller Pool; BP, Buyer Pool.

The desired gains of LNGs are not without the possibility of potential negative consequences. For example, a frequently used argument against LNGs is that they run a high risk of becoming a buyer's cartel, where the members collude to agree to pay a maximum royalty well below the FRAND rate, which would not give a SEP holder a reasonable reward for its contributions to a standard and reduce incentives to participate in standardization. Similarly, it is postulated that LNGs could facilitate a collective holdout strategy with the goal to ultimately pay lower than FRAND royalties or no royalties at all.[11] Furthermore, the anticompetitive risks of LNGs have been said to be too great in relation to their potential reduction of transaction costs, which can more effectively be managed through existing patent pool models.[12]

While it is correct that these risks internal to an LNG may exist, that should not be a reason to simply reject the concept of LNGs as an undertaking that is anticompetitive per se. In this respect, a parallel with patent pools can be drawn, since analogous arguments can be made against patent pools, which could facilitate the formation of a seller's cartel to capture supra-FRAND royalties (that is, a collective patent holdup strategy).[13]

It is also important to acknowledge that both patent pools and LNGs, as well as other collective licensing organizations (CLOs), are part of a broader multilayered, open innovation ecosystem of privately ordered market governance.[14] The first layer of collective action includes the open, consensus standardization process of SDOs, which is widely accepted as pro-competitive. The second layer of collective action includes the FRAND-based intellectual property (IP) policies that incentivize (1) investment in upstream R&D and contribution of technology, and (2) investment in the production and distribution of standard-enabled products and services. As both

[11] Igor Nikolic, *Licensing Negotiation Groups for SEPs – Collusive Technology Buyers Arrangements: Pitfalls and Reasonable Alternatives*, LES NOUVELLES, Dec. 2021, at 351, https://ssrn.com/abstract=3926650.

[12] Jonathan M. Barnett, *The Economic Case against Licensing Negotiation Groups in the Internet of Things*, 10 J. ANTITRUST ENFORCEMENT 518 (2022).

[13] Erik Hovenkamp & Herbert Hovenkamp, *Patent Pools and Related Technology Sharing*, *in* THE CAMBRIDGE HANDBOOK OF ANTITRUST, INTELLECTUAL PROPERTY, AND HIGH TECH 358–76 (Roger D. Blair & D. Daniel Sokol eds., 2017).

[14] Bowman Heiden & Justus Baron, *A Policy Governance Framework for SEP Licensing: Assessing Private versus Public Market Interventions* (2021), https://ssrn.com/abstract=3872493.

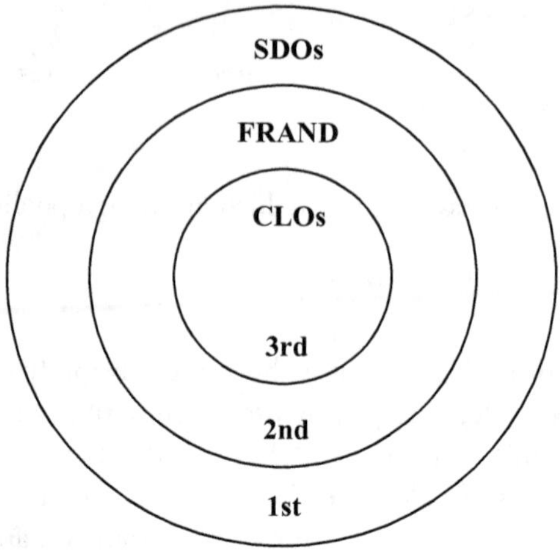


FIGURE 7.2. Multilayered collective action in SEP-enabled standardized markets

these investments are sunk costs, the FRAND commitment is a critical mechanism necessary to balance the interests and reduce the overall financial risks shared between SEP holders and implementers. Therefore, CLOs are born within this already highly collaborative ecosystem of vertical and horizontal competitors, operating as a third-order means of collective action to solve the remaining SEP licensing challenges emanating from the expansion of connectivity into new IoT markets. Figure 7.2 illustrates the levels of collective action in which the norms of LNGs are embedded and through which the norms of antitrust must be interpreted.

The historically developed norms of collective action within the cellular ecosystem have incentivized private firms to invest tens of billions of dollars in fundamental cellular R&D and contribute millions of person-hours in joint standard development, which has resulted in open standards implemented across industries that have enabled trillions of dollars of economic impact.[15] Creating a Pareto improvement to this ecosystem is a humbling task given the complex interaction of increasing technological functionality, expanding industry use-cases, and converging market norms. The remaining sections of this chapter are a first attempt to design an LNG that increases the efficiency of SEP licensing for both licensors and licensees while reducing the relevant antitrust risks.

The structure of the chapter is the following. Section II describes the general antitrust treatment of seller and buyer collaborations and then focuses on specific antitrust concerns and safe harbors related to collective action in the SEP licensing context. Section III describes the *Huawei v. ZTE* licensing framework and how it may apply to LNGs. Section IV discusses how LNGs could be implemented in compliance with antitrust safeguards, sketches the internal governance rules of

[15] Heiden & Petit, *supra* note 8.

LNGs, and provides some examples from practice where collective implementer groups were used to facilitate SEP licensing.

## II. ANTITRUST, COLLECTIVE ACTION, AND SEP LICENSING

### A. *General Antitrust Principles for Collective Action among Competitors*

Antitrust laws have been traditionally suspicious about collaborations among competitors. Two main concerns are associated with competitor collaborations. The first is that cooperation may result in a cartel, where companies would discuss and exchange sensitive commercial information and agree on prices, output, quality, or innovation.[16] The second concern is that collaboration among competitors may increase collective market power and harm competition by increasing the ability and incentive of companies to raise prices above competitive levels or reduce output.

On the other hand, in some instances, cooperation among competitors may lead to pro-competitive benefits. Competitor collaboration may enable companies to offer new or cheaper products or services. The key is the combination of *complementary* activities, skills, or assets.[17] For instance, companies may combine their complementary research and development activities to produce new and improved products or combine complementary assets and skills to achieve economies of scale or scope. In contrast, the combination of substitutes will normally raise antitrust concerns.[18]

Competitor collaborations are regulated by special antitrust guidelines of the European Commission in the European Union and the Department of Justice (DOJ) and the Federal Trade Commission (FTC) in the United States. Guidelines provide a general "safe harbor" or "safety zone" for agreements that are not hidden cartels and do not exceed a certain market share threshold. In the United States, a safety zone is established for such agreements among competitors, which do not collectively constitute more than 20% of the relevant market,[19] while in the European

[16] Eur. Union Comm'n, Guidelines on The Applicability of Article 101 of The Treaty on the Functioning of the European Union to Horizontal Co-Operation Agreements, 11 O.J. 1, para. 3 (2011) [hereinafter Guidelines on Horizontal Cooperation Agreements]; Fed. Trade Comm'n & US Dep't of Just., Antitrust Guidelines for Collaborations among Competitors (Apr. 2000), p. 3 [hereinafter Competitor Collaboration Guidelines].

[17] Guidelines on Horizontal Cooperation Agreements, *supra* note 16, para. 2, 51; Competitor Collaboration Guidelines, *supra* note 16, p. 6.

[18] Guidelines on Horizontal Cooperation Agreements, *supra* note 16, para. 52 ("Horizontal co-operation agreements that do not involve the combination of complementary skills or assets are less likely to lead to efficiency gains that benefit consumers. Such agreements may reduce duplication of certain costs, for instance because certain fixed costs can be eliminated. However, fixed cost savings are, in general, less likely to result in benefits to consumers than savings in, for instance, variable or marginal costs.").

[19] Competitor Collaboration Guidelines, *supra* note 16, p. 26.

Union, the collective market share threshold is 15%.[20] Additionally, there are special rules for assessing specific types of competitor collaborations, such as joint research and development, specialization, production, purchasing, commercialization, and standardization agreements.

The same rules in principle apply to both seller and buyer collaboration. However, seller cooperation is regulated more extensively, as it may include a broad spectrum of activities from joint R&D production to joint commercialization of products or services. Buyer collaboration covers joint purchasing, enabling buyers to negotiate better terms and conditions with sellers, leading to lower prices for consumers. The negative aspects of buyer collaboration may, in contrast, be felt by consumers or purchasers. Suppose buyer collaboration results in significant buyer power. In that case, buyers may decide not to pass on lower purchase prices to final consumers or collectively reduce the purchase price below the competitive level, harming sellers and reducing their incentives to innovate and, as a result, lowering the quality or output supplied by sellers.[21]

### B. *Collective Action and SEP Licensing*

Competitor collaboration exists in relation to SEP licensing as well. To date, we have largely seen collaboration on the seller side, where SEP owners form patent pools to jointly license SEPs to third parties. Antitrust authorities have adopted specialized rules for assessing the formation of patent pools and licensing-out of SEPs from a patent pool to third parties.[22] In contrast, LNGs for implementers are a new phenomenon, and there are currently no specialized rules for their assessment.

[20] Guidelines on Horizontal Cooperation Agreements, *supra* note 16, para. 208, 240; *but see* Eur. Union Comm'n, Regulation (EU) 1217/2010 of December 14, 2010 on the application of Article 101 (3) of the Treaty on the Functioning of the European Union to Certain Categories of Research and Development Agreements, 335 O.J. 36, art. 4 (2010) (combined market share threshold of 25% for joint R&D agreements); Eur. Union Comm'n, Commission Regulation 1218/2010 of December 14, 2010 on the application of Article 101(3) of the Treaty on the Functioning of the European Union to Certain Categories of Specialisation Agreements, 335 O.J. 43, art. 3 (2011) (combined market share threshold of 20% for horizontal specialisation agreements).

[21] German Organisation for Economic Co-operation and Development, *Monopsony and Buyer Power*, 80 DAF/COMP 1, 11–12 (2008).

[22] In the European Union, rules are contained in Eur. Union Comm'n, Guidelines on the application of Article 101 of the Treaty on the Functioning of the European Union to technology transfer agreements, 89 O.J. 3, para. 244–73 (2014) [hereinafter Technology Transfer Guidelines]. In the United States, rules are contained in numerous business review letters of the DOJ and the joint antitrust guidelines for the licensing of IP rights by the DOJ and the FTC; see Dep't of Just., Letter from Joel Klein, Assistant Att'y Gen. to Garrard Beeney (June 26, 1997) (MPEG-LA Business Review Letter); Dep't of Just., Letter from Joel Klein, Assistant Attorney General to Garrad Beeney (Dec. 16, 1998) (DVD3 Business Review Letter); Dep't of Just., Letter from Joel Klein, Assistant Att'y Gen. to Carey Ramos (June 10, 1999) (DVD6 Business Review Letter); Dep't of Just., Letter from Charles James, Assistant Att'y Gen. to Ky Ewing (Nov. 12, 2002) (3GPP Partnership Business Review Letter); Dep't of Just., Letter from Makan Delrahim, Assistant Att'y Gen. to Mark Hamer (July 28, 2020) (Avanci Business

TABLE 7.1. *Anticompetitive concerns with pools and joint purchasing agreements*

| Patent pools | Joint purchasing agreements |
|---|---|
| Cartelization concerns:<br>• If the pool consists of substitute patents, it amounts to a price-fixing cartel.<br>• Coordination of downstream prices, output, and markets.<br>• Exchange of sensitive business information may lead to collusion. | Cartelization concerns:<br>• If a joint purchasing agreement is a façade for a disguised cartel (exchange of sensitive business information, price-fixing, market allocation, output limitation, etc.) |
| Market power concerns:<br>• Foreclosure of alternative technologies (for example existence of a standard-related pool or a pool consisting of non-essential complementary patents may reduce innovation and make it more difficult for new and improved technologies to enter the market.) | Market power concerns:<br>• If substantial parts of products are purchased collectively, the incentives for price competition in the downstream market may be reduced.<br>• Cost-savings from reduced wholesale prices may not be passed on to consumers but retained by members.<br>• Suppliers may be hurt by lower prices and, as a result, may reduce quality or output, lessening innovation incentives. |

*Sources*: Eur. Union Comm'n, Guidelines on the application of Article 101 of the Treaty on the Functioning of the European Union to technology transfer agreements, 89 O.J. 3, para. 246 (2014); US Dep't of Just. & Fed. Trade Comm'n, Antitrust Guidelines for the Licensing of Intellectual Property (Jan. 12, 2017), pp. 30–31; Eur. Union Comm'n, Guidelines on The Applicability of Article 101 of The Treaty on the Functioning of the European Union to Horizontal Co-Operation Agreements, 11 O.J. 1, para. 200–06 (2011); Fed. Trade Comm'n & US Dep't of Just., Antitrust Guidelines for Collaborations among Competitors (Apr. 2000), p. 14.

Thus, general rules on assessing horizontal joint purchasing agreements could be used by way of analogy.

Pools and joint purchasing agreements share the same antitrust concerns. The main risk is that the aggregation of substitute products or services would constitute a price-fixing cartel and amount to a "per se" restriction (US) or a restriction of competition "by object" (EU). Additionally, horizontal cooperation among sellers or buyers risks exchanging confidential business information that may lead to direct or indirect collusion and cartelization on downstream product markets and upstream technology markets. The increased market power of such horizontal cooperation is another concern, which may lead to foreclosure of alternative technologies or harm to sellers manifesting in reduced innovation and product quality. Table 7.1 summarizes antitrust-related concerns of patent pools and horizontal joint purchasing agreements.

Review Letter); US Dep't of Just. & Fed. Trade Comm'n, Antitrust Guidelines for the Licensing of Intellectual Property (Jan. 12, 2017) p. 30 [hereinafter IP Licensing Guidelines].

TABLE 7.2. *Current antitrust safe harbors for pools and joint purchasing agreements*

| Patent pools | Joint purchasing agreements |
|---|---|
| • voluntary and open membership<br>• independently verified essentiality of pooled patents<br>• safeguards to limit the exchange of sensitive commercial information<br>• FRAND licensing terms and conditions<br>• freedom of members to license independently<br>• freedom to challenge the validity and essentiality of pooled patents by licensees<br>• freedom to develop competing products and technologies | • not by object (EU) or per se restrictions (US)<br>• Combined market share thresholds of all parties:<br>○ European Union: ≤15%<br>○ United States: ≤20%. |

*Sources*: Eur. Union Comm'n, Guidelines on the application of Article 101 of the Treaty on the Functioning of the European Union to technology transfer agreements, 89 O.J. 3, para. 246 (2014); US Dep't of Just. & Fed. Trade Comm'n, Antitrust Guidelines for the Licensing of Intellectual Property (Jan. 12, 2017), pp. 30–31; Eur. Union Comm'n, Guidelines on The Applicability of Article 101 of The Treaty on the Functioning of the European Union to Horizontal Co-Operation Agreements, 11 O.J. 1, para. 208–12 (2011); Fed. Trade Comm'n & US Dep't of Just., Antitrust Guidelines for Collaborations among Competitors (Apr. 2000), p. 26.

That said, antitrust authorities have produced guidelines for assessing joint purchasing agreements and have identified conditions when such competitor collaborations would not raise antitrust concerns. The rules for patent pools are much more elaborate than for joint purchasing agreements. There may be several reasons for such differentiated treatment. Joint purchasing agreements are primarily associated with the procurement of physical goods and services, while patent pools relate to the joint selling of technology (that is, intangible assets). Guidelines for joint purchasing agreements rely on market share thresholds to control the potentially negative effects of the increased market power of its members. On the other hand, patent pools are more efficient when they aggregate as much of the selling side of the market as possible. Thus, safeguards are focused not on the market share thresholds of pool members, but on mechanisms to control and negate the market power of a pool, for example, by only allowing the pooling of complementary patents, the freedom to license outside of the pool, and FRAND licensing terms for pooled patents.

Table 7.2 provides an overview of current antitrust safe harbors related to patent pools and joint purchasing agreements.

Considering that LNGs are a new phenomenon consisting of buyer collaboration in technology markets, some antitrust principles for LNGs could be borrowed both from antitrust safeguards related to joint purchasing agreements and from those related to patent pools.

## III. LNGS AND THE *HUAWEI V. ZTE* NEGOTIATION FRAMEWORK

In July 2015, the European Court of Justice (CJEU) issued its opinion with respect to certain questions that the German Federal Court of Düsseldorf referred to the CJEU in a SEP infringement case between Huawei and ZTE.[23] The opinion of the CJEU clarified the conditions under which a SEP holder could seek an injunction against an infringer without violating European competition laws by abusing its dominant position by virtue of holding a SEP.

The judgment of the CJEU created the clarity desired by SEP holders, implementers, and the national courts in EU member countries. The CJEU judgment provided guidance for how SEP licensors and implementers should behave in SEP licensing negotiations. It sets out several steps that SEP licensors and implementers should follow for a SEP licensor to be considered a willing licensor and an implementer to be regarded as a willing licensee. Although it is not mandatory for a SEP licensor to follow these steps and it can demonstrate that it is a "willing licensor" in other ways, a SEP licensor following these steps can seek an injunction against an implementer who does not conform to these guidelines, without risking a violation of the competition laws.

Under the Intellectual Property Rights Policy of ETSI (ETSI IPR Policy), a leading SDO in wireless communications, a SEP holder must undertake to license its SEPs under FRAND terms to implementers of the relevant standard. A SEP holder that is not willing to license its SEPs to an implementer seeking a license would breach that undertaking in the ETSI IPR Policy. Under French contract law, which governs the ETSI IPR policy, many courts have deemed implementers as third-party beneficiaries of the FRAND commitment in the ETSI IPR Policy.[24] However, any dispute about whether terms offered by a SEP licensor are indeed FRAND is a matter that national courts should judge under their national laws.

Under the CJEU guidelines, a SEP holder seeking to license its SEPs to an implementer must, as a first step, make the implementer aware of the alleged infringement by notifying them in writing of these SEPs and the relevant infringing products. In response, the implementer must, as a second step, express its willingness to conclude a license under these SEPs on FRAND terms. If the implementer does not do so in a timely manner, it may be considered an unwilling licensee, opening up the path for a SEP holder to seek an injunction. After the implementer has indicated its willingness to enter into a FRAND license, the SEP holder must, as a third step, make an offer in writing to the implementer specifying the royalty and how they determined that royalty. In return, the implementer must, as a fourth step,

[23] Case C-170/13, Huawei Techs. Co. Ltd. v. ZTE Corp. [2015] 5 C.M.L.R. 779.

[24] Unwired Planet Int'l Ltd. v. Huawei Techs. (UK) Co. Ltd. [2017] EWHC 2988 (Pat), at 146; Microsoft v. Motorola, 854 F. Supp. 2d 993 (W.D. Wash. 2012).

diligently respond to that offer by either accepting it or promptly providing a counteroffer in writing to the SEP holder, which the implementer believes is FRAND. In making a counteroffer, the implementer must also provide security in the form of a bank guarantee or put into escrow an amount equivalent to the royalties for its past sales, if any, based on his counteroffer. The implementer must supplement this amount to reflect estimated royalties on future sales.

An important aspect to mention here is that an implementer may challenge the essentiality, validity, or infringement of the asserted SEPs during the negotiations with the SEP licensor and may even do so after concluding a license agreement.

After the CJEU published its judgment in 2015, national courts further refined the various steps of the Huawei–ZTE negotiation framework in several infringement cases in the years thereafter. For example, national courts have specified the requirements for a written notice of infringement to the implementer,[25] the conditions under which an implementer is not considered to be a willing licensee,[26] and the conditions for providing security.[27] Courts have also clarified that a SEP holder must provide the reasons why it considers its proposed royalty rate to be FRAND.[28] Sometimes, different courts in the same jurisdiction reached different conclusions on the same topic.[29] Nevertheless, SEP licensors have mostly conducted licensing negotiations following the steps of the Huawei–ZTE negotiation framework to safeguard the ability to seek an injunction in case an implementer does not conform to this framework. Likewise, implementers have followed the steps of the framework to avoid being considered an unwilling licensee and risking an injunction.

### A. *Implications for LNGs*

Now turning to LNGs, various questions may be raised when considering how the Huawei–ZTE negotiation framework should be applied to LNGs or, stated differently, how LNGs should conduct licensing negotiations in line with this framework. The different aspects of applying the Huawei–ZTE negotiation framework to LNGs will be addressed in the remainder of this section, following the subsequent steps of the framework. For this purpose, it is assumed that the LNG has been established so that it can be considered as lawfully representing its members in any interaction with the SEP licensor.

[25] Philips v. Archos, Case 7 O 209/15 (Regional Court Mannheim 2016); Sisvel v. ZTE, Case 4a O 16/16 (Regional Court Düsseldorf 2017); Intellectual Ventures v. Vodafone, Case 4c O 81/17 (Regional Court Düsseldorf 2018).

[26] *See id.*

[27] *See id.*

[28] *See id. See also* Sisvel v. Haier, Case 15 U 66/15 (Higher Regional Court Düsseldorf 2016).

[29] Saint Lawrence Commc'ns v. Deutsche Telekom, Case 2 O 106/14 (Regional Court Mannheim 2015); Saint Lawrence Commc'ns v. Vodafone, Case 15 U 36/16 (Regional Court Düsseldorf 2016).

If the SEP licensor notifies the LNG in writing of alleged infringement of its SEPs by LNG members, it seems reasonable that the SEP licensor must only specify the category of products considered to be infringing and does not have to indicate at least one specific product of each LNG member. This category of products is also what creates the common interest of the members in having this SEP licensing matter addressed through the LNG.

Another interesting aspect to consider is whether the SEP licensor is entitled to exclude one (or more) members of the LNG for purposes of resolving a particular infringement dispute, because it prefers to have bilateral negotiations with this member. This may be the case if, for example, the SEP licensor and that member have multiple overlapping business activities and the SEP licensor wants to enter into negotiations with that member covering a broader scope than the specific products for which the LNG will negotiate licenses. It may also be the case that the SEP licensor and the member already have a license agreement in place covering part of the SEPs it is offering to license to the LNG. This is the mirror situation of a scenario in which a patent pool approaches an implementer for a pool license, and the implementer prefers to have bilateral negotiations with licensors in the pool, because, for example, the implementer already has a (cross-) license agreement in place with one or more pool licensors and covering the products licensed by the pool. A licensor cannot refuse to enter into bilateral license negotiations with an implementer that makes such a request. Following the same approach for LNGs, an LNG member should not be allowed to refuse a request for bilateral negotiations from a SEP licensor.

In response to the written notification of the SEP licensor, the LNG must communicate that its members are willing to enter into a license agreement on FRAND terms with the SEP holder. This willingness should be unconditional in the sense that this willingness should not depend on whether, and the extent to which, individual members accept the outcome of the negotiations between the LNG and the SEP licensor. Negotiations between a SEP licensor and an LNG are likely to fail without a firm commitment by members of the LNG to accept the outcome of the negotiations and enter into a license agreement with the SEP licensor on that basis. Suppose the LNG and the SEP licensors reach an agreement on the FRAND terms of a license, and the LNG members are free to accept or decline this outcome and thus can freely determine whether or not to enter into a license agreement on these terms with the SEP holder. In that case, the SEP holder may find it unattractive to enter into license negotiations with the LNG at all. It will create an incentive for at least some LNG members to request bilateral negotiations with the SEP licensor and take the outcome of the negotiations with the LNG as the starting point for their own bilateral negotiations, hoping to negotiate separately even better licensing terms. The freedom to decline the agreed outcome of the negotiations between the SEP holder and the LNG may also be used as part of a holdout strategy by an implementer, who can defer paying royalties that licensed

LNG members are already paying. Members should commit to entering into a license agreement with the SEP holder if the LNG and the SEP holder have reached an agreement approved by the members. This aspect will be addressed in greater detail when discussing the governance of LNGs in Section IV.

A key factor in the success of LNGs relies on members' commitment to enter into a license agreement with the SEP licensor based on the approved outcome of negotiations between the LNG and the SEP licensor. Suppose, despite such a commitment, a particular member does not enter into a license agreement with the licensor within a reasonable period after the agreement is reached. In that case, it seems reasonable that this member should be deemed an unwilling licensee and the licensor should be entitled to seek an injunction against the implementer without being accused of misusing a dominant position. If the SEP holder initiates litigation against a recalcitrant LNG member and the member, in the face of a likely injunction, elects to accept the FRAND terms negotiated with the LNG, it seems unfair that the implementer would still be entitled to a license under the same FRAND terms. Even if a court finds that the implementer is still entitled to a FRAND license, the implementer should be required to pay a penalty on top of the agreed FRAND terms (for example, in the form of a higher royalty for infringing sales made prior to entering into the license agreement with the SEP licensor).

Suppose a SEP licensor submits a FRAND license offer specifying the royalty and giving the reasons why it considers this royalty to be FRAND. In that case, the LNG must diligently respond to that offer without undue delay. If the LNG does not accept the license offer, it should timely make a FRAND counteroffer in writing. In bilateral negotiations, courts have held that a response time of three months is not timely.[30] In contrast, in another case, a five-month period was not regarded as timely.[31] In the case of LNGs, courts may take into consideration that it may take more time for the members of an LNG to agree internally and deliver a counteroffer, but even in this case, a court may find that making a counteroffer after more than five months is not timely.

If that situation would arise, would all LNG members be considered unwilling licensees? It is difficult to predict how a court would determine this point. Nonetheless, the mere fact that an LNG is negotiating with a SEP licensor does not release that LNG from its obligation to provide a timely FRAND counteroffer after having rejected a licensor's offer. Additionally, there may be a different response time that a licensor is willing to accept before taking legal action. The SEP licensor may prefer the prospect of getting all members of an LNG licensed through a single negotiation over litigation, and may therefore be willing to accept a

30 Saint Lawrence Commc'ns v. Deutsche Telekom, Case 2 O 106/14 (Regional Court Mannheim 2015).

31 Saint Lawrence Commc'ns v. Vodafone, Case 15 U 36/16 (Regional Court Düsseldorf 2016).

longer period for the LNG to respond to its FRAND offer than would be the case in a bilateral negotiation.

The same question arises if, in response to an offer by a SEP licensor, the LNG responds with a very low counteroffer that is clearly non-FRAND. Will all LNG members be considered unwilling licensees and potentially risk an injunction if the SEP licensor commences infringement actions against individual LNG members? In this case, all LNG members should again be considered unwilling licensees. Under the Huawei–ZTE negotiation framework, a group of licensees should not behave and be treated differently than any individual licensee.

Moreover, suppose the LNG does not negotiate in good faith by a delayed response to the SEP holder's offer and then making a clear non-FRAND counteroffer. In that case, the SEP licensor may accuse the LNG members of engaging in a group holdout, which may be the basis for an antitrust complaint by the SEP licensor.

Under the Huawei–ZTE negotiation framework, if the SEP licensor rejects a counteroffer from an implementer, the implementer must provide a bank guarantee or put in escrow an amount for the royalties on its past sales based on the rate in its counteroffer. Applied to the LNG context, this implies that the LNG must provide a collective bank guarantee for all its members or put an amount in escrow for the royalties on the collective past sales of all its members, or alternatively, each individual member would have to provide a bank guarantee or put money into an escrow account for the royalties on its past sales. In both cases, the members would have to provide an accounting of these past sales. In the first case, it may be difficult for the LNG to get a bank guarantee, since it may not have funds to support this guarantee, and it will not be an easy task to ensure that all members place the appropriate amounts in escrow.

In negotiating a SEP license, an LNG must follow the Huawei–ZTE negotiation framework in the same manner as an individual company. In theory, since LNGs are likely formed with the goal of reducing transaction costs for their members, to negotiate better FRAND royalty terms than each individual member could negotiate, and to level the playing field among their members, they should be incentivized to negotiate in good faith, especially if its behavior could result in all its members being deemed unwilling licensees, which would entitle the SEP holder to seek injunctive relief against LNG members. This accountability should also encourage the LNG to timely respond to written notifications (infringement letters and/or FRAND offers).

## IV. IMPLEMENTATION OF LNGS

### A. A *Safe Harbor for LNGs*

As discussed in Section II.B, a set of guidelines similar to those that have been developed for patent pools is needed to create a safe harbor for LNGs. Currently, no request for review of an LNG for licensing purposes has been submitted to the DOJ or EC (or other competition authority). Nevertheless, it is possible to identify a number of conditions that LNGs must satisfy to avoid likely conflicts with antitrust laws.

One key question is whether the members of an LNG should be allowed to have a collective market share that could enable the LNG to exercise market power. Generally, a collective market share of less than 15–20% is considered acceptable in the case of other types of collaborative buying groups.[32] However, greater market shares may be deemed pro-competitive given the potential to reduce transaction costs and limit patent holdout, especially as the LNG is bound to a FRAND rate and, in Europe, the negotiation guidelines of Huawei–ZTE. However, as the market share of an LNG increases, additional antitrust scrutiny is justified, concerning both the SEP licensing market and the possibility for collusion among LNG members in the downstream product market. Thus, it will likely be necessary to assess the maximum collective market share for LNGs on a case-by-case basis across different industry verticals. Taking the smartphone market as an example, an LNG including Apple, Samsung, Xiaomi, Oppo, and Vivo with roughly a 70% collective market share would be heavily scrutinized. On the other hand, companies in the tail of the smartphone market, each of which has less than a few percent market share, should be allowed to form an LNG without significant inquiry, since they do not have a strong market position, do not collectively hold significant market power, and the LNG will have better negotiation capabilities than each individual member. If successful, the LNG will not harm competition or innovation in the smartphone market.

LNGs will tend to be more successful if members are situated at approximately the same level of the value chain. In the example just discussed earlier, the smaller companies in the tail of the market are all similarly situated. This would likely help solve the collective action problem (that is, patent holdout) introduced in Section I by facilitating a level playing field among similarly situated competitors. It would be more difficult if the LNG has members operating at different levels in a value chain that make and sell different products (components versus end products). In that situation, conflicting interests among LNG members may make it unattractive for a SEP licensor to start licensing discussions with the LNG.

LNG members could be allowed to share information within the group about the essentiality, validity, and infringement of the relevant SEPs. However, they should not share any opinions or conclusions regarding this information, as this may be considered collusion. On the other hand, the LNG could be allowed to act on behalf of its members by seeking outside counsel's opinion regarding certain matters, since this would likely reduce the costs borne by each member, consistent with a key purpose of an LNG.

[32] Guidelines on Horizontal Cooperation Agreements, *supra* note 16, para. 208; Competitor Collaboration Guidelines, *supra* note 16, p. 26.

During their meetings, LNG members should not be allowed to discuss prices of products, profit margins, or market shares. The same principle applies in meetings of SEP licensors in the context of patent pools. To ensure that this rule is upheld, it would be advisable for an antitrust lawyer to attend all their meetings and remind the participants at the start of each meeting about the members' duties to operate in accordance with the antitrust laws and about the subjects they must not discuss. This external counsel should also intervene if any member raises a topic that should not be addressed.

LNG members should be allowed to disclose their position on royalty rates only with the LNG representative negotiating with the SEP licensor and not between LNG members themselves. This alleviates the concerns about possible coordination of royalties by LNG members, as only the LNG negotiator will have all the information on royalties collected from members. And only the LNG negotiator needs to know the pricing position of all LNG members in order to attempt to arrive at the most acceptable royalty level in negotiations with SEP owners.

Moreover, it should be realized that the LNG and the SEP licensor are bound by the Huawei–ZTE negotiation framework, which reduces the risk that the LNG will be able to negotiate or dictate a sub-FRAND royalty. Concluding licenses with a group of companies in a single negotiation through an LNG reduces the cost of licensing for a SEP licensor. The SEP licensor may share part of these benefits with the LNG members by accepting a lower royalty, but in doing so, the SEP licensor will still have to consider the nondiscrimination prong of its FRAND obligations toward other similarly situated licensees outside the LNG.

These proposed steps for conducting negotiations between SEP licensors and LNGs and other pertinent steps should be formulated into guidelines that aim to create a safe harbor for LNGs and which, if followed, would greatly reduce the most problematic anticompetitive risks. As mentioned, LNGs seem particularly interesting for similarly situated companies in the tail of markets, which collectively do not possess a market share that gives them market power. Given the increasing use of connectivity standards in various IoT verticals and the increasing number of companies that require licenses under SEPs from licensors for these standards for a wide variety of products, LNGs become increasingly attractive for both SEP licensors and similarly situated implementers given lower transaction costs and the opportunity for smaller implementers to secure more competitive royalty rates and enter the product market on a level playing field. This is even more relevant, as implementers in emerging IoT verticals may be less familiar with standards and SEP licensing than implementers in the telecom sector.

### B. *Governance of LNGs*

In the previous sections, the organization, operation, and decision-making of an LNG have not been addressed. However, the governance of an LNG is one of the decisive elements in determining its success. Member selection, member commitments to enter into license agreements with a SEP licensor(s) on agreed terms, voting rules, and providing clear mandates for the negotiators are principal elements that must be included in the governance of LNGs. This section will describe the various governance elements of an LNG in more detail.

An LNG can be set up in different ways, from purely formal to informal. LNG members can establish a legal entity specifically for this purpose, which is jointly controlled by the members. This may be attractive in a setting where the members are already members of a professional organization or association in their industry and where they will use this entity for handling negotiations with other SEP licensors for the same and other standards that they may use in their products. They can hire one or more licensing experts and other staff required to do the actual negotiations with a SEP licensor. The LNG members need to agree among themselves and with the legal licensing entity what mandate is given to the licensing entity and how that entity should interact with the LNG members.

Alternatively, the LNG members could simply contract several experts or a law firm to handle the negotiations on behalf of the LNG members on a project basis. In this case, the relationship between LNG members and the contractor must be formalized so that the negotiators have a clear mandate. In a more loosely controlled arrangement, the members of an LNG could elect a number of their representatives as the negotiators on behalf of the members and potentially hire a law firm or licensing experts to support them in the negotiations with the SEP licensor. Whatever the setup of the actual group or entity for handling the negotiations, it is important that negotiators have a clear mandate from the LNG. In particular, the process of communicating with the LNG during the negotiations should be clear and fully transparent, as it may otherwise frustrate not only the relationship between the LNG entity and LNG members but also the relationship between the entity and the SEP licensor(s).

When handling a specific SEP licensing opportunity, relevant companies need to determine individually whether they want to become members of the LNG or wish to opt out, because they prefer to negotiate with the relevant SEP licensor(s) bilaterally. As discussed in Section III, giving members the opportunity to opt out once the results of the LNG negotiations are known incentivizes individual members to use the outcome of the negotiations as a starting point for separate bilateral negotiations, aiming to get a better deal for themselves. This could lead to a disparity in royalty rates among LNG members. As a member of an LNG, a company enjoys the benefits of lowering its transaction costs and potentially obtaining better FRAND terms than it would be able to negotiate itself. However, it also runs a risk that it will not be satisfied with the terms negotiated with the SEP licensor and approved by the LNG members. Each company should assess this risk and determine whether it wants to opt out at the beginning of the negotiations with the SEP licensor. Also, allowing LNG members to opt out at the end of the process could induce certain implementers to hold out as long as possible based on the expectation that the SEP licensor may not be willing to litigate against an individual member, since the litigation cost may be higher than the revenues that can be collected from that implementer. Therefore, to prevent use of the LNG as a vehicle for holdout, the LNG should at the start of the process request that members commit to the agreed outcome of the negotiations and enter into a license with the SEP licensor within a predetermined period after approval of the result of the negotiations. In this way, SEP implementers can either negotiate collectively in good faith or opt out from the beginning, where the latter is simply the status quo.

Suppose an LNG would include companies that operate at different levels of a value chain, for example, members who operate at the downstream end product level and other members who operate at the component level. In that case, the LNG should be formed only by members operating at the same level in the value chain. The appropriate membership for the LNG will depend on the level in the value chain that the SEP holder targets for its licensing program. This should be clear from the assertion letter from the SEP licensor, in which it should indicate the devices alleged to infringe its SEPs. If LNG members occupy different levels of the value chain, the likelihood of a successful outcome to the negotiations is lower.

LNG members will also need to conclude a nondisclosure agreement to keep the information regarding the negotiations confidential and not disclose it to others both inside or outside the LNG. It is recommended that the LNG involves an outside counsel who can advise the LNG members on the obligations of the LNG to operate in line with the Huawei–ZTE negotiation framework and also to remind members at every meeting that members should not discuss prices, profits, market shares, and other information that would run afoul of antitrust guidelines. Moreover, this counsel should guide the members on what types of information they can share concerning the essentiality, validity, and infringement of the SEPs asserted by the licensor. Again, the goal is to reduce transaction costs while managing the risk of antitrust behavior.

Another important governance aspect of an LNG is how decisions are taken and thus what voting rules and procedures are put in place for various categories of topics that the LNG members may have to decide upon. Requiring full consensus is not recommended, as this gives individual members a veto right to block important proposals that are acceptable to all other members. Voting by a supermajority on major issues may provide a better approach for LNGs to facilitate broad market impact in the shortest timeframe (that is, reducing patent holdout).

Once the SEP licensor has provided the LNG with its FRAND offer, the LNG members will need to conduct a consultative process with their representatives concerning the terms of the counteroffer to the licensor. This will require taking into account all information regarding the licensor's SEP portfolio, including any opinions from outside counsel. The negotiators must be given clear, approved instructions (in accordance with the voting rules) about the counteroffer and a clear mandate for the subsequent negotiations with the SEP licensor that may ensue.

Suppose the negotiators can reach an agreement with the SEP licensor about the FRAND terms for a SEP license, and the LNG members approve the results according to their voting rules. In that case, each LNG member should enter into a license agreement with the SEP licensor within a predetermined time (for example, within six months in accordance with their commitment at the start of the process). It is in the interest of all LNG members that every member honors its licensing commitment, which creates a level playing field among the LNG members and avoids the collective action problem that can induce patent holdout. Any LNG who fails to enter into a license on a timely basis may be deemed an unwilling licensee and is therefore running the risk that the SEP licensor may seek an injunction against that member. This incentive structure is critical to maintaining the level of accountability and commitment required to make an LNG a viable SEP licensing mechanism.

Having a clear and transparent governance structure will enhance the ability of an LNG to achieve its goals of reducing members' transaction costs and obtaining better FRAND licensing terms than each member can negotiate individually. Additionally, implementing proper rules of governance can ensure that LNG members act in good faith toward both licensors and LNG members and, as a result, overcome the collective action problem that can result in systemic holdout.

## C. *Example of an LNG*

Although LNGs exist in other fields, such as joint purchasing groups for physical goods, LNGs in the field of licensing are mostly uncharted waters, although some precursor organizations and activities do exist. For example, defensive aggregators, such as RPX and AST, aggregate buyers of patents to facilitate licensing and reduce transaction costs, which is somewhat similar to the proposed role of LNGs. The recent syndication deal between RPX and Sisvel regarding SEPs for the Wi-Fi standard illustrates the transaction cost savings obtained by "pooling" both buyers and sellers in a single transaction as shown earlier in Figure 7.1.[33] Additionally, patent licensing platforms such as Avanci exemplify how linking licensors and licensees through a single platform can potentially enhance the efficiency of SEP licensing markets.[34] One-Blue is possibly the best example of a successful negotiation between groups of SEP licensors and licensees. In this section, the experience with an actual LNG in the One-Blue context will be discussed further.

The SEP licensor, in this case, was the One-Blue patent pool, which included a large majority of all the licensors holding SEPs for the Blu-ray standard and holding

[33] *Sisvel and RPX Conclude Licensing Agreement for Wi-Fi Standard Essential Patents*, RPX (Jan. 10, 2019), www.rpxcorp.com/about/news/sisvel-and-rpx-conclude-licensing-agreement-for-wi-fi-standard-essential-patents/.

[34] Avanci, www.avanci.com (last visited Mar. 18, 2022).

a significant part of the relevant SEPs for the standard for Blu-ray Disc™ products.[35] The Blu-ray standard is the successor to the DVD standard, offering consumers higher video quality with interactive features. Blu-ray players and recorders are backward-compatible with the various DVD and CD playback and recordable/rewritable standards. For the previous standards, joint licensing programs had been formed on a standard-by-standard basis, which required securing licenses from one or more patent pools, as well as from several individual licensors. Had this approach been followed for the Blu-ray Disc standards, the number of licenses necessary to manufacture and sell Blu-ray Disc players and recorders would have been so large that it would have discouraged manufacturers from developing Blu-ray-compatible products. To promote widespread use of this standard, the three originators of this standard, Philips, Sony, and Panasonic, developed the concept of a product pool or "pool of pools" for Blu-ray products, where all the SEPs for the Blu-ray standards and the backward-compatible standards were included in one licensing package offered to potential licensees. This provided a one-stop-shop licensing mechanism that yielded a reduced aggregate royalty rate compared to the sum of the royalties that would have been paid if each standard had been licensed separately.

Based on experiences in licensing older-generation formats, it was known that licensing manufacturers in one particular major country had taken longer and had required more efforts than in many other countries. Since the purpose was to stimulate and develop the market for Blu-ray products as quickly as possible, a different approach was chosen. While the basic framework of the patent pool and its licensing program were being established, licensing discussions were commenced with the industry association in that country for the relevant type of consumer audio/video products. All the relevant manufacturers were members of this industry association. Several meetings were held, in which the basics, including the royalty structure of the new licensing concept, were explained, questions were answered, and feedback on the various elements of the licensing program was sought, which were taken into account in finalizing the patent pool program.

After announcing the patent pool licensing program, meetings between the patent pool administrator and the industry association representing their member companies continued to discuss the collective licensing of these companies. Multiple meetings took place to discuss the benefits of reduced transaction costs for these members, including the elimination of the risk of ending up in costly litigation, the prospect of ensuring a level playing field among the association members, and the efforts that the pool would undertake to license other implementers of the Blu-ray standard. This holistic value proposition incentivized the industry association to agree with the patent pool administrator to advise their relevant members to sign a standard license agreement with the pool. To incentivize members to sign up with the patent pool within a six-month period, the industry

35 ONE-BLUE, www.one-blue.com (last visited Mar. 18, 2022).

association also agreed that any member not signing up in this period would not be given access to certain additional services provided by the association as long as they remained unlicensed.

Due to the close and constructive cooperation between the industry association acting as the representative of the relevant group of manufacturers (that is, as an LNG), it was possible for the patent pool to sign up a group of 15 midsize manufacturers within six months. Achieving the same result conventionally through bilateral negotiations may have taken years and required one or more litigations. This example demonstrates that licensing groups of implementers in a single licensing negotiation process through an LNG can offer significant benefits to both SEP licensors and implementers, not only reducing transaction costs but also eliminating holdout through collective action.

## V. CONCLUSION

Using a combined set of legal, economic, and managerial tools, LNGs can be designed to accomplish many different tasks. These tools include (1) proper guidelines to create a safe harbor, in which LNGs can operate without risking antitrust liability, (2) appropriate governance of LNGs' internal operations, and (3) conducting negotiations between LNGs and SEP licensors in accordance with the Huawei–ZTE negotiation framework for SEP licensing. Through careful institutional design, LNGs can facilitate SEP licensing efficiencies through reduced transaction costs for both licensees and licensors. Moreover, LNGs can create a level playing field among similarly situated implementers, who, as direct competitors, are rationally unwilling to take a license until everyone is licensed. LNGs can potentially solve this collective action problem and reduce the threat of patent holdout, which in turn could increase the leverage toward unlicensed companies in a virtuous cycle. Although it remains an empirical question whether LNGs can be designed to address antitrust concerns and then successfully implemented to facilitate increased SEP licensing on FRAND terms and at lower transaction costs, the substantial benefits LNGs may create for both SEP licensors and implementers make them worthwhile to explore.

# 8

# How to Create a Smoother SEP Licensing Ecosystem for IoT

*Ruud Peters, Fabian Hoffmann, and Nikolaus Thumm*

## I. STANDARDIZATION AND STANDARD ESSENTIAL PATENTS (SEPS)

We may not always realize that we live in a world where standardized devices and services are ubiquitous. We use them both professionally and privately, and our activities would largely come to a halt if these devices and services would not be available. Many high-volume products and services use one or more standardized technologies. Products like PCs, TV sets, DVD/Blu-ray Disc players, and streaming services like Netflix and Amazon Prime use various audio and video compression standards, and smartphones, probably the highest-selling tech device of all time, use several connectivity and audio and video compression standards.

The market success of these products is to a large extent determined by the interoperability that standards provide between products and systems of different suppliers, ensuring customers that they can buy and use products from different vendors, that all will operate in the same way in combination with other parts of the system, and that consumers can enjoy the same services on products from different vendors. A person can use their smartphone, tablet, or laptop of whatever brand to view content on the networks of different operators.

Standardization can be considered as one of the most successful examples of precompetitive open innovation, where commercial entities of different sizes, research institutes, universities, nonprofit organizations, and government bodies collaborate in standards developing organizations (SDOs) or ad hoc consortia to create technical standards that meet the needs of the market in a specific domain. Participants invest in the development of relevant technologies to which they are willing to contribute and from which the best technical solutions are selected to be incorporated into the standard. The SDOs (and consortia) set these standards with the aim to have them used as widely as possible.

Entities participating in standard-setting and making technical contributions often file patents on inventive elements in their proposals. When these proposals are adopted into the standard, these patents may become standard-essential patents (SEPs), which are necessarily infringed when implementing a standard. Participants need to be incentivized to invest in research and development to develop the technologies and contribute these to standards so SDOs can develop the best possible standards from a technical perspective on a continuing basis. Licensing their SEPs to implementers of standards provides technology contributors with such an incentive. SDOs have developed intellectual property rights (IP rights) policies for how to deal with SEPs. These policies seek to balance, on the one hand, the interest of SDOs in stimulating the widespread use of standards, and on the other hand, the interest of technology contributors in securing an appropriate return for making their technologies available for incorporation into standards.

The standards likely to be most widely used in the broad field of Internet of Things (IoT) are cellular standards (3G, 4G, 5G) developed by the 3GPP,[1] a partnership of seven SDOs, and a number of different wireless standards, including Wi-Fi standards developed by IEEE,[2] and a number of standards developed by ad hoc standard groupings, including the Zigbee, Lora, and Bluetooth standards. In this chapter, we will focus on cellular standards, for which the SEPs are governed by the European Telecommunications Standards Institute (ETSI) IP Rights Policy.

Under this policy, ETSI members participating in the standard-setting process have an obligation to disclose in a timely manner any patent or patent application that may be or may become essential to a standard. ETSI maintains a publicly accessible database of these declared SEPs. Also, the members holding SEPs for a standard have to be willing to license under fair, reasonable, and nondiscriminatory (FRAND) terms to third parties interested in implementing that standard. The ETSI IP Rights Policy does not provide any further information about what FRAND means, and ETSI does not want to become involved in any commercial discussions. They leave it to SEP licensors and implementers to negotiate an acceptable FRAND royalty.

## II. SEP LICENSING CHALLENGES

If SEP licensors and implementers do not succeed in negotiating a license, they have to turn to courts or arbitration to get a decision on their dispute. In the last 15 years, we have seen many SEP litigations relating to smartphones. In the period 2010–2015, litigation was used as a weapon in the platform battle between the mobile phone operating systems, Apple's iOS and Google's Android, which ended after Apple and Google entered into a patent truce in 2014. Most other litigation should just be seen as financial disputes between the various parties, where the SEP holder could be a commercial entity or a licensing company. Originally most cases were in the United States but over time also increasingly in Europe, in particular Germany

[1] *3rd Generation Partnership Project*, 3GPP, www.3gpp.org/ (last visited Mar. 19, 2022).
[2] *Institute of Electrical and Electronics Engineers*, IEEE, www.ieee.org/ (last visited Mar. 19, 2022).

and the United Kingdom, and more recently also increasingly in China. Litigation is initiated by both SEP licensors and implementers, in most cases because the parties could not come to an agreement on the royalty rate. To support their case, implementers mostly argue that the asserted SEPs are not truly essential, not infringed, or invalid and that the royalty offered is non-FRAND, whereas SEP holders argue the opposite.

In 2015, the Court of Justice of the European Union introduced the Huawei–ZTE negotiation framework[3] that provided guidance for SEP licensors and implementers on how to behave during licensing negotiations. A SEP licensor can seek an injunction against an implementer that is an unwilling licensee without violating competition laws, and implementers can show that they are a willing licensee and avoid an injunction if they follow the relevant steps of this framework. Although parties negotiating SEP licenses generally follow this framework, it has not led to a significant reduction in SEP litigation. Since the introduction of the Huawei–ZTE framework in 2015, more than 65 court cases have been decided in European countries, including more than 40 in Germany alone.[4]

Courts in various countries, and also the Patent Trial and Appeal Board (PTAB),[5] in the United States hold SEPs invalid or partly invalid in a majority of the cases where validity is challenged. Generally, willing implementers face little risk of an injunction when challenging essentiality, validity, and royalties as non-FRAND, because even if they are unsuccessful in litigation, they will still likely wind up paying only a FRAND royalty. Moreover, implementers might benefit from a hold-out or delaying strategy since SEP holders are often willing to give discounts on past sales when negotiating a license retrospectively. The longer the past sales period, often the higher the benefit from such discounts.

Given this situation, SEP litigation rates are unlikely to decline in the years to come. To the contrary, due to the increasing use of connectivity standards in the various IoT verticals, the number of companies having to take SEP licenses for these standards for widely different products will rapidly grow, and the same is likely to be true of SEP litigation. Companies in these IoT verticals may be less familiar with standards and SEP licensing, which may create additional difficulties in SEP licensing. The European Commission (EC) has recognized that this may slow down the development of digital and sustainable technologies and related markets in Europe. As announced in its 2020 IP Action Plan,[6] the EC is considering steps to create a more transparent and predictable SEP licensing ecosystem. Realizing that SEP licensing is frequently done at a global level, the EC will promote its SEP

3 Case C-170/13, Huawei Techs. Co. Ltd. v. ZTE Corp. [2015] 5 C.M.L.R. 779.

4 Marie Barani et al., *Case Law Post CJEU Ruling Huawei v. ZTE*, 4IP Council, https://caselaw.4ipcouncil.com (last visited Mar. 19, 2022).

5 *Patent Trial and Appeal Board*, USPTO, www.uspto.gov/patents/ptab (last visited Mar. 19, 2022).

6 Eur. Comm'n, *Commission Communication for an Intellectual Property Action Plan to Support the EU's Recovery and Resilience*, COM (2020) 760 final (Nov. 25, 2020).

licensing principles to, and cooperate with, other countries and regions, including Japan and the United States.

The EC will focus on three policy pillars to introduce new regulations or guidelines: (i) enhancing transparency on SEPs; (ii) providing clarity on various aspects of FRAND; and (iii) improving the effectiveness and efficiency of enforcement. Since this is still a work in progress, it is not known yet which specific measures the EC will take. (This discussion was finalized prior to, and therefore does not address, the EC's announcement of new proposed SEP regulations in April 2023.) However, we believe that creating a smoother and more efficient SEP licensing system leading to less litigation requires a holistic approach that considers all elements of the SEP licensing process that trigger litigation or are mostly used in litigation to secure royalty terms that are more favorable than the SEP licensor is offering or than the implementer is willing to accept. By addressing only some elements, parties in SEP negotiations will likely focus on other elements to get better financial terms, and these elements may again be triggers of litigation.

In the end-to-end licensing process, we think that five elements are the main reasons for disputes and litigation in SEP licensing negotiations: (i) lack of SEP transparency;[7] (ii) low confidence in the validity of SEPs; (iii) inability to assess a reasonable aggregate royalty; (iv) lack of incentives to seek licenses; and (v) concerns about an unlevel playing field.

In the following sections, we will go deeper into these issues and propose solutions for each of them. We want to emphasize that these solutions should not be considered in isolation, but rather integrally as a single solution for the total SEP licensing process. Each individual part of the solution may give rise to obligations that seem to fall more heavily on SEP licensors rather than implementers, or *vice versa*. However, when considering the integral solution as a whole, we believe that it achieves a fair balance between SEP licensors and implementers.

The solutions presented in this chapter are based on some of the mostly unrelated proposals described in the EC Expert Group report on SEP Licensing and Valuation,[8] and are presented here for the first time as a holistic solution. In this chapter, we have put a set of proposals together that in combination reduce the main causes of licensing disputes and litigation in a fairly balanced way for SEP licensors and implementers.

## III. SEP TRANSPARENCY

The ETSI database of declared SEPs[9] was established for the purpose of recording patents that are or may become standard essential and are submitted by members in

7 SEP transparency: clarity about which patents can be deemed essential based on independent essentiality assessments.

8 Justus Baron et al., *Group of Experts on Licensing and Valuation of Standard Essential Patents (E03600)*, Eur. Union Comm'n (Jan. 2021), https://ec.europa.eu/transparency/expert-groups-register/screen/expert-groups/consult?lang=en&groupID=3600.

9 ETSI IPR Online Database, ETSI, https://ipr.etsi.org/ (last visited Mar. 19, 2022).

accordance with their disclosure requirements under the ETSI IP Rights Policy. Due to over-declaration to safeguard compliance with the IP Rights Policy or for strategic reasons, only an estimated 25–40% of the patents in this database are actually essential.[10] This database is therefore not a reliable source of information for implementers to identify which companies have SEPs, to assess the size of each company's SEP portfolio, to assess what licenses may be required to produce a standard-compliant product, and to estimate the aggregate royalty for those products. Additionally, it is difficult for SEP licensors to determine a FRAND royalty for their SEP portfolios absent reliable information about the estimated total number of true SEPs for the relevant standard.

The EC 2020 IP Action Plan indicates that the Commission will seek to improve the transparency and predictability of SEP licensing. In particular, the Commission will explore the creation of an independent system of third-party essentiality checks in view of improving legal certainty and reducing litigation costs. Patent pools have shown that large-scale essentiality checks can be done. The EC Pilot Project for Essentiality Assessment of Standard Essential Patents has confirmed the technical and organizational feasibility of such essentiality checks.[11]

An essentiality check system needs to be designed and implemented in an efficient and cost-efficient manner. Patent examiners from patent offices, like the EPO, or attorneys from law firms doing evaluations for patent pools are well positioned to do these checks. Guidelines must be formulated ensuring that essentiality checks are done based on clear and transparent criteria. A supervising body (new or existing) should monitor compliance with these guidelines by the evaluators. This body should also arrange for certification of any entity or person that wants to perform these essentiality checks.

Essentiality findings should be treated as expert opinions, which could be appealed by patent holders and which could also be challenged by implementers and licensees in a fast and cost-effective challenge procedure (for example, within six months is considered feasible) based on a "loser pays" principle. Of course, a party may still bring its case to court, but it is expected that if the independent body does essentiality checks consistently with high quality and courts generally do not come to different conclusions, parties will likely increasingly rely on this body. If a party files an action in court and loses, without having first used the less expensive and shorter essentiality check procedure, the opposing party should be awarded its reasonable legal fees and other costs to be paid by the party bringing suit.

It is often argued that doing essentiality checks for all declared SEPs would take too much time and resources as well as cost too much. This is based on the misunderstanding that checks would be needed for all declared SEPs. However, the essentiality check process should be based on claim charts prepared by the patent holder and submitted to

[10] Baron, *supra* note 8, at 35.

[11] Rudi Bekkers et al., Pilot Study for Essentiality Assessment of Standard Essential Patents, EUR 30111 EN (Nikolaus Thumm ed., 2020).

the independent evaluator to start the process. A company will only submit those declared SEPs for evaluation, for which it has sufficient confidence in its claim charts; it will not be willing to pay the evaluation cost (estimated average cost around €5,000 per patent)[12] for patents with deficient claim charts. This will already eliminate an estimated 50–70% of all declared SEPs.

The cost of essentiality checks can be limited by checking only one member of a patent family in a major jurisdiction (including at least China, European Union, or the United States) and certification by the patent holder that the specified other members of that family include a claim that is substantially similar as the claim found essential in the checked patent. We believe that it is appropriate for SEP licensors rather than implementers to bear these costs, as licensors will benefit the most from having their SEPs checked for essentiality. The reasoning is as follows. In the first phase of SEP licensing negotiations, the licensor and implementer usually discuss the SEP portfolio as presented by the licensor, and the implementer may dispute the essentiality of one or more of the patents. These discussions can take considerable time and may even end up in litigation. By having the licensor's SEPs checked by an independent, trusted body, any discussions about whether or not presented SEPs are truly SEPs can be avoided. This will save time and effort both for the SEP licensor and the implementer. Since the SEP licensor can avoid this phase of the discussions with all implementers in all different IoT verticals and the implementer can avoid this phase only with the relevant SEP licensor, the total savings for the SEP licensor are higher than for the implementer, so that it seems justified that the SEP licensor should bear the cost for the essentiality checks of its patents. Since essential checks will likely reduce negotiation time and time to agreement, a licensor will also likely receive revenues earlier. Also, the practical complications of allocating these costs among an unknown number of implementers of unknown sizes recommend allocating these costs to the SEP licensors. Moreover, the SEP licensor is likely to earn a "return" on its investment in essentiality checks through the cost savings from more efficient licensing negotiations.

Another important aspect often not addressed is the timing of essentiality checks. Delaying these checks until years after the market for certain standard-compliant products has developed will result in little improvement in the licensing ecosystem. By that time SEP licensors and implementers will already have negotiated and concluded licenses, disputes will already have arisen, and litigation initiated, settled, or adjudicated. Checks need to be done as soon as possible in the early stages of development of the market for a category of standard-compliant products, which will allow the checks to take place before licensors and implementers start their negotiations.

[12] Nikolaus Thumm & Ruud Peters, A *Six-Point Plan for a New Approach to Assessing SEP Essentiality*, IAM (Feb. 3, 2021), www.iam-media.com/article/new-approach-assessing-sep-essentiality.

For each new product category, the relevant SEPs need to be identified. As the various products for different IoT verticals will be launched at different points in time after the adoption of the standard, licensors' investments in essential checks will also be spread out over time. Moreover, it should be realized that the 5G standard comprises a baseline component (New Radio/Network Core-NR/NC) and additional components for the different use cases related to different IoT verticals, which will similarly spread out essentiality checks over time.

Checks must be done for granted patents only. Currently, on average, more than 50% of all declared SEP families have a granted patent in one of the major market countries at the time of publication of a standard, but with significant time lag in granted patents across companies due to different filing routes.[13] The percentage of granted patents will grow over time. For standard-compliant product categories that enter the market several years after the adoption of the standard, the percentage of granted patents will have increased significantly, and it will be large enough to give a reliable picture of the size of licensors' SEP portfolios and thus also their share in the total stack of SEPs for those product categories. Since the first product category enters the market relatively shortly after the publication of a standard (for cellular standards, these are usually smartphones), the percentage of granted patents is still relatively low, and the distribution of granted patents across companies is skewed. This may make the picture of the SEP landscape less reliable. Many SEP holders already make use of accelerated patent examination procedures, and this should be further encouraged to allow the percentage of granted patents to increase more rapidly after publication of a standard.

Additional measures could be taken to further stimulate companies to take steps to have their patents granted quickly (or at least, not to delay the process) as well as to have their patents evaluated quickly after grant. For example, these practices can be encouraged by adopting rules that companies may only assert SEPs that have been confirmed to be essential after a check by a certified body. Alternatively, rules could be adopted that companies can only collect royalties after the date they have submitted their alleged SEPs for an essentiality check.

Essentiality checks by an independent body, based on agreed guidelines and supervised by an authority to ensure that they are done consistently and with high quality, is a first and important step in promoting a smoother licensing environment for SEPs. On the one hand, essentiality checks will assist SEP holders in estimating their SEP share in the total stack of SEPs for a category of standard-compliant products and use that information as an input to determine the royalty for their SEP portfolio, taking into account a reasonable aggregate royalty for the total stack. On the other hand, essentiality checks will assist implementers in identifying the companies from which they may need to take licenses for their products, to estimate the aggregate royalty for their products, and to take those considerations into

[13] Based on nonpublic input from IPlytics GmbH, a patent data analytics company.

account in their business plans. In the aggregate, these steps will result in more efficient licensing negotiations and fewer disputes and litigation concerning essentiality.

## IV. IMPROVING ON VALIDITY

### A. *Validity Rates*

Licensing negotiations tend to follow a rather fixed pattern. In the first phase, the implementer presents its arguments why one or more of the asserted SEPs are believed to be non-essential. In the second phase, the implementer makes arguments why one or more SEPs are believed to be invalid. As discussions about validity involve judgments about whether or not a patented invention is obvious, it might not be easy to reach agreement on validity. The objective of implementers is to try to undermine the SEP position of the licensor by advancing claims that the royalty offered by the licensor is too high and not FRAND. In cases where the parties are not able to reach an agreement and proceed to litigation, the implementer will in many cases contest the validity of the SEPs being asserted against it.

Since the introduction in the United States in 2012 of *inter partes* review (IPR),[14] implementers faced with SEP patent assertions have used IPRs in an effort to invalidate the SEPs. Today many large implementers file IPR petitions as a response to a SEP assertion letter while, at the same time, ensuring that they take those steps in line with the Huawei–ZTE negotiation framework and, as a result, are likely to be viewed as a willing licensee who is negotiating in good faith. In some cases, large implementers file multiple IPR petitions, which may place financial and resource pressure on the SEP licensor. This strategy may discourage smaller SEP licensors from asserting their SEP portfolios, which may have a negative impact on their investments in innovation and willingness to participate in standard-setting processes in the future.

Based on various reports, the PTAB invalidates about 65% of the challenged patent claims in accepted cases (in the term used by the PTAB, "instituted petitions").[15] Also, courts in Germany have declared 33% of all litigated patents in the period 2018–2020 fully invalid and 41% partially invalid.[16] These rates are more or less in line with the results of opposition proceedings against European patents before the European Patent Office.[17] These figures apply to all patents and not only

[14] *Patent Trial and Appeal Board*, *supra* note 5.

[15] Clark A. Jablon, *Is the Sky Falling in the U.S. Patent Industry?*, 36 Info. Display 37 (2020).

[16] Dr. Tilman Müller-Stoy et al., *Aktuelle Vernichtungsquoten im deutschen Patentnichtigkeitsverfahren* [*Current Cancellation Rates in German Patent Nullity Proceedings*], GRUR 142 (2022).

[17] Barker Brettell Intell. Prop., *Opposition Proceedings at the EPO* (2018), www.euro-ip.com/content/uploads/2018/08/Opposition-Proceedings-at-the-EPO.pdf.

SEPs, but it can be assumed that invalidation rates for SEPs will not be lower than for non-SEPs. Any SEP invalidity determinations reached through these adjudicative processes will impact a licensor's SEP position toward not only the implementer involved in each proceeding but also all other implementers. These invalidation rates show that implementers have a substantial likelihood of success in contesting the validity of SEP patent claims in litigation, IPRs, or oppositions. An implementer can use the risk of invalidation to try to secure better SEP royalty terms through settlement prior to adjudication. On the other hand, SEP licensors might already price into their royalty rates the likelihood that roughly half of their SEPs may be declared invalid in litigation, IPR, or oppositions.

It is not expected that this situation will change any time soon. As long as major patent offices continue to examine all patent applications with approximately the same degree of scrutiny, the percentage of invalidated claims of granted patents that are used in SEP licensing is not likely to change. Implementers will continue to contest the validity of asserted SEP patents, and SEP licensors will continue to be faced with invalidations of patents in their SEP portfolios offered for a license. Nonetheless, it still makes sense to consider mechanisms that may provide a reasonable estimate of whether the patent will be upheld or invalidated, quickly and easily, right at the beginning of negotiations.

### B. *In-Depth Prior Art Searches*

SEP licensors could undertake in-depth prior art searches on their SEPs prior to submitting them for an essentiality check or even prior to starting to prepare claim charts. The quality of state-of-the-art semantic search engines, often with additional artificial intelligence/machine learning (AI/ML) functionalities, has improved in recent years, especially for application in the field of information and communication technologies. These engines could be used to conduct fast, low-cost full-text searches against patent databases without limitations on technical classes in the Cooperative Patent Classification (CPC) system or other classification systems. Companies offering these search engines as a commercial service are readily available. The searches may reveal relevant prior art that has not yet been considered in patent examination procedures. Based on this prior art, a patent holder could decide that the patent in its current form is not likely to stand a validity test in court (or in an IPR) and consequently that it hardly makes sense to spend money on having it checked for essentiality. Patent holders may also opt to use these search engines during the examination procedure and bring any relevant prior art to the attention of patent examiners so they can take this prior art into account when evaluating the patentability of the claimed inventions. If used pre-grant, these prior art searches would contribute to reducing the likelihood that SEPs will be declared invalid when scrutinized in court or in IPRs. The post-grant use of such searches would make it possible to predict the answer to this question with reasonable certainty. Both the pre-grant and the post-grant use may reduce litigation based on invalidity claims of litigated SEPs. However, if such prior art searches show no indication of invalidity, an implementer would still have the right to claim the invalidity of SEPs in court. Litigation costs considerable time and money for both parties, and, moreover, it may take years before a final decision is made about the validity of a SEP, and clarity is achieved not only for the parties involved in litigation but also for other potential licensees.

## C. *Validity Challenges*

It is desirable to achieve the clarity described in the preceding section in an early phase of the development of a standard-compliant product market. This could be achieved if implementers could challenge the validity of asserted SEPs in an out-of-court challenge procedure before panels of independent patent experts. These panels could be selected from a pool of experienced and qualified patent experts certified by an independent body that facilitates and supervises these panels. This body could, for example, be the same body supervising the essentiality checks as described in Section III.

The challenge procedure should be relatively fast and inexpensive. It seems feasible that with a strict process where parties bring their arguments and counter-arguments in a limited number of rebuttals, panels should be able to produce valid opinions in about six to seven months. This should also keep the cost relatively low and well below the average cost of IPRs, which are estimated between $300,000 and $600,000.[18]

These panels would issue opinions about the likelihood that a patent will withstand a validity challenge when scrutinized in court (or in an IPR). They could not invalidate a patent, as this can only be done by a court. The parties could agree to accept the opinion of such a panel, or a party not accepting the opinion of a panel could elect to go to court. If an implementer went directly to court to claim invalidity without first using the faster and less expensive validity challenge panel and the patent's validity is upheld in court, the implementer should be ordered to pay the licensor's reasonable out-of-pocket costs. The same should apply to a SEP licensor who commences litigation without having completed the challenge procedure, provided the implementer initiated the challenge to the SEP's validity in a timely manner. This would create an incentive for both licensors and implementers to use the validity challenge procedure before going to court. This would also counter any hold-up or hold-out strategies.

If the panels produce high-quality opinions and courts generally do not come to different conclusions, the parties will increasingly rely on such opinions and will

[18] *Cost of Inter Partes Review: Everything You Need to Know*, UpCounsel, www.upcounsel.com/cost-of-inter-partes-review.

tend not to bring such cases to court. This would reduce the number of litigations based on claims that asserted SEP patents are invalid.

## V. INCENTIVIZING IMPLEMENTERS TO SEEK LICENSES

### A. *Publishing Standard License Terms*

Even when the SEP owners and the size of their SEP portfolios are known in the case of a particular standard, it is unlikely that implementers will approach the relevant licensors for their standard-compliant products. A SEP licensor will still have to identify the implementers that commercialize standard-compliant products using their SEPs and assert their SEPs against these implementers. This wait-and-see approach may mean that an implementer is approached by SEP licensors years after they started to commercialize standard-compliant products. Without information about the estimated aggregate royalty for these products, many (or even most) implementers would not take an estimated aggregate royalty into account in their business plans and would not make provisions for the royalties they will have to pay. In the meantime, these implementers may have considerable liability exposure to royalties owing on sales made prior to being approached by a SEP licensor. This liability exposure will increase even further, as licensing negotiations may also take considerable time (easily 18–36 months). Although SEP licensors are usually willing to give discounts on royalties for past-use sales, the outstanding past sales amount may create such a financial burden for the implementer that this may prolong negotiations even further.

The aforementioned situation could be avoided if licensors with confirmed SEPs make their standard license terms (or, alternatively, their standard license agreements) for a standard-compliant product publicly available through the relevant SDO, for example, ETSI. This could be done by recording those terms and conditions in the ETSI database or alternatively by showing a link to the website of the relevant SEP licensor where these terms and conditions are listed, which ensures that the latest terms and conditions are shown.

It should be appreciated that the standard license terms will apply only to specifically identified standard-compliant product(s). There may be different standard-compliant products within one application field – for example, infrastructure equipment and smartphones in the telecoms area, but also different products in various IoT verticals. A SEP licensor may publish different license terms for these different products since they may use different SEP families and the value that these patented technologies add to these products may be different. A SEP licensor does not need to publish the terms for all these different products at the same time. Rather, it could publish terms when the markets for the relevant compliant products start to develop.

### B. *Implementers Proactively Seeking Licenses*

By publicly disclosing its standard terms for a license under its confirmed SEPs for specific standard-compliant products, the SEP licensor would not be required to take the initiative to approach an implementer and should be assumed to have fulfilled both the first step of asserting its SEPs against an implementer selling such products and the third step of making a FRAND offer for a SEP license for these products according to the Huawei–ZTE negotiating framework.

The licensor's disclosure of its license terms should obligate the implementer to proactively seek a license from the SEP licensor, as required by the Huawei–ZTE negotiations framework. The implementer should fulfill the second step of this framework by expressing its willingness to take a license on FRAND terms and also should fulfill the fourth, fifth, and sixth steps, which obligate the implementer to respond diligently to the SEP offer without delay, to promptly make a counteroffer if it does not accept the published offer of the SEP licensor as being FRAND, and to provide security for the payment of past and future royalties based on its counter-offer, respectively. Moreover, an implementer would be required to take these steps prior to the commercialization of the relevant standard-compliant products.

This extension of the Huawei–ZTE negotiation framework will create incentives both for SEP licensors with true SEPs to publicly disclose their licensing terms and conditions before, or as early as possible after, the market for the relevant standard-compliant products starts to develop and for implementers to proactively seek licenses from such licensors prior to commercialization.

If implementers can access the published standard licensing terms of a SEP licensor, this will promote a more level playing field among implementers. Despite the nondiscrimination obligation under the FRAND undertaking, implementers have regularly expressed hesitancy to take licenses out of concerns that different terms may be offered to their competitors.

Under the extended negotiation framework, SEP licensors would be able to conclude licenses more rapidly and earlier in the commercialization process as compared to the current negotiation framework. When a licensor's SEPs are confirmed to be true SEPs, it can bypass the usual negotiation phase of discussing with each potential licensee whether or not its SEPs are true SEPs, saving both licensors and implementers time and effort and shortening the time to negotiate license agreements. Moreover, when implementers proactively seek licenses, SEP licensors could negotiate licenses with implementers not only more rapidly but also at an earlier point in time, which will enable them to receive licensing revenues earlier than otherwise would be the case. SEP licensors would be faced with an increased workload by having to negotiate licenses with several implementers in a more condensed period of time. However, greater efficiency in the licensing process seems to be needed in any case in light of the increasing number of implementers due to the increasing use of connectivity standards in the various IoT verticals.

By having to proactively seek licenses from a SEP licensor with confirmed SEPs and published standard license terms, an implementer would also avoid payment of a substantial amount of royalties for past sales that could form a significant obstacle to reaching a SEP license agreement under the current negotiation framework. This scenario is likely to arise when, as described previously, implementers wait until a SEP licensor asserts its patents to start negotiations.

We emphasize that a SEP licensor would have the choice to determine whether or not it wishes to publicly disclose its license terms for its SEPs and for which products. If a SEP licensor declines to disclose its license terms, the current Huawei–ZTE negotiation framework would continue to apply. This means that a SEP licensor has to assert its SEPs against an implementer as a first step to start the negotiation process. Additionally, the implementer would only be obligated to proactively seek a license if the licensing terms are also published in a database of the relevant SDO.

If the SEP holder takes the steps described previously and an implementer either does not proactively seek a license or fails to make a substantiated FRAND-counter-offer in a timely manner, the SEP holder would be entitled to seek an injunction against the unwilling implementer.

If the license offered by the SEP holder relates to one component of a complex product incorporating many different technologies, including several standardized technologies, an injunction may be too harsh a penalty for the implementer, who may have already invested in the development and manufacture of this product. On the other hand, an injunction may be too weak a sanction if the implementer can avoid it by submitting a counteroffer in the course of litigation initiated by the SEP licensor. Therefore, it may be more appropriate in this case to impose a penalty in the form of increased royalties on the implementer's past-use sales prior to entering into a license agreement. The size of this penalty should depend on the time elapsed between the implementer's response and the licensor's offer. This type of penalty is necessary to create sufficient incentives to induce the implementer to submit a counteroffer in a timely manner and present any other arguments it wishes to present against the assertions of the SEP licensor. In addition to this penalty, a SEP licensor may claim damages to the extent permitted under relevant national laws.

This sanction would also promote a more level playing field among implementers. If most and even all implementers seek to obtain a SEP license in a timely manner, they will also include the royalties in the calculation of the price of their product from the outset. Given an implementer's exposure to increased royalties in the absence of a timely response to a published license offer from a licensor, implementers would incorporate the expected royalties in the price of their standard-compliant products.

To the same extent, this sanction would also counteract hold-out tactics. If the implementer fails to make a timely response to a standard license offer, then the increased royalty should apply to all sales made by the implementer until a license agreement is reached, either by settlement or judicial determination. Therefore, an implementer would have an incentive to respond in a timely manner to a standard license offer or, in the absence of such a response, to limit the prospective penalty by negotiating expeditiously a license agreement with the SEP holder.

## VI. A REASONABLE AGGREGATE ROYALTY

A SEP licensor must determine what would be an appropriate FRAND royalty for its SEP portfolio applicable to certain standard-compliant products, taking into account a reasonable aggregate royalty for the total SEP stack for those products. We do not believe it is helpful to explain to SEP licensors and implementers what FRAND means or to provide guiding principles for FRAND license negotiations. It will probably raise more questions and trigger more litigation than it would avoid. Rather, it seems more constructive to focus on what practical methods could be applied to assess the reasonable aggregate royalty for the total SEP stack for a certain standard-compliant product. As is known, this aggregate royalty is not a single figure but a range of figures. We will outline in this section a three-layered approach, with each layer bringing an additional level of refinement in attaining a reasonable aggregate royalty.

The first layer makes use of the results of the essentiality checks as described in Section III. Where the estimated share of each SEP holder in the total SEP stack for a certain standard-compliant product is known, a SEP licensor can assess the estimated aggregate royalty based on its proposed royalty rate by "grossing up" its royalty by applying the licensor's royalty rate across the entire SEP stack. If royalty rates of other SEP licensors are known (for 5G smartphones, several SEP licensors, including Qualcomm, Nokia, Ericsson, and Huawei have published their maximum royalty rates), then the estimated aggregate royalty based on their royalties can be determined. The more datapoints, the better a SEP licensor can assess whether its royalty rate is in the right ballpark. This approach could lead to a kind of self-regulation of FRAND royalty rates for individual portfolios based on what is considered to be a reasonable aggregate royalty range.

A SEP licensor could argue that its SEP portfolio is more valuable than that of other SEP holders, and therefore its royalty may be higher than those of others. It is indeed reasonable to assume that some SEP holders have several SEPs that have a higher value than average, and some have several SEPs that are of lower-than-average value. But across all SEP holders, this will likely balance out in assessing the aggregate royalty range.

It should be realized that the estimated aggregate royalty rate based on the total SEP stack will be higher than the actual aggregate royalty will be in practice because certain SEP holders will not actively license their patents but use them only defensively (like Samsung and Apple, who based on their large market shares, and thus large SEP exposure basically pursue a defensive SEP strategy).

In the second layer, it is assessed whether the aggregate royalty rate range resulting from "grossing up" royalty rates for individual SEP portfolios reflects the added value of the patented technologies to the relevant standard-compliant product. This added value may differ from product to product (for example, compare the value of the 5G standard to a self-driving car and an offshore energy turbine). Various methods can be applied to estimate this added value. Hedonic price regression, choice modeling, and demand modeling approaches[19] can be used but may be too complex to be used on large scale, and in many cases, the outcomes may be less reliable for standard-compliant product categories for which the market is in the very early stage of development. These methods can be appropriate in litigations that take place at a point in time when there is an established market. In such cases, usually sufficient money and thus resources are available for these methods to be applied by economic specialists familiar with the relevant technology area and standard-compliant products, but even then, different experts may reach different conclusions.

Using comparable license agreements as a reference point for the estimated aggregate royalty rate seems to be a somewhat easier and more practical approach. Comparable licenses are usually considered to be licenses concluded with implementers that are similarly situated in the relevant product market. People may differ in what they consider to be similarly situated, but competing for customers with the same products in the same market seems to be a good description. It is unlikely that many comparable license agreements will have been concluded at an early stage of the development of the market for a new compliant product category. In that case, one can examine comparable license agreements for another complaint product category that may have been introduced to the market earlier or, if that is not the case, SEP license agreements negotiated for a previous generation standard. The available license agreements must be broken down into various elements. For example, some may be based on a running royalty and others on a lump-sum basis. They may also apply to a different royalty base – for example, an end product or a component. The different agreements may have to be weighted to take into account the level of similarity between these agreements. In case licenses for a previous generation standard are used, "scale up" factors may need to be applied to reflect the ratio between the number of SEPs for the related products for each generation and other factors, including price and performance ratios for the two generations. If sufficient licenses are available, the estimated aggregate royalty rate based on announced or known royalty rates for individual SEP portfolios can be compared with the aggregate royalty rate based on the comparable licenses.

A problem in applying this approach is that most license agreements are confidential, which may result in an insufficient number of license agreements being available to determine a reasonably reliable aggregate royalty estimate. This brings

[19] Baron, *supra* note 8, at 110–11 and Annex 6.

us to the third layer, which aims to ensure that a sufficient number of comparable license agreements will be available.

Parties concluding SEP licenses should be obligated through regulation or other means to submit their SEP license agreements to a market transparency office under the supervision of a governmental authority. Given the sensitive nature of many of these agreements, this office must keep them strictly secret. This office should be staffed with experienced licensing, patent, standards, and economic experts to be able to categorize, analyze, normalize, weigh, and scale up the different agreements to make them reasonably comparable. Based on this work, they could regularly publish reports with aggregate royalty rates (or, preferably, ranges) for the different standard-compliant products governed by the different relevant standards, together with the applied methods used to arrive at their figures. The estimated aggregate royalties for these products based on grossed-up individual royalties can be checked against these published aggregate royalties.

This third layer approach would support licensors in setting FRAND royalty rates for their SEP portfolios taking into account the reasonable aggregate royalty ranges for the relevant products. In case all or at least the major SEP licensors establish a patent pool, it can also make use of this aggregate royalty information to set the royalty rates for the products licensed by the pool. Usually, patent pools offer discounted rates by sharing part of the lower transaction cost with their licensees.

Additionally, implementers would be able to better assess the estimated aggregate royalty for the SEP licenses they need for their standard-compliant products. They could include these royalties as costs in their business plans and could make provisions for the payment of these royalties for the period that they have not yet negotiated the required licenses. This will avoid exposure to large unpaid royalties for past-use sales, which can discourage an implementer from entering into a license agreement with a SEP licensor.

We believe this framework for determining the aggregate royalty for various standard-compliant products will facilitate licensing negotiations between individual SEP licensors and implementers and result in fewer FRAND-royalty disputes and thus less litigation.

## VII. A BETTER LEVEL PLAYING FIELD

Implementers frequently express concerns that their competitors may be paying lower royalties than they are paying or are being offered. They also fear that a SEP licensor may not undertake sufficient efforts to license all of an implementer's competitors or that later licensees will benefit from higher discounts or discounts over longer past-sales periods. Moreover, they worry that larger, more powerful players in their market get much better terms than themselves, which makes it difficult to grow their market share for most products in highly competitive markets.

Due to the lack of transparency in SEP licensing, the nondiscrimination prong of FRAND is one of the more difficult issues to overcome.

In the current SEP licensing climate, large, financially powerful implementers are frequently only prepared to take licenses if they are forced to do so after litigation. They might put a lot of pressure on a licensor by counterclaiming non-essentiality and invalidity of their SEPs, including filing many costly IPRs in the United States (filing 15–20 and even more IPRs are not unusual today). Under the financial pressure of the huge litigation costs (including IPRs) and lacking the prospect of collecting royalty payments soon, SEP licensors might tend to make large royalty concessions. They may give large discounts on royalties for past sales, substantial volume discounts for future sales, or discounted lump sums on highly "de-risked" or worst-case sales projections. The effects are sometimes masked by including other elements in the deal, such as including non-SEPs or announcing a technical or other collaboration between the parties, to avoid any accusations of discriminatory licensing practices. The resulting effective royalty rates may be significantly lower than agreed with other smaller, less powerful players operating in the same market. Competing and growing market share for these smaller players may become more difficult, and it could also create a barrier for new companies to enter the market. This will ultimately lead to less choice and higher prices for consumers.

Frequently the argument is used that these larger, financially powerful companies with large market shares are not similarly situated as the smaller players since they are operating mostly in the premium segment of the market, whereas the smaller players are mostly operating in the budget segment of the market. If this argument is accepted, it would put these large companies in a separate league in the market, making the bigger players even bigger, and forcing smaller players to stay small. It all depends on whether or not all companies competing for customers for their products in the same market are considered similarly situated, even in a situation where they have widely different market shares.

These discrimination concerns can be mitigated by the various steps proposed in this chapter: increased SEP transparency based on the introduction of SEP checks; the availability of market information about aggregate royalty ranges for standard-compliant products; the obligation of implementers to seek licenses if SEP licensors have published their standard license terms for these products as described in the previous sections; and patent pools and licensing negotiation groups (as addressed in Chapter 7).

Nevertheless, the nondiscrimination concerns could be reduced even further if the market transparency office would not only publish the aggregate royalty rates per product but also upon the request of the competition authorities, and if courts would also aggregate royalties per company so that they could investigate whether there is any discrimination that would lead to hampering competition in the relevant product market. Providing access to this information through these mechanisms will likely have a positive effect on avoiding unnecessary discrimination. If there are clear indications of discrimination, then the competition authorities can start a full investigation and courts would make a determination.

## VIII. CONCLUSION

Taken together, the practical solutions described in this chapter could promote a more efficient SEP licensing ecosystem, where SEP licensors and implementers would have greater incentives to negotiate license agreements, rather than to litigate over their differences. These solutions address the major reasons behind most SEP litigation, including disputes about essentiality, validity, reasonable aggregate royalties for SEPs, hold-out behavior, and discrimination concerns. These solutions as a whole are designed to balance costs and benefits for both SEP licensors and implementers.

# PART V

# Patent Enforcement, Wireless Markets, and Global Competitiveness

# 9

# The Geopolitical Implications of Patent Holdout and the Ensuing Race to the Home Court

*Jorge Padilla and Andrew Tuffin*

## I. INTRODUCTION

During the last few decades, patent holders and implementers participating in standards developing organizations (SDOs) have successfully cooperated to develop new wireless standards that have benefited consumers all around the world. Thanks to such collaborative efforts, consumers around the world can communicate with each other, play games, watch movies, and enjoy various other activities using wireless networks.[1]

The standardization process has been successful because it has made *all* its participants – innovators, implementers, governments, and consumers – better off. The future of collaborative standards hinges on ensuring that this remains to be the case. That requires two *necessary conditions*: Implementers must have access to the best technological solutions under terms and conditions that allow them to profitably commercialize the products embedding those standards, and innovators must receive fair compensation for their research efforts.[2]

Much of the policy debate during the last 10 years has focused on how to reduce patent holders' leverage in negotiations with implementers that aim to license their technologies to avoid the risk of patent holdup and, therefore, ensure that the *first necessary condition* holds (that manufacturers can profitably commercialize products embedding standardized technology). Several authors have repeatedly warned about the risk of patent holdup in the licensing of standard-essential patents (SEPs).[3]

[1] Jorge Padilla, John Davies, & Aleksandra Boutin, *Economic Impact of Technology Standards: The Past and the Road Ahead* (Sept. 24, 2017), www.compasslexecon.com/wp-content/uploads/2018/04/CL_Economic_Impact_of_Technology_Standards_Report_FINAL.pdf.

[2] Justus Baron et al., *Contribution to the Debate on SEPs (E03600)*, Eur. Comm'n – Internal Mkt., Indus., Entrepreneurship & SMEs – Indus. Pol'y: Standard Essential Pats. (2021), https://ssrn.com/abstract=3778166.

[3] Mark A. Lemley & Carl Shapiro, *Patent Holdup and Royalty Stacking*, 85 Tex. L. Rev. 1991, 2049 (2007).

Holdup is a classic problem in economics; it arises in circumstances when firms negotiate trading terms after they have made costly, relationship-specific investments.[4] Since the costs of these investments are sunk when trading terms are negotiated, they are not factored into the agreed terms. As a result, depending on the relative bargaining power of the firms, the investments made by the weaker party may be undercompensated. In the context of SEPs, patent holdup would arise if SEP owners were able to take advantage of the essentiality of their patents to charge excessive royalties to manufacturers of products reading on those patents that made irreversible investments in the standard.

After years of heated debate, however, there is *no* consensus about whether holdup exists. Some argue that there is no evidence of holdup in practice.[5] If patent holdup were a significant problem, manufacturers would anticipate that their investments would be expropriated and would thus decide not to invest in the first place. But end product manufacturers have invested considerable amounts in standardized technologies. Others claim that while investment is indeed observed, actual investment levels are "necessarily" below those that would be observed in the absence of holdup. They allege that, since that counterfactual scenario is not observable, it is not surprising that more than 15 years after the patent holdup hypothesis was first proposed, empirical evidence of its existence is still lacking.[6]

The *second necessary condition* for the proper functioning of the standardization process, namely that patent holders be properly compensated, has received much less attention from scholars and policymakers. As Epstein and Noroozi explain:

> By "patent holdout" we mean [. . .] that an implementer refuses to negotiate in good faith with an innovator for a license to valid patent(s) that the implementer infringes, and instead forces the innovator to either undertake significant litigation costs and time delays to extract a licensing payment through court order, or else to simply drop the matter because the licensing game is no longer worth the candle.[7]

Arguably, the possibility of patent holdout is especially relevant in the standardization context. As SEP owners that made a commitment to license on fair, reasonable, and nondiscriminatory (FRAND) terms are typically limited in their ability to request an injunction in case of patent infringement, they have little or no

[4] Oliver Williamson, Markets and Hierarchies: Analysis and Antitrust Implications (1975).

[5] Alexander Galetovic & Stephen Haber, *The Fallacies of Patent Holdup Theory*, 13 J. Competition L. & Econ. 1, 44 (2017).

[6] Rebecca Kelly Slaughter, Comm'r, Fed. Trade Comm'n, SEPs, Antitrust, and the FTC, Remarks Prepared for Delivery at the ANSI World Standards Week: Intellectual Property Rights Policy Advisory Group Meeting (Oct. 29, 2021). *See also* Carl Shapiro & Mark A. Lemley, *The Role of Antitrust in Preventing Patent Holdup*, 168 U. Pa. L. Rev. 2019, 2060 (2020).

[7] Richard A. Epstein & Kayvan B. Noroozi, *Why Incentives for "Patent Holdout" Threaten to Dismantle FRAND, and Why It Matters*, 32 Berkeley Tech. L.J. 1381, 1384 (2017).

leverage when negotiating a licensing deal. The very same restrictions that limit the bargaining power of licensors to deal with the alleged risk of holdup aggravate the risk of patent holdout and the likelihood of undercompensating innovation. Furthermore, the risk of holdout is more significant for SEP owners with many complementary patents reading across jurisdictions. Patentees with large and global patent portfolios naturally seek to license their portfolio of SEPs at once to minimize transaction costs. Yet, some implementers refuse to negotiate in this way and choose to challenge the validity and/or essentiality of the SEP portfolio "patent-by-patent" and/or "jurisdiction-by-jurisdiction."[8] This strategy involves excessive litigation costs and is, therefore, inefficient. It may also lead to excessively low royalties and undercompensation.[9]

While patent holdout concerns have attracted the attention of the leadership of the US Department of Justice (DOJ) and the United States Patent and Trademark Office (USPTO) in the recent past,[10] some authors have rejected them as relatively immaterial.[11] However, the risk of holdout is not a mere theoretical *curiosum*. Heiden and Petit empirically document that some implementers do engage in patent holdout by ignoring correspondence, postponing negotiations, or simply by making counteroffers that are inconsistent with industry practice.[12] Other strategies include trying to affect the policies of SSOs or appealing to competition authorities. Of course, by delaying and stalling negotiations, potential licensees aim to obtain better licensing terms.

8 Baron et al., *supra* note 2.

9 In the Unwired Planet v. Huawei case [2017] EWHC 2988 (Pat), Judge Birss asked "[W]hat sort of license for Unwired Planet's portfolio would be FRAND in terms of its geographical scope when applied to a multinational licensee like Huawei? I will start by asking what a willing licensor and a willing licensee with more or less global sales would do. There is only one answer. Unwired Planet's portfolio today is (and in 2014 it was) sufficiently large and has sufficiently wide geographical scope that a licensor and licensee acting reasonably and on a willing basis would agree on a worldwide license. They would regard country by country licensing as madness. A worldwide license would be far more efficient."

10 Makan Delrahim, Assistant Att'y Gen., "Telegraph Road": Incentivizing Innovation at the Intersection of Patent and Antitrust Law, Remarks at the 19th Annual Berkeley-Stanford Advanced Patent Law Institute (Dec. 7, 2018), www.justice.gov/opa/speech/assistant-attorney-general-makan-delrahim-delivers-remarks-19th-annual-berkeley-stanford.

11 Letter from 77 former government enforcement officials and professors of law, economics, and business to Makan Delrahim, Att'y Gen. (May 17, 2018), www.competitionpolicyinternational.com/wp-content/uploads/2018/05/DOJ-patent-holdup-letter.pdf.

12 Bowman Heiden & Nicolas Petit, *Patent "Trespass" and the Royalty Gap: Exploring the Nature and Impact of Patent Holdout*, 34 Santa Clara High Tech. L.J. 179 (2018); Brian J. Love & Christian Helmers, *An Empirical Test of Patent Hold-Out Theory: Evidence from Litigation of Standard Essential Patents*, Santa Clara Univ. Legal Stud. Rsch. Paper (2021), https://ssrn.com/abstract=3950060. The authors find some evidence of an association between in-litigation holdout and both SEP portfolio size and patent quality; however, they find no evidence associating pre- or in-litigation holdout with the international breadth of SEP rights.

Indeed, the extant debate about patent holdout is not about whether implementers engage in so-called efficient[13] infringement; they do.[14] Heiden and Petit argue that the delay and the costs associated to patent holdout may also be related to the significant decrease in licensing coverage in the mobile phone industry, which has dropped from 73% to 36% between 2006 and 2016. Rather, what some scholars, such as Shapiro and Lemley,[15] claim is that the patent holdout concern is a theoretical and groundless "chimera," which at most affects only the distribution of surplus from innovation, stating that, in any case, it could be addressed through ex post court-mandated damages. FTC Commissioner Rebecca Slaughter maintains a similar position:

> Holdout, as long as it is unilateral and not done collusively among licensees, fits squarely into the box of problems that have patent law solutions. If a potential licensee has engaged in willful infringement, the patent holder has remedies in patent law, including the potential for enhanced damages. Unilateral holdout does not involve the abuse of market power to stymie consumer choice that holdup does, and therefore does not trigger antitrust concerns in the same way.[16]

In plain English, their claims are that when a licensee takes actions to stop paying licensors for the patents, it matters to the *licensors*, but (a) it should not concern consumers, because the latter's slice of the pie is unchanged, (b) it should not bother the licensees' competitors that pay religiously for the use of the innovator's technologies, and, in any event, (c) the licensor can always be compensated in court.

However phrased, these claims are wrong. In a recent paper, Llobet and Padilla show that patent holdout can engender significant social-welfare losses under a wide range of realistic circumstances.[17] The implications of patent holdout are not merely distributional. They find that implementers may have the incentive to incur significant costs to litigate SEPs sequentially (that is, patent-by-patent and/or jurisdiction-by-jurisdiction) even when this is socially inefficient. Such a strategy leads to lower royalty payments and may result in undercompensation of innovation. Furthermore,

[13] Efficient in a private, self-interested sense, but not in the collective interest, of course.

[14] The former head of patent licensing at Apple, Boris Teksler, explained that in his opinion "'efficient infringement,' where the benefits outweigh the legal costs of defending against a suit, could almost be viewed as a 'fiduciary responsibility,' at least for cash-rich firms that can afford to litigate without end." *The Trouble with Patent-Troll Hunting*, ECONOMIST (Dec. 14, 2019), www.economist.com/business/2019/12/14/the-trouble-with-patent-troll-hunting.

[15] According to the authors, "[p]atent advocates have sought to deflect concerns about patent holdup not only by denying its existence but by concocting a supposedly parallel story of 'patent holdout.'" They claim that "[p]atent holdout is incoherent as a theoretical matter and rejected as an empirical matter" and conclude that "[t]hose who express concerns about patent holdout seem to want to increase the returns to patent holders whose inventions add little or no incremental value, possibly because they advise SEP owners." Shapiro & Lemley, *supra* note 6, at 2047 n.91.

[16] Slaughter, *supra* note 6.

[17] Gerard Llobet & Jorge Padilla, *A Theory of Socially Inefficient Patent Holdout* (2021), https://papers.ssrn.com/sol3/papers.cfm?abstract_id=4021461.

it is likely to cause the dissipation of social surplus, as it leads to excessive litigation and leads to the exclusion of other implementers that, due to their smaller size or because of their start-up nature, cannot afford to engage in a similar litigation strategy.

In addition, there are powerful reasons to conclude that ex post court-mandated damages are likely to be insufficient to deter such a socially costly holdout strategy. First, it is obvious that if the cost of a patent holdout strategy is payment of reasonable royalties ex post, then (rational) implementers will have no incentive to pay early, given that a dollar today is worth more than a dollar in the future.[18] Second, when all that the SEP holder can recover in adjudication is cash royalties, not the other terms and conditions (for example, a cross license) that it would have been able to obtain during good faith bilateral negotiations, then an injunction is strictly needed to make the SEP holder whole. Likewise, when a delay in payment causes the SEP holder's bankruptcy or undermines its ability to fund valuable R&D, either by exhausting its internal funds or weakening its credit relative to third-party investors, then an injunction may also be strictly needed.[19] This is especially likely because holdout tends to occur in cascades: Once a major licensee engages in holdout, all others, insofar as they compete with the former, have an incentive to shirk on their payments too. As we subsequently explain, not even *enhanced* damages may be able to address the problems identified.

In Llobet and Padilla's paper, inefficient patent holdout can be prevented in a global court or when a local court is empowered to determine the validity of patents across all jurisdictions. However, this finding rests on some strong (and unrealistic) assumptions. Most importantly, local courts typically lack the authority to adjudicate with respect to the validity and infringement of foreign patents.[20] In addition, Llobet and Padilla assume that local courts approach patent disputes based on a similar legal framework – statutes and case law – and possess the same level of technical competency, and their decisions are unbiased (that is, based exclusively on objective information about the patent portfolio and, possibly, the outcome of previous trials). Finally, they implicitly assume that if a court with global jurisdiction were created de novo, it would also apply a similar legal framework and be unbiased.

[18] This could be addressed through enhanced damages. Jonathan M. Barnett & David J. Kappos, *Restoring Deterrence: The Case for Enhanced Damages in a No-Injunction Patent System*, in this book.

[19] David Goldman, *Qualcomm Made a Deal with Apple. Its Stock Has Soared 40%*, CNN Bus. (Apr. 17, 2019), https://edition.cnn.com/2019/04/17/tech/qualcomm-stock/index.html ("With the company no longer at risk of losing one of its most important sources of revenue, Qualcomm's stock has soared 40% to a 5-year high since it announced Tuesday it had settled all litigation with Apple. Qualcomm will continue charging Apple royalties for its patents, and Apple will pay Qualcomm a substantial fee as part of the agreement.").

[20] Jorge L. Contreras, *Global Rate-Setting: A Solution for Standards Essential Patents?*, 94 Wash. L. Rev. 701, 757 (2019).

As of today, we are not aware of any realistic initiative to create a multilateral institution with authority to resolve SEP disputes globally. Instead, we observe courts in various jurisdictions (for example, the United Kingdom and China) attributing to themselves the right to decide global license terms.[21] We also see how licensors and licensees file anti-suit, anti-anti-suit injunctions, and anti-anti-anti-suit injunctions seeking to influence which court ends up setting global terms. These maneuvers only make sense if courts are heterogenous, whether their differences are driven by differences in legal statutes, case law, speed, or objectivity.

In this chapter, we explore the implications of these developments for the future of the standardization process. Specifically, we consider the implications of extraterritoriality when licensors and licensees are located in different jurisdictions and local courts may be biased in favor of local litigants.

The pursuit of domestic industrial policy objectives through the biased enforcement of the law is likely to backfire and generate negative effects for everyone. Yet countries and their companies may face a prisoners' dilemma in which all litigants strive to get their disputes resolved by their local courts. This prisoners' dilemma may undermine the creation of global standards that, in the past, have contributed to the development and diffusion of technologies, such as mobile telephony, so successfully. It may cause the fragmentation of global standards along geopolitical lines: US firms would contribute with technologies covered by US patents to standards with a US-only geographic scope; EU firms would contribute with technologies covered by EU patents to standards with an EU-only geographic scope; and Chinese firms would contribute with technologies covered by Chinese patents to standards with a Chinese-only geographic scope; and so forth. This fragmented landscape will result in delayed innovation and result in worse and more expensive end products around the world due to lost economies of scale and scope.

The remainder of this chapter is organized as follows. In Section II, we detail the conditions under which patent holdout is socially inefficient. In Section III, we explain why ex post damages, even if somewhat enhanced, are likely to be insufficient to deter willful infringement. In Section IV, we explain how a global dispute resolution mechanism – a global court or mandatory arbitration tribunal – could eliminate the incentives to engage in socially inefficient patent holdout litigation

[21] Richard Lloyd, *UK Supreme Court Hands Unwired Planet and Conversant Victory in Key SEP FRAND Dispute*, IAM (Aug. 26, 2020), www.iam-media.com/frandseps/breaking-uk-supreme-court-hands-unwired-planet-and-conversant-victory-in-key-sep-frand-dispute; Bing Zhao & Jacob Schindler, *Inside Samsung's Wuhan Anti-suit Injunction against Ericsson*, IAM (Jan. 6, 2021), www.iam-media.com/frandseps/more-details-emerge-wuhan-anti-suit-ruling; Bing Zhao, *Chinese Judges Can Set Global SEP Rates and License Terms, Supreme People's Court Confirms*, IAM (Sept. 2, 2021), www.iam-media.com/frandseps/chinese-courts-can-set-global-sep-rate-and-licensing-terms-spc-confirms; and Jacob Schindler, *Sharp-Oppo Patent Dispute Ends with Cross-License Deal*, IAM (Oct. 8, 2021), www.iam-media.com/frandseps/sharp-oppo-patent-dispute-ends-cross-licence-deal.

strategies and discuss the institutional framework that would make that possible. In Section V, we document that, in the absence of a global dispute resolution mechanism, local courts around the world are moving to set global license terms and explain the risks and challenges posed by these developments. Most importantly, we expose the risk of fragmentation of otherwise global standards. Section VI concludes with a discussion of alternative ways of dealing with such risks.

## II. SOCIALLY INEFFICIENT PATENT HOLDOUT

As noted in the Introduction, the real debate about patent holdout concerns two issues: (a) whether it only affects the distribution of surplus from innovation, and (b) whether it can be addressed through ex post court-mandated damages. In this section, we explain that patent holdout's implications are not merely distributional; rather, patent holdout is socially inefficient under realistic conditions. Then, in the next section, we show that patent holdout's adverse consequences are unlikely to be effectively addressed in the absence of injunctions – for example, through the award of ex post damages.

### A. *Conditions for Inefficient Patent Holdout*

Llobet and Padilla model the negotiation between a licensor owning a SEP portfolio with patents in two jurisdictions and a global implementer that needs access to the patented technology to develop its products.[22] Due to its commitment to license on FRAND terms, the innovator is constrained to set the same royalty in both jurisdictions (to the extent that those jurisdictions are similarly situated) and to honor the offer made prior to litigation even after it is successful on validity in court.

The theoretical model rests on the following realistic assumptions. First, the SEP owner possesses many complementary patents and therefore seeks to license its whole portfolio at once to minimize transaction costs. Second, because standardized products are sold globally and the SEP portfolio at issue includes patents from different jurisdictions, the global implementer can challenge the validity of patents in that portfolio in different national courts (that is, "jurisdiction by jurisdiction").[23] Third, the implementer has the option to challenge the validity of these patents simultaneously (for example, globally) or sequentially (for example, patent by patent or jurisdiction by jurisdiction). Fourth, in sequential lawsuits, the result of a trial affects the probability that each party wins the following one. That is, if the implementer wins the first trial, it has a higher probability to win the second, as a first victory may uncover information about the validity of other patents that relate to

[22] Llobet & Padilla, *supra* note 17.

[23] Which in the context of the Llobet & Padilla paper is equivalent to "patent by patent" litigation since the licensor in their model owns a patent per jurisdiction.

the same type of innovation, which will be less likely to be upheld in court. Fifth, the impact of a validity challenge on royalty payments is asymmetric: Payments are reduced to zero if the patent is found to be invalid but are not increased if it is found valid (and infringed). This last assumption is consistent with the commitment to license on FRAND terms, whereby the innovator is constrained to set the same royalty across jurisdictions and to honor the offer made even after it is successful in court.

Llobet and Padilla show that the features of the legal system described in the last three assumptions can be strategically used by the manufacturer to reduce the compensation received by SEP owners even when that strategy entails a significant social cost. This result does not rely on the differential legal costs that global and local litigation might entail but, rather, on informational spillovers across jurisdictions.

The intuition is as follows. Suppose, for the sake of argument, that the innovator sets a royalty for each patent for which, in the simultaneous trial case, the implementer would be indifferent between settlement and litigation.[24] Under sequential litigation, however, the implementer may be willing to challenge a patent because of the gain in a future trial. This is due to the asymmetric effects that winning or losing the second trial has on the royalty rate that the implementer will have to pay. If the implementer wins the first trial, so that the first patent is invalidated, its probability of winning the second one increases, which means that the innovator is likely to settle for a lower royalty for the second patent or see both patents invalidated in court. In the opposite case, if the innovator wins the first trial, so that the second is also likely to be unfavorable to the implementer, the latter always has the option to pay the original royalty rate and avoid the second trial. In other words, the possibility that the implementer might be able to negotiate the royalty rate downward after a victory in the first trial, without the risk of it being increased in case of a defeat, fosters sequential litigation and results in lower royalties than the simultaneous litigation of all patents would produce.

When the innovation has a moderate value, the implementer's sequential litigation strategy forces the patent holder to lower its royalty to avoid being dragged from court to court. In contrast, a patent holder with a high-value innovation might decide to increase its royalty even if that generates inefficient litigation. When the patent is highly valuable and the informational spillovers between jurisdictions are sufficiently strong, raising the royalty, rather than decreasing it, might be profitable for the patent holder. Its success in court in the second jurisdiction is very likely upon success in the first one, and this implies that the downstream producer would settle even if the royalty were high. In that case, the patent holder trades off the losses

[24] That rate will necessarily be below the incremental value of the licensor's technology, which is the level of the royalty at which the licensee would be indifferent between using the patented technology or not, since exiting the market is less desirable than engaging in costly litigation.

from the initial litigation with the higher royalty payment in the second jurisdiction after an initial success.

When the value of the innovation is moderate, the implementer clearly prefers to litigate sequentially since that leads to a lower royalty. Yet, it also prefers to do so when the innovation is highly valuable even if that means that legal costs are incurred, and the royalty is higher. In this case, since the patent holder chooses a high royalty rate, litigation will take place in the first jurisdiction whether litigation is sequential or not. But, sequential litigation, by making the success probabilities in the second jurisdiction more extreme, always discourages one of the parties from going to court again, which, since litigation is costly, makes both parties better off.

### B. *Patent Holdout and SEPs*

The risk of socially inefficient holdout, while being applicable to any portfolio that includes patents the validity of which is related, becomes more significant in the context of SEPs for the following reasons. The first is the difficulty of SEP holders in adjusting their royalties upward after the first successful trial, as it might be considered a breach of their FRAND commitments. Indeed, we find that while a sequential litigation strategy may prove socially inefficient when the patent holder can revise the royalty upward, the distortion is more likely and more severe when the royalty initially chosen by the patent holder cannot be revised upward after a success in court.

The second is that, following recent intellectual property (IP) and competition law litigation in the United States,[25] the European Union,[26] and other jurisdictions, SEP owners are restricted in their ability to seek injunctions even in case of willful infringement.[27] By increasing the cost of holdout, injunctions curtail the incentives for the downstream producer to engage in sequential litigation and can help restore efficiency. However, while the threat of injunction mitigates the incentive to litigate sequentially and, therefore, excessively (that is, even when such litigation reduces social welfare), Llobet and Padilla demonstrate that it is unlikely to eliminate it.

The third reason is that patent holdout may undermine the standardization process. A recent Draft Statement published by the DOJ, USPTO, and the National Institute of Standards and Technology (NIST), states that:

> At the same time, when standards implementers are unwilling to accept a F/RAND license or delay licensing negotiations in bad faith, these strategies can lessen patent holders' incentives to participate in the development process or contribute technologies to standards voluntarily. Without adequate incentives to contribute to a

[25] eBay Inc. v. MercExchange, L.L.C., 547 U.S. 388 (2006).

[26] Case C-170/13, Huawei Techs. Co. Ltd. v. ZTE Corp., ECLI:EU:C:2015:477 (July 16, 2015), https://eur-lex.europa.eu/legal-content/EN/ALL/?uri=CELEX%3A62013CJ0170.

[27] Barnett & Kappos, *supra* note 18.

> consensus-based process, patent holders may opt for closed, proprietary standards that do not offer the same benefits of interoperability and enhanced consumer choice.[28]

We agree that if potential licensees are allowed to delay licensing negotiations in bad faith, this may lessen patent holders' incentives to participate in SDOs or contribute technologies to standards voluntarily.[29] But that is not the only, or even the most important, cost of such nefarious strategies. As explained earlier, patent holdout often leads to socially excessive litigation and causes innovators to be undercompensated, thus hurting innovation. Patent holdout strategies may also distort competition in markets where implementers engaging in such strategies compete if there are asymmetries in the ability to engage in costly litigation (see Section II.C). Furthermore, and most importantly, such strategies risk undermining the integrity and efficiency of the patent system. As a result, far fewer innovations will be developed in the first place, irrespective of whether they end up being standardized or not.

As explained by Haber and Lamoreaux,[30] patents are valuable because the right to exclude that they confer protects innovators against the free riding of their ideas, and because that right takes the form of a temporary property right that can be sold, licensed, and traded. This is of fundamental importance, since many innovators are just not good at running businesses and often prefer to transfer the task of commercialization to others. As these authors state,

> [t]he temporary property right that comes with a patent grant provides the requisite assurance [that their ideas will not be stolen by the licensees], facilitating a division of labor in which innovators can specialize in what they do best.[31]

Policy interventions that weaken the bargaining position of patent holders vis-à-vis unwilling licensees will discourage innovation by specialized firms, which depend for their existence on the proper functioning of markets for technology where they can license their technologies.[32] Such misguided interventions may force implementers to divert their own R&D resources to address the gap, which may

[28] Press Release, U.S. Dep't of Just., U.S. Pat. & Trademark Off. & Nat'l Inst. of Standards and Tech., Draft Policy Statement on Licensing Negotiations and Remedies for Standards-Essential Patents Subject to Voluntary F/Rand Commitments (Dec. 6, 2021), www.justice.gov/opa/pr/public-comments-welcome-draft-policy-statement-licensing-negotiations-and-remedies-standards.

[29] Anne Layne-Farrar, Gerard Llobet, & Jorge Padilla, *Payments and Participation: The Incentives to Join Cooperative Standard Setting Efforts*, 23 J. Econ. & Mgmt. Strategy 24 (2014).

[30] Stephen H. Haber & Naomi R. Lamoreaux, *The Battle over Patents: History and Politics of Innovation* (Nat'l Bureau of Econ. Rsch., Working Paper No. 28774, 2021), www.nber.org/papers/w28774.

[31] *Id.* at 8.

[32] "In sum, a market for technology refers to transactions for the use or creation of technology. It includes transactions ranging from full technology packages (patents and other intellectual property, along with know-how and services) to bare-bones patent licensing." Ashish Arora &

limit their ability in terms of simultaneously developing new products or services. Moreover, they may allow large, vertically integrated firms, mostly relying on secrecy, which is socially inefficient in a dynamic sense,[33] to protect their discoveries and capture all returns from innovation. Policy measures that place markets for technology at risk cannot constitute appropriate public policy.

Geradin *et al.* explain that:

> the effects of patents in the hands of upstream specialists are far more complex than is recognized in much of the policy debate, by the lower courts, by some competition officials, or in segments of the academic literature. In fact, patents held by NPEs can offer a number of pro-competitive benefits. First, IPRs, and especially patents, assist the entry of specialists into a market, which has direct implications for the level of competition and therefore the prices that consumer pay. Second, as is well recognized, specialization can mean higher quality. This is no less a factor in IP contexts. Third, when it is upstream, specializing can also translate into more innovation, as rival firms are pushed to innovate in order to remain competitive in the market. These many positive effects must be weighed against the negatives presented by blocking patents and opportunistic ex post licensing.[34]

Standardization enables smaller and non-vertically integrated innovators to collaborate to create valuable technologies that rival the proprietary solutions in the control of a handful of vertically integrated companies. Such pure or horizontal innovators deserve an appropriate return on their investments, which may not occur unless implementers are required, or at least incentivized, to negotiate in good faith.

### C. *Patent Holdout and Antitrust*

Llobet and Padilla find a second motivation for the holdout strategy: business stealing. Global implementers may engage in inefficient patent holdout – that is, litigate excessively – to gain a valuable cost advantage over their competitors, especially those who may not be able to afford such a costly strategy because they are relatively small, are financially constrained (as many startups are), or have a local dimension. Global implementers may prefer to litigate, even when litigation costs are so large that it would be preferable for society to avoid litigation, because their royalty burden may be reduced both in absolute terms and, in particular, relative to the royalty burden for its rivals if successful in litigation (while it would not go up if the patents are found valid). This business stealing incentive will result in

Alfonso Gambardella, *The Market for Technology*, *in* HANDBOOK OF THE ECON. OF INNOVATION 646 (Bronwyn H. Hall & Nathan Rosenberg eds., 2010).

33 Klaus Kultti, Tuomas Takalo, & Juuso Toikka, *Simultaneous Model of Innovation, Secrecy, and Patent Policy*, 96 AM. ECON. REV. 82 (2006).

34 Damien Geradin, Anne Layne-Farrar, & Jorge Padilla, *Elves or Trolls? The Role of Nonpracticing Patent Owners in the Innovation Economy*, 21 INDUS. & CORP. CHANGE 73, 90 (2012).

undercompensation of innovators, but, importantly, it may also result in the anticompetitive foreclosure of more efficient downstream competitors.

Llobet and Padilla consider a scenario in which a large implementer with the ability to fund protracted litigation competes in a downstream market with a competitive fringe, comprising small firms for which litigation is not an option. In this scenario, the large manufacturer may choose to litigate to force the innovator to settle on a low royalty. The large manufacturer exploits the asymmetry with its defenseless small rivals to reduce its (relative) IP costs. In some jurisdictions, it may also exploit yet another asymmetry in the legal system to achieve an even larger cost advantage. If both the large manufacturer and the innovator choose to litigate and the former wins, the patent is invalidated, and the large manufacturer avoids paying royalties altogether. Whether this confers a comparative advantage on the large manufacturer depends on whether the invalidation results in the immediate termination of all other existing licenses or not. If not, then an additional competitive advantage is obtained.

## III. ON THE INADEQUACY OF EX POST COMPENSATORY DAMAGES

It is often argued that monetary damages will usually be adequate to fully compensate a SEP holder in cases of strategic patent holdout. For example, Shapiro and Lemley state that:

> [w]hile courts may have difficulty calculating those damages, they tend to err on the side of paying patent owners too much, not too little. Plus, a defendant deliberately infringing a patent must also pay punitive damages for willful infringement, and often attorneys' fees as well. Some companies may try to "hold out" by infringing a patent and refusing to pay reasonable royalties, but the law can and does call them to account for it. Patent holdout might be a worry if we did not have a patent system, but that system by design prevents patent holdout.[35]

This proposition is incorrect. Absent injunctions, remedies available at law are inadequate to compensate for willful infringement. That is, reasonable royalties may be insufficient to deter patent holdout. Even enhanced damages may prove insufficient.

As explained by Vincenzo Denicolò and coauthors,[36] licensing negotiations are multidimensional, typically encompassing all IP issues between two companies.[37] Cross-licensing can be a part of the negotiations even for non-practicing entities (NPEs), for example, when follow-on research relies in part on complementary

35 Shapiro & Lemley, *supra* note 6. (Footnotes omitted.)

36 Vincenzo Denicolò et al., *Revisiting Injunctive Relief: Interpreting eBay in High-Tech Industries with Non-practicing Patent Holders*, 4 J. Competition L. & Econ. 571 (2008).

37 Ashish Arora, *Contracting for Tacit Knowledge: The Provision of Technical Services in Technology Licensing Contracts*, 50 J. Dev. Econ. 233 (1996).

patents held by others. Courts, however, do not have the authority to order an infringer to grant a cross-license of the infringer's patents to the successful patent holder plaintiff, nor to impose any other terms. Thus, all that the patent holder can recover in adjudication is cash royalties, not the other terms and conditions it would have been able to obtain during good faith bilateral negotiations. If shifting bargaining power reduces parties' ability to reach agreement on these terms, patent holders cannot be made whole through reasonable royalty awards alone.

Furthermore, because patents have expiration dates, timing issues must be considered when assessing the adequacy of monetary damages.[38] Patent holders face substantial delays in receiving payment, delays that might jeopardize their operations. If court proceedings moved at a quick pace, ignoring delays might be reasonable, but, in reality, patent infringement cases can take years to wend their way through the courts. Any delay in payment benefits the infringer and harms the patent holder, since a dollar today is always worth more than a dollar tomorrow. This is especially true for R&D-focused NPEs,[39] which rely on licensing for their revenues. Thus, infringers tend to have strong incentives to drag out proceedings, while patent holders generally have incentives to settle.

Even when courts finally enforce payment, patent holders face considerable dangers. If a court sets royalties too low, it will not only cost the patent holder in that one transaction but also will hinder its future negotiations with other potential licensees, as no other party will pay more than the judicially determined royalty rate. The opposite is not true, since it is not necessarily in the patent holder's benefit to enforce a ruling involving very high royalties (given the adverse volume effects).[40] This dynamic may reinforce patent holders' incentives to settle on a license, even when it appears that they will win a court case, just to avoid judicially determined rates, and conversely strengthens the incentives of infringers to engage in delaying tactics.

Finally, while SEPs are usually licensed on a portfolio basis, they are generally litigated on a patent-by-patent and jurisdiction-by-jurisdiction basis, even when that is socially inefficient. Damages awarded on this basis are therefore inadequate (by an order of magnitude) when compared with the trespass on intellectual property and the losses incurred.

For all these reasons, not even enhanced damages may be able to address the problems we have identified, though, of course, that depends on the magnitude and nature of the penalty imposed.

[38] The literature has incorrectly assumed away timing issues. For instance, Lemley and Shapiro argues that "[i]t is true that stays will allow the infringing party to keep infringing for some period after the patent is found valid and infringed, but we do not see this as terribly unfair to the patent holder, since the infringing party will owe reasonable royalties for those infringing sales, so any adverse impact on the patent holder is no greater than the impact caused by the infringement during the pendency of litigation." Lemley & Shapiro, *supra* note 3, at 2041.

[39] As opposed to patent assertion entities that conduct no R&D of their own.

[40] Denicolò et al., *supra* note 36.

## IV. GLOBAL DISPUTE RESOLUTION IN A NO-INJUNCTION WORLD

So far, we have established that patent holdout is a real-world problem with significant efficiency effects and potential antitrust implications. We have also shown that the incentives to engage in patent holdout are unlikely to be addressed effectively by means of ex post damages awards, even if the damages awarded exceed compensatory royalties. We have also discussed that injunctions, if available, would mitigate the problem, but are also unlikely to eliminate all incentives to infringe in cases involving SEPs. So, is there any more effective solution?

### A. *Mandating Global Litigation*

In Llobet and Padilla, implementers litigate jurisdiction by jurisdiction (and/or patent by patent), even when that strategy entails socially wasteful litigation, to force licensors to set lower royalties or, more generally, to reduce their expected royalty burden. In this model, the way to defeat such a strategy is to compel patent holders and licensees to accept a global jurisdiction where the validity of all patents, irrespective of their geographic scope, is determined by one of the local courts. That is, inefficient patent holdout can be prevented when patent validity across jurisdictions is decided either by a global court or a local court making extraterritorial determinations. Llobet and Padilla find that global litigation is more efficient than a system where each patent is independently tried in each jurisdiction even if the legal costs of global litigation are higher than the costs of litigating in multiple jurisdictions (that is, even in the absence of economies of scale in the legal process).

This conclusion holds because Llobet and Padilla assume that local courts have the authority to adjudicate with respect to the validity and infringement of foreign patents, approach patent disputes based on a similar legal framework – statutes and case law – and possess the same level of technical competency, and their decisions are unbiased (that is, based exclusively on objective information about the patent portfolio and possibly the outcome of previous trials rather than the identities of defendant and plaintiff). They also assume that if a court with global jurisdiction were created de novo, it would adopt a similar legal framework and be unbiased. These are all very strong assumptions. In particular, local courts typically lack the authority to adjudicate on validity and infringement with respect to foreign patents.[41]

### B. *Mandatory v. Voluntary Arbitration*

While *mandatory* global arbitration would produce similar outcomes to global litigation, it seems unclear how to create such an obligation in practice. This is

[41] Contreras, *Global Rate-Setting*, *supra* note 20.

important because, unfortunately, *voluntary* global arbitration cannot play a similar role. Licensees unwilling to pay the royalty rate proposed by a patent holder could, in principle, voluntarily submit their pledge to an arbitrator that would produce a globally binding ruling. Yet they have no incentive to do so. Since there is no commitment to arbitration before the royalty is chosen, the option to arbitrate does not affect the implementer's incentives to engage in sequential litigation. The implementer will engage in socially costly sequential litigation (holdout) under the very same circumstances in which it did so in the absence of the arbitration alternative.

### C. *Global Rate Setting*

In any event, as of today, we are not aware of any realistic initiative to create a multilateral institution with authority to resolve patent validity disputes globally, whether this is a court of justice or an arbitration tribunal. Yet, as noted by Contreras:

> Courts adjudicating FRAND disputes face a dilemma. On one hand, patents are issued under national law and, by definition, have legal effect only in the issuing jurisdiction. On the other hand, the parties to FRAND disputes are often multinational corporations with operations (and patents) in jurisdictions around the world. Moreover, many of these parties privately negotiate worldwide license agreements to cover their global operations, without regard for the particular patents issued in any given country.[42]

In the absence of a global rate setting body, as the one proposed by Contreras, we observe courts in various jurisdictions (for example, the United Kingdom and China) asserting the right to decide global royalty terms.[43] Because FRAND disputes are essentially contractual disputes, national courts may have the jurisdictional authority to determine a global rate for the portfolio licensed under the agreement at issue.[44]

The move toward global rate setting started in 2017, when the UK High Court for Patents ruled in *Unwired Planet v. Huawei* that it was authorized to set the terms of a global FRAND license between the parties, covering not only the SEP holder's UK patents but also foreign patents covered by its FRAND commitment.[45] The court concluded that a licensor and licensee acting reasonably and on a willing basis

[42] Jorge L. Contreras, *Anti-suit Injunctions and Jurisdictional Competition in Global FRAND Litigation: The Case for Judicial Restraint*, 11 N.Y.U. J. Intell. Prop. & Ent. L. 171, 172 (2021).

[43] Contreras, *Global Rate-Setting*, *supra* note 20.

[44] Jorge L. Contreras et al., *The Effect of FRAND Commitments on Patent Remedies*, in Patent Remedies and Complex Products: Toward a Global Consensus 160, 163 (C. Bradford Biddle et al. eds., 2019).

[45] Unwired Planet Int'l Ltd. v. Huawei Techs. Co. Ltd. [2017] UKSC 37, aff'd [2020] EWHC (Pat) 711 (Eng.).

would agree on a worldwide license, since country-by-country licensing would be highly inefficient. A similar approach was taken by the US District Court for the Central District of California in *TCL v. Ericsson*.[46] Most recently, courts in China have also moved to adopt global FRAND rates.[47]

## V. SETTING GLOBAL TERMS BY BIASED DOMESTIC COURTS

These developments have given rise to what Contreras characterizes as a "race to court,"[48] where licensors and licensees have been filing anti-suit injunctions (ASI),[49] anti-anti-suit injunctions (AASI),[50] and even anti-anti-anti suit injunctions (AAASI), seeking to influence which court ends up setting global royalty rates. Of course, these maneuvers only make sense because courts are heterogenous, with asymmetries driven by differences in legal statutes, case law, speed, or objectivity. Leaving aside differences in substantive law and procedure, these strategic races to the courthouse appear to be motivated by actual or perceived institutional capture and domestic favoritism. In short, foreign litigants may be trying to avoid Chinese courts because of fear of bias and vice versa. The race to court is in practice a race to the "home court."

### A. *The Race to the Home Court*

Like other regulatory instruments, such as antitrust or merger control,[51] IP law and contract law enforcement might be used in unorthodox ways to favor domestic firms

46 TCL Commc'n Tech. Holdings, Ltd. v. Telefonaktiebolaget LM Ericsson, 955 F.3d 1317 (Fed. Cir. 2020). The same is true of the N.D. Cal. ruling (later overturned) in FTC v. Qualcomm. FTC v. Qualcomm Inc., 411 F.Supp.3d 658 (N.D. Cal. 2019), *rev'd*, 969 F.3d 974 (Fed. Cir. 2020).

47 Huawei Techs. Corp. Ltd. v. Conversant Wireless Licensing S.A.R.L., case ID: 2019 Zui Gao Fa Zhi Min Zhong No. 732, 733, 734 Part I (Sup. People's Ct., Aug. 28, 2020). An unofficial translation is available at https://patentlyo.com/media/2020/10/Huawei-V.-Conversant-judgment-translated-10-17-2020.pdf. Xiaomi Commc'n Tech. Co., Ltd. v. InterDigital Inc., case ID: 2020 E 01 Zhi Min Chu 169 (Wuhan Intermediate People's Ct., Hubai Province, Sept. 23, 2020). An unofficial translation is available at https://patentlyo.com/media/2020/10/Xiaomi-v.-InterDigital-decision-trans-10-17-2020.pdf; Guangdong OPPO Mobile Telecommunications Corp. Ltd. v. Sharp Corporation, case ID 2020 Yue 03 Min Chu No. 689-1 (Intermediate People's Court of Shenzhen City of Guangdong Province, Dec. 3, 2020). The Supreme People's Court upheld the Shenzhen ruling on Sept. 2, 2021.

48 Contreras, *Anti-suit Injunctions and Jurisdictional Competition*, *supra* note 42.

49 An anti-suit injunction (ASI) is an interlocutory *in personam* remedy issued by a court in one jurisdiction to prohibit a litigant from initiating or continuing parallel litigation in another jurisdiction. While an ASI can bind a party to litigation, it has no binding effect on a foreign court.

50 AASI operates *in personam* prohibiting a litigant from taking a particular action (seeking or enforcing an ASI), rather than purporting to restrain the authority of a foreign court.

51 Mario Mariniello, Damien Neven, & Jorge Padilla, *Antitrust, Regulatory Capture and Economic Integration*, Bruegel Pol'y Contribution (July 2015), www.bruegel.org/wp-content/uploads/imported/publications/pc_2015_11_.pdf.

competing in global markets at the expense of foreign competitors and economic integration. A small country, or a country in which new technologies are mostly used as inputs by domestic manufacturers, might find it optimal to adopt a pro-implementer bias and, hence, use the law to reduce the cost of IP of its domestic firms. Lower IP prices need not have negative effects on that country's economy if the incentives of high-tech multinationals supplying domestic manufacturers to develop new technologies are warranted by bigger markets in other countries.

A country could, for example, use antitrust policy and, in particular, the laws against abusive conduct by dominant firms, opportunistically. Companies licensing their valuable IP to domestic manufacturers may be accused of charging excessive prices and compelled to license their IP at rates that are disproportionately low to grant domestic manufacturers a competitive advantage over their foreign counterparts, both domestically and in foreign markets.

These concerns are real. In February 2015, for example, the US chipmaker Qualcomm paid $975 million to Chinese authorities to end a 14-month antitrust investigation into its patent licensing practices.[52] The fine was then the largest fine in China's corporate history. The settlement required Qualcomm to reduce the royalty rates on its standard-essential patents applied to sales of mobile phone made in China by Chinese smartphone makers, such as Xiaomi, Lenovo, and Huawei. We do not intend to discuss whether the decision was justified or not. We simply note that the move must have helped Chinese manufacturers to compete against market leaders Apple and Samsung in the growing Chinese mobile phone market and, possibly, elsewhere.

More generally, research points out that domestic bias in law enforcement is pervasive – whether developed or developing countries, centers of innovation or centers of manufacturing, or other differences in industrial policy. Bhattacharya and coauthors, for example, show that there is a lower probability of adverse US court judgments for US domestic companies compared to foreign companies.[53] This could reflect a conscious bias (an explicit tool of industrial policy) or unconscious bias (that is, courts ideologically sympathetic to a particular position or domestic companies just know their way around the local legal system better).

### B. *The Global Costs of Biased Domestic Enforcement*

Regulatory capture in the enforcement of competition, contract, and IP law could cause significant distributional effects, shifting rents from efficient and innovative foreign firms to less efficient domestic companies, to the ultimate detriment of local

[52] *Qualcomm Settlement with China's NDRC Removes Major Speedbump*, FORBES (Feb. 10, 2015), www.forbes.com/sites/patrickmoorhead/2015/02/10/qualcomm-settlement-with-chinas-ndrc-removes-major-speedbump/?sh=27342c24431a.

[53] Utpal Bhattacharya, Neal Galpin, & Bruce Haslem, *The Home Court Advantage in International Corporate Litigation*, 50 J.L. & ECON. 625 (2007).

and foreign consumers. This reallocation of rents could distort incentives to invest and innovate and hence reduce the overall growth potential of the global economy. The risk of domestic bias also creates regulatory and legal uncertainty, thus reducing the incentives of foreign companies to invest both domestically and overall.

One can distinguish between four channels through which investment decisions can be affected by domestic bias in law enforcement. First, capture increases uncertainty: Political interference in the enforcement process makes it more difficult to predict the final outcome of a court case. Regardless of what that outcome could be, the mere inability to anticipate it reduces the incentive to invest.[54]

The second channel is through direct distorting effects: These arise if the main objective of political intervention is to protect domestic companies. Political intervention that biases enforcement in favor of local players might have effects on the competitiveness of foreign companies already present in the domestic market, undermining their competitive position in both domestic and international markets. Foreign companies may be forced to revise downward their expectations about future profits from innovation, which would reduce their incentives to invest and innovate.

The third channel is through indirect effects if the distortions introduced by political interference in law enforcement affect domestic markets in such a way that it is less appealing for foreign investors to produce or invest in that country.

Finally, there are potential dynamic effects. Strategic trade theory suggests that the more leeway countries have in using the law to pursue protectionist goals, the greater the risk that penalized companies' countries of origin will retaliate by implementing equally distorting measures. The end result is a reduction of the inflow and outflow of trade for all jurisdictions involved.[55]

## C. *The Prisoners' Dilemma of Biased Domestic Enforcement*

In the long term, the pursuit of domestic industrial policy objectives through the (possibly unconscious) biased enforcement of the law is likely to backfire and generate negative effects for everyone. Any short-term advantages conferred on domestic firms by the strategic use of the domestic laws will evaporate once trading partners respond to those abuses and retaliate. A well-functioning global economy

[54] Brandon Julio & Youngsuk Yook, *Political Uncertainty and Corporate Investment Cycles*, 67 J. FIN. 45 (2012). The authors investigate the relationship between cross-border capital flows and host economies' political uncertainty. They find that the capital flow from US companies to their foreign affiliates drops by 12% during election years in host economies. Investment is lower when investors find it more difficult to anticipate future government policy.

[55] Michal S. Gal & Jorge Padilla, *The Follower Phenomenon: Implications for the Design of Monopolization Rules in a Global Economy*, 76 ANTITRUST L.J. 899 (2010).

requires laws designed and enforced without bias. Yet, countries and their companies may face a prisoners' dilemma: All of them would benefit if local courts seeking to adjudicate on global royalty terms acted objectively so that no litigant benefited from a home court advantage, but because all have the incentive to act opportunistically, they are likely to end up in a world where all litigants strive to get their disputes resolved by their local courts, making use of ASIs, AASIs, and AAASIs or whatever needed to secure and protect that advantage.

The impact of this prisoners' dilemma on royalty payments and, therefore, on the balance of interests between innovators and implementers is unclear. Assuming local courts adopt the objectives of their local constituencies, we may have pro-licensor rulings in jurisdictions where (pure or horizontal) innovators hold sway, pro-implementer rulings where implementers dominate, and unbiased decisions when neither group has greater political clout. However, irrespective of the direction of the bias, the uncertainty associated with its existence is what causes the problem.

Ultimately, our main concern is that this prisoners' dilemma may undermine the creation of the sort of global standards that have contributed so successfully in the past to the development and diffusion of technologies such as mobile telephony. Firstly, biased courts may shift rents away from innovators (respectively, implementers) if rates are decided by local courts biased in favor of local implementers (respectively, local innovators). Secondly, they may cause innovators (or implementers) to be under- or overcompensated depending on their nationality, irrespective of the incremental contributions of their technologies. Thirdly, biased adjudication may lead to conflicting legal determinations across jurisdictions and, therefore, to enhanced business uncertainty and protracted conflict. Lastly, and most importantly, for all these reasons, the decisions of domestically biased courts may cause the regional fragmentation of global standards into, for example, the United States, the European Union, and Chinese zones. These competing standards may compete outside their respective home bases, as GSM and CDMA did in the past.

Standards' success depends on their ability to take advantage of economies of scale and scope, which would be lost if standards become geographically fragmented. Competition across standards may result in wasted duplication of R&D expenses, limit the scope for specialization, and, ultimately, and perhaps more importantly, cause technological and economic divergence and raise new geopolitical tensions.

## VI. SOLVING THE PRISONERS' DILEMMA

Contreras suggests that the solution to this dilemma could be that:

> while international bodies develop a more comprehensive, efficient and transparent methodology for assessing global "fair, reasonable and nondiscriminatory"

(FRAND) royalty rates, national courts voluntarily "stand down" from assessing global FRAND rates and instead limit their assessments to FRAND royalty rates applicable to patents within their own jurisdictions.[56]

While we understand the logic of this proposal, we do not believe it to be the solution to the problem. The global adjudication of SEP disputes has a logic that cannot be denied: It makes no sense to resolve disputes involving SEP "global" portfolios on a "jurisdiction by jurisdiction" basis. The voluntary moratoria proposed earlier could perpetuate indefinitely. The solution is radical action: the creation of an impartial global FRAND rate setting tribunal.[57] SDOs' IP policies would require that SEP holders and implementers resolve their license disputes through that impartial body.

This is not a new solution. SDOs overcame self-interest from companies and countries when developing computing and communications standards. They need to do it again for remunerating contributions to developing and implementing those standards.

[56] Contreras, *Anti-suit Injunctions and Jurisdictional Competition*, *supra* note 42.

[57] One option would be to rely on the WIPO Arbitration and Mediation Center. Based in Geneva, Switzerland, with a further office in Singapore, the WIPO Arbitration and Mediation Center was established in 1994 to offer Alternative Dispute Resolution (ADR) options for the resolution of international commercial disputes between private parties. The subject matter of these disputes includes both contractual disputes (for example, patent and software licenses, trademark coexistence agreements, distribution agreements for pharmaceutical products, and research and development agreements) and noncontractual disputes (for example, patent infringement), including court referrals. WIPO disputes have involved parties based in different jurisdictions, including Austria, China, France, Germany, Hungary, India, Ireland, Israel, Italy, Japan, the Netherlands, Panama, Spain, Switzerland, the United Kingdom, and the United States of America. *WIPO Arbitration and Mediation Center*, WIPO, www.wipo.int/amc/en/center/background.html (last visited May 19, 2022).

# 10

# China's Practice of Anti-suit Injunctions in Standard-Essential Patent Litigation

## *Transplant or False Friend?*

*Mark A. Cohen*

## I. INTRODUCTION TO THE ANTI-SUIT INJUNCTION TRANSPLANT

Based on published data from various sources, in late 2020, China emerged, at least temporarily, as the largest grantor of anti-suit injunctions (ASIs) against overseas assertions of standard-essential patents (SEPs) in the world with the issuance of five ASIs.[1] This unexpected development was attributed to China "transplanting" Chinese ASI practices from other countries. A legal transplant is "the borrowing and transmissibility of rules from one society or system to another."[2] In a typical ASI, a Chinese court will order parties appearing before it to refrain from pursuing litigation on the same matter in one or more foreign countries on the basis that the foreign litigation would have a substantially negative impact on Chinese litigation. It may also involve imposing penalties on a litigant who seeks an ASI in another jurisdiction. If the order seeks to prohibit the litigant from filing an anti-suit injunction in another country, it is more precisely known as an "anti-anti-suit injunction" or AASI. ASIs entered Chinese legal practice in maritime disputes[3] and then quickly emerged as a tool for litigants to seek control over non-Chinese patent litigation. In Chinese civil litigation practice, an ASI is accomplished by a Chinese court granting a Behavior Preservation Order, which is similar to a preliminary injunction.

[1] King Fung Tsang & Jyh-An Lee, *The Ping-Pong Olympics in Antisuit Injunction in FRAND Litigation*, 28 Mich. Tech L. Rev. 305 (2022) [hereinafter *Ping-Pong Olympics*].

[2] Alan Watson, Legal Transplants: An Approach to Comparative Law (2d ed. 1993), cited by Peter K. Yu, Jorge L. Contreras, & Yang Yu, *Transplanting Anti-suit Injunctions*, 71 Am. U.L. Rev. 1537, 1545 n.36 (2022) [hereinafter *Transplanting ASIs*].

[3] John Liu & Minli Tang, *The First Official Shot from Chinese Court against an Anti-suit Injunction*, AllBright Law Offices (July 27, 2017), www.lexology.com/library/detail.aspx?g=cf8d3bde-2c31-4526-9848-eab60842f016; *see also* Zhang Weiping, *The Construction and Implementation of My Country's Anti-suit Injunctions* (我国禁诉令的建构与实施), Guangzhou Maritime Court of the PRC (Apr. 24, 2022), www.gzhsfy.gov.cn/web/content?gid=93982&lmdm=1029 [hereinafter *Zhang Weiping Article*].

Chinese ASIs have not yet extended into other areas of law, including in international commercial arbitration.[4] Prior to their adoption by Chinese courts, they were long in use in common law countries, including the United States, particularly in matters where courts of equity conflicted with the law courts or ecclesiastical courts.[5]

ASIs are the newest of several Chinese tools to undermine foreign parallel intellectual property (IP) litigation. As with any legal transplant, questions may be raised concerning the ability of China to "customiz[e] and assimilate[e] the imported standards based on local needs, interests, conditions, and priorities."[6] In view of its motives and implementation, an alternative perspective is to view China's ASIs as a type of "false friend." A "false friend" is a linguistic term referring to words in a foreign language bearing resemblance to words in one's own language but having different meanings. A Chinese saying describes this legal phenomenon: "similar appearance, different spirit" (形似神异).[7] In legal terms, a false friend may manifest itself as a legal transplant that has significantly departed from the original purposes of its "exporting" country. In this sense, a false friend may present itself as a more extreme form of the challenges in legal and cultural adaption of a transplant, which are often themselves "endemically riddled with value laden, open-ended notions" of a country's legal culture.[8] However, when a system is transplanted but is widely divergent (or unstable), characterization as a transplant, which is inherently value-neutral, may also lead to undue acceptance of borrowed practices that may be fundamentally different, or even experimental in nature.[9]

As an example of prior efforts to normalize an IP transplant, China on two separate occasions sought to justify an expansion of its administrative system for IP protection on the basis that the United States had a similar system administered by the United States International Trade Commission (USITC). The first such effort occurred in 2004. At that time, China legislated a procedure to exclude infringing imports under China's Foreign Trade Law modeled on US law. However, Chinese Customs already had the authority to exclude infringing imports without the

4 *Zhang Weiping Article*, *supra* note 3.

5 S. I. Strong, *Anti-suit Injunctions in Judicial and Arbitral Procedures in the United States*, 66 Am. J. Comp. L. 153 (2018) [hereinafter *S. I. Strong Article*].

6 *Transplanting ASIs*, *supra* note 2, at 1549.

7 Mark Cohen, David Kappos, & Randall Rader, *Faux Amis: US-China Administrative Enforcement Comparison*, 4 China Patents & Trademarks 33 (2016) [hereinafter *Faux Amis Article*].

8 Paul Edward Geller, *Legal Transplants in International Copyright: Some Problems of Method*, 13 UCLA Pac. Basin L.J. 199, 210 (1994).

9 *Faux Amis Article*, *supra* note 7. For a different view of how developing countries such as China may support IP false friends without understanding their commercial consequences, see Miranda Forsyth & Blayne Haggart, *The False Friends Problem for Foreign Norm Transplantation in Developing Countries*, 6 Hague J. Rule L. 202 (2014). In this chapter, I take a different approach by suggesting that calling a legislative "import" a transplant may be part of an intentional effort to make the receiving country's conduct less offensive to other countries.

necessity of extensive and costly USITC-style administrative proceedings.[10] As far as I know, this system has never been implemented. Another effort at transplanting USITC practice occurred in 2015, when China sought to justify a 23-fold increase in its vast domestic administration for patent enforcement over a six-year period on the basis that the USITC had a similar procedure. As a consequence, China had nearly 1,000 times more cases than the USITC by 2015. In fact, China's administrative system did not replicate the USITC remedy, an in rem enforcement remedy used solely to address the challenges of infringing imports, based on US civil procedure, and the subject of a published docket. In a sense, USITC remedies were transplanted twice, and not at all.[11] China's efforts to transplant USITC practice also shared a common theme of legal experimentation with ASIs, which can make it especially difficult to anticipate how a legal transplant can develop in China over time.

Unlike other recent foreign transplants in China's IP regime, such as the recent introduction into China's civil practice of punitive damages, burden of proof reversals, and "patent linkage" in pharmaceutical IP disputes,[12] ASIs do not address domestic lawsuits but are directed exclusively to overseas cases. One scholarly Chinese commentator has explicitly disavowed the possibility of a domestic ASI regime, by noting that Chinese ASIs "can only be used in international or extraterritorial parallel litigation situations" and that they are a legal tool to deal with "differences arising from different legal systems." This scholar has contrasted the Chinese system with other countries, such as the United States, where ASIs can also affect federal/state judicial relationships.[13]

Put another way, there does not appear to be a Chinese interest in a domestic counterpart to China's practice of ASIs. This is not because it is unneeded. Conflicts can arise from China's extensive multiple-track domestic IP litigation system, including overlapping civil, criminal, administrative, and customs remedies and conflicts; delays attributed to separation of IP validity and infringement proceedings; and conflicts that may arise between national and local government authorities.[14] The Chinese government has attempted to mitigate those differences through

[10] *Faux Amis Article*, *supra* note 7.

[11] *Id.*

[12] Mark Cohen, *Synthesizing Developments on Linkage from the July 15 Berkeley Program*, China IPR (July 19, 2021), https://chinaipr.com/2021/07/19/synthesizing-developments-on-linkage-from-the-july-15-berkeley-program/.

[13] *Zhang Weiping Article*, *supra* note 3; see also *S. I. Strong Article*, *supra* note 5.

[14] Mark Cohen, *Draft Judicial Interpretation on Patent Linkage Released by SPC*, China IPR (Oct. 29, 2020), https://chinaipr.com/2020/10/29/draft-judicial-interpretation-on-patent-linkage-released-by-spc/; *see also Outline for a Strong IP Country (2021–2035)* (中共中央 国务院印发《知识产权强国建设纲要》(2021–2035)), (Sept. 22, 2021), www.gov.cn/zhengce/2021-09/22/content_5638714.htm (Item 10 states that in order to address the complex enforcement roles of different Chinese institutions, China will "clarify the responsibilities, powers and jurisdictions of administrative organs and judicial organs, improve the connection mechanism between administrative protection of intellectual property rights and judicial protection, and form a joint protection force.").

national and local coordinating agencies and through policies mandating cooperative interactions. However, due to the lack of a separation of powers doctrine and limited opportunity for judicial review of administrative action, Chinese courts have no general authority to countermand the action of another domestic court or administrative agency.[15] Chinese ASIs stand out in part because they are, in the words of Judge Zhu Jianjun, who has adjudicated many key SEP decisions, intended to help assist China "to build the main battlefield for foreign-related dispute resolution."[16]

This outwardly facing approach of issuing ASIs also differs from prior approaches of the Chinese courts and governmental institutions toward transnational parallel litigation. In the past, China typically used domestically focused civil strategies to ensure that foreign litigation did not impact Chinese domestic litigation. Judge Zhu described this approach: "China's traditional legal system mainly focuses on resolving domestic disputes, which is coupled with a relatively inward-facing traditional culture (including legal culture). Chinese laws mainly deal with foreign-related disputes based on the rules and experience of handling domestic disputes."[17]

Arguably, all courts rely on domestic law in the treatment of cases. What I believe instead distinguished China's prior approach from the common law system in handling parallel litigation was something more narrow: the judicial attitude of "not formally recognizing or enforcing" ASIs.[18] There are, indeed, arguments to be made that China's new system of extraterritorial ASIs is now instead a part of a tradition within Chinese law of enacting separate institutions, procedures, and rules for dealing with foreign-related matters. These differences today include a separate tribunal for foreign-related civil cases at China's Supreme People's Court (the Number 4 Civil Tribunal), separate time frames for the resolution of foreign-related cases pursuant to China's Civil Procedure Law (CPL) (Art. 277), an inability of foreigners to sit for the bar and an inability of foreign law firms to appear before Chinese courts,[19] and a greater likelihood for intervention in a case by an

[15] Shen Kui, *Administrative "Self-Regulation" and Rule of Administrative Law in China*, 13 U. PA. ASIAN L. REV. 72 (2018); Ian Johnson, *China Grants Courts Greater Autonomy on Limited Matters*, N.Y. TIMES, Jan. 3, 2016, www.nytimes.com/2016/01/03/world/asia/china-grants-courts-greater-autonomy-on-limited-matters.html.

[16] Mark Cohen, *Unwired Planet and the Role of Chinese Courts: A Perspective from Shenzhen*, CHINA IPR (Jan. 18, 2021), https://chinaipr.com/2021/01/18/unwired-planet-and-the-role-of-chinese-courts-a-perspective-from-shenzhen/.

[17] Zhu Jianjun, *Conflicts and Responses in Issuing SEP ASIs and AASIs and How to Counter Them* (标准必要专利禁诉令与反禁诉令颁发的冲突及应对), S. CHINA INST. INT'L INTELL. PROP. (Aug. 8, 2021), https://sciiip.gdufs.edu.cn/info/1026/1760.htm [hereinafter *Zhu Jianjun Article*].

[18] *Zhang Weiping Article*, *supra* note 3; *see also* Vivienne Bath, *Overlapping Jurisdictions and the Resolution of Disputes before Chinese and Foreign Courts*, SYDNEY L. SCH. RSCH. PAPER NO. 16/102, Dec. 5, 2016, https://ssrn.com/abstract=2880942 [hereinafter *Vivienne Bath Article*].

[19] Mark A. Cohen, *International Law Firms in China: Market Access and Ethical Risks*, 80 FORDHAM L. REV. 2569 (2012).

"Adjudication Committee," rather than the judges listening to the case, in resolving a dispute involving a foreigner.[20] While the utility and necessity of disparate treatment of foreigners in China's civil IP system is a general topic that is beyond the scope of this chapter, differential treatment of foreigners, in fact, has a long history in China's legal system and could be considered a distinguishing feature of China's ASI regime.

China's historical approaches to foreign ASIs were also based on judicial practices that generally respected the territoriality of IP rights by not considering validity, infringement, or damages in overseas patent litigation, unless there was consent by the parties to consider extraterritorial issues. China instead relied on certain advantages of its legal proceedings to exert some influence over parallel proceedings, including an expedited court docket, expert judges, and the near-automatic granting of injunctive relief to stop the manufacturing or sales of infringing products within China. Unlike US patent litigation, but like continental legal systems, Chinese injunctions are granted nearly 100% of the time when requested by a successful patent litigant who is suing on patents that have not yet expired.[21] Injunctions have traditionally been important for litigants in China due to traditionally low damages for patent infringement. Chinese injunctions have also become increasingly valuable with the growth of China's manufacturing prowess and the expansion of its domestic consumer markets. Injunctions may also provide a significant incentive toward settlement of a global SEP litigation.

China has also taken other nonjudicial steps to insulate itself from foreign parallel cases, such as by limiting possibilities for dismissal of cases on grounds of forum non conveniens[22] or limiting exposure to enforcing foreign SEP judgments through mutual legal assistance agreements. One narrowly tailored example of the latter is found in the "Arrangement on Reciprocal Recognition and Enforcement of Judgments in Civil and Commercial Matters by the Courts of the Mainland and of the Hong Kong Special Administrative Region," which specifically excludes "cases on the confirmation of the license fee rate of a standard-essential patent heard by a court of the Mainland or a court of the Hong Kong Special Administrative Region."[23]

A turning point in China's approach toward ASIs occurred about the time of the *Huawei v. Samsung* litigation that resulted in a US court issuing an ASI against

[20] *Dialogue – Issue 39: Only in China: "Adjudication Committees" Serve Judicial System*, Dui Hua Foundation, https://duihua.org/dialogue-issue-39-only-in-china-adjudication-committees-serve-judicial-system/ (last visited on June 11, 2022).

[21] Bian Renjun, *Patent Litigation in China: Challenging Conventional Wisdom*, 33 Berkeley Tech L.J. 413, 436 (2018).

[22] *Vivienne Bath Article*, *supra* note 18, at 12–17.

[23] Agreement on Reciprocal Recognition and Enforcement of Judgments in Civil and Commercial Matters by Courts of the Mainland and of the Hong Kong Special Administrative Region, Hong Kong Dep't of Just. (Jan. 18, 2019), at Art. 3, www.doj.gov.hk/en/mainland_and_macao/pdf/Doc3_477379e.pdf.

Huawei.[24] Huawei had simultaneously filed parallel rate setting proceedings in the Northern District of California and Shenzhen, China. Judge Zhu described China developing its own ASIs as a necessary response to this type of ASI:

> The extraterritorial court issued ASIs to Chinese wireless communication companies that were seriously affected and hindered in the civil litigation rights enjoyed by Chinese civil litigants in accordance with Chinese law. These ASIs may further affect and hinder civil substantive rights. At the same time, although the object of an ASI issued by an extraterritorial court is a party to a civil lawsuit in China, it will hinder the normal development of civil lawsuits in my country, and even lead to the termination of civil lawsuits already underway in my country or unenforceable judgments. The extraterritorial ASI directly or indirectly affects the exercise of judicial jurisdiction by Chinese courts over SEP disputes and interferes with and undermines China's judicial sovereignty.[25]

Judge Zhu's reference to an extraterritorial court issuing an ASI may suggest the negative pregnant that there are Chinese territorial courts issuing domestic ASIs. As noted, this does not appear to be the case in China. His focus on "sovereignty" is also not atypical in Chinese discussions of ASIs. As Professor Vivienne Bath has noted, "Chinese cases and judicial documents dealing with international legal matters tend to refer both to the important concept of judicial sovereignty ... and to the more general idea of reciprocity. The phrase 'judicial sovereignty' is used in connection with the protection from foreign encroachment of the jurisdiction and autonomy of Chinese courts."[26] Judge Zhu also does not account for the fact that Huawei was the plaintiff in both the US and Chinese *Huawei v. Samsung* cases. According to Judge Orrick in the California dispute, Huawei had requested a global rate determination from the US court.[27] *Huawei v. Samsung* may not, therefore, be a good example of why China should develop its own ASI system, as the offending party was Huawei when it initiated duplicative lawsuits in two jurisdictions. Indeed, the ASI decisions subsequent to *Huawei v. Samsung*, such as *Samsung v. Ericsson*,[28] have looked to the minimization of inefficient duplicative litigation as a reason for granting an ASI and rejected comity arguments to defer to an overseas court. Arguably, a Chinese court could also have considered Huawei estopped by subsequently pursuing a contrary position in its parallel litigation in Shenzhen. This

[24] Huawei Techs. Co. v. Samsung Elecs. Co., No. 3:16-cv-02787-WHO, 2018 U.S. Dist. LEXIS 63052 (N.D. Cal. Apr. 13, 2018).

[25] *Zhu Jianjun Article*, *supra* note 17.

[26] *Vivienne Bath Article*, *supra* note 18, at 25.

[27] Huawei Techs. Co. v. Samsung Elecs. Co., No. 3:16-cv-02787-WHO, 2018 U.S. Dist. LEXIS 63052, at *11 (N.D. Cal. Apr. 13, 2018).

[28] Wuhan Intermediate People's Court, *Samsung v. Ericsson, Civil Ruling of PRC Wuhan Intermediate People's Court (2020) E 01 Zhi Min Chu No. 743*, at pp. 10–11 (Dec. 25, 2020), www.ipeconomy.cn/index.php/mobile/news/magazine_details/id/2148.html (last visited June 26, 2022).

position also appears to be supported by Article 100 (now Art. 103) of China's CPL regarding "Behavior Preservation," which Chinese courts have utilized as the legislative basis for granting extraterritorial ASIs. In its current form, it primarily seeks to address damages caused to a litigation based on behavior of the "other party."

Article 100 was based in part on China's experience in granting preliminary injunctions in IP matters, as was required by its accession to the WTO. It was a rarely used IP remedy, that has since been made broadly available in all civil disputes:

> **Chapter IX Preservation and Preliminary Execution**
>
> Article 100 In the event that the judgment on the case may become impossible to enforce or such judgment may cause damage to a party *because of the behavior of the other party* to the case or because of any other reason, the people's court may, upon the request of the said party, order the preservation of the property of the other party, specific performance or injunction; in the absence of such request, the people's court may, where it deems necessary, also order property preservation measures.
>
> When a people's court adopts any preservation measure, it may order the applicant to provide security; where the party refuses to provide such security, the court shall reject the application.
>
> When a people's court receives an application for preservation in an emergency, it shall decide within 48 hours after the receipt of the application; if the court accepts the application, such measures shall come into force immediately.[29]

Article 100 was also an odd basis on which to consider the situation posed in SEP cases such as *Huawei v. Samsung*. As Zhang Weiping, a noted scholar of China's CPL, has pointed out, Article 100 was drafted with "the understanding that its significance did not include ASIs."[30] Moreover, by its own terms, it is primarily intended to "resolve domestic disputes."[31] It gives no guidance concerning its potential extraterritorial application. "Chinese law," one commentator has noted, "does not explicitly permit the courts to issue anti-suit or anti-arbitration injunctions."[32] Chinese academics and others have justified these actions as responses to similar actions taken by foreign courts on SEP-related litigation

[29] *Civil Procedure Law of the People's Republic of China*, CHINA INT'L COM. CT. (June 29, 2017) (emphasis added), http://cicc.court.gov.cn/html/1/219/199/200/644.html.

[30] *Zhang Weiping Article*, *supra* note 3.

[31] *Id.*

[32] Sophia Tang, *Anti-suit Injunction Issued in China: Comity, Pragmatism and Rule of Law*, CONFLICT OF LAWS (Sept. 27, 2020), https://conflictoflaws.net/2020/anti-suit-injunction-issued-in-china-comity-pragmatism-and-rule-of-law/.

involving China,[33] either more generally on the basis that this was a "legal transplant" that was part of the "ping-pong"[34] of conflicting ASIs from different jurisdictions, or the "gaming"[35] by litigants in pursuit of optimal judicial fora, which China was fully in its rights to enter. Not surprisingly, in light of its domestic orientation, Article 100 does not explicitly consider the impact of an ASI on a foreign jurisdiction, nor is it reflected in a Judicial Interpretation of the Supreme People's Court regarding behavior preservation orders in IP-related cases that are intended to provide more granular guidance under China's CPL.[36] Comity was subsequently introduced as a consideration in the landmark *Huawei v. Conversant* decision and has thereafter been followed in other ASI cases.[37] In that decision, the Chinese court ordered an ASI directed to the pendency of a potentially conflicting German court decision regarding a patent royalty rate that was significantly higher than the calculation of a Chinese court. By contrast, courts in Wuhan have since issued global ASIs precluding lawsuits anywhere in the world that might interfere with their efforts to set global FRAND rates, as did the Shenzhen court in *Oppo v. Sharp*.

Descriptions of Chinese procedures as a "transplant" based on outward similarities might also be understood as part of a broader effort to normalize novel actions by Chinese courts toward other sovereign courts. Such euphemistic nomenclature downplays any deficiencies in a domestic court's practices. One misleading aspect of that description is that it does not account for changes that the transplant "receiver" must make to adjust to a new legal institution. China's experience with ASIs to date shows that ASIs have required reinterpretation of China's CPL and adoption of other measures. Judge Zhu has noted that Article 100 did not explicitly contemplate the complexities entailed in granting an ASI compared to other types of provisional measures called for under Article 100.[38] Other changes have also occurred to accommodate this more aggressive posture of the Chinese courts, including global FRAND rate setting,[39] judicial jurisdiction based on the *situs* of negotiations,[40]

33 Cheng Zhongren, *The Chinese Supreme Court Affirms Chinese Courts' Jurisdiction over Global Royalty Rates of Standards Essential Patents*, Berkeley Tech. L.J. Blog (Jan. 3, 2022), https://btlj.org/2022/01/the-chinese-supreme-court-affirms-chinese-courts-jurisdiction-over-global-royalty-rates-of-standard-essential-patents-sharp-v-oppo [hereinafter *BTLJ Blog*].

34 *Ping-Pong Olympics*, *supra* note 1.

35 Guan Yuying, *ASIs: China's Attitude towards Responding to Global IP Dispute Jurisdiction Gaming* (禁诉令：应对全球知识产权纠纷司法管辖权博弈的中国态度), Intell. Prop. Ct. Supreme People's Ct. (Feb. 26, 2021), https://ipc.court.gov.cn/zh-cn/news/view-1060.html.

36 SPC, Provisions of the Supreme People's Court on Several Issues Concerning the Application of Law in Examining Act Preservation Cases in Intellectual Property Disputes, Art. 7 (issued Dec. 12, 2018, effective Jan. 1, 2019).

37 Huawei v. Conversant, Zuigaofa Zhimin Zhong (最高法知民中) (Aug. 28, 2020), translation available at https://patentlyo.com/media/2020/10/Huawei-V.-Conversant-judgment-translated-10-17-2020.pdf.

38 *Zhu Jianjun Article*, *supra* note 17.

39 *BTLJ Blog*, supra note 33.

40 *Id.*

imposition of daily recurring penalties for continuous violation of the ASI despite a Judicial Interpretation that had provided otherwise,[41] and creation of a new *sui generis* cause of action for a FRAND rate setting.[42]

Another possible indication of the unique challenges posed by this transplant is the conflicts with foreign countries that have arisen from a Chinese court's granting of ASIs. Foreign courts have issued preemptive ASIs forbidding parties from seeking or enforcing ASIs in a Chinese court. Judge Gilstrap in *Ericsson v. Samsung* imposed an indemnity on Samsung for any fine imposed by a Chinese court for Ericsson seeking relief in a US court.[43] In March 2022, the Defending American Courts Act was introduced in Congress.[44] It could impose penalties upon foreign litigants seeking ASIs involving US court proceedings. As the Chinese decisions directly impact foreign courts, the lack of transparency over the cases has also been troubling. The European Union filed a WTO "transparency" request pursuant to Article 63 of the TRIPS Agreement during the summer of 2021, which has since been followed by a request for WTO consultations filed by the European Union on February 18, 2022.[45] The United States, Canada, and Japan have all requested to join these consultations.[46]

Other indications that a rush to normalize ASIs may be premature are the attendant instabilities of China's ASI practice as it seeks to address the challenges just noted. Judge Zhu and others have called for China to further clarify how China should grant ASIs, including harmonizing varying local Chinese judicial approaches to granting ASIs, incorporating the experience of the Conversant decision, and carefully considering the experience of foreign countries in handling ASIs.[47] As one indication of a possible change in direction, there appear to have been no new published decisions on ASIs since the initial spate of late 2020 when China emerged as the global leader in granting ASIs. Another indication of that possible reconsideration is the decision to publish the *Lenovo v. Nokia* case in March 2022 on the website of the IP Court of the Supreme People's Court, about

[41] SPC, Interpretations on Application of Civil Procedure Law, Art. 184 (2020).

[42] *Rule on Civil Procedure Cases Causes of Action* (最高人民法院印发修改后的"民事案件案由规定), Sup. People's Ct. China (Dec. 30, 2020), www.court.gov.cn/shenpan-xiangqing-282031.html.

[43] Ericsson Inc. v. Samsung Elecs. Co., No. 2:20-CV-00380-JRG, 2021 U.S. Dist. LEXIS 4392, at *23-24 (E.D. Tex. Jan. 11, 2021).

[44] Defending American Courts Act, S. 3772, 117th Cong. (2021–2022).

[45] Mark Cohen, *EU Files Request for Consultations on Chinese Judicial SEP Practices*, China IPR (Feb. 18, 2022), https://chinaipr.com/2022/02/18/eu-files-request-for-consultations-on-chinese-judicial-sep-practices/.

[46] Council for Trade-Related Aspects of Intellectual Property Rights, Request for Information Pursuant to Article 63.3 of the TRIPS Agreement, Communication from the European Union to China, WTO Doc. IP/C/W/682 (June 7, 2021).

[47] *Zhu Jianjun Article*, *supra* note 17; *Zhang Weiping Article*, *supra* note 3.

14 months after its decision date.[48] In that case, a Chinese plaintiff was denied an ASI in circumstances that seemed quite similar to other cases where ASIs were granted.[49] Judge Zhu was a member of the judicial panel deciding that case. This is an important decision, as it may be the first Chinese case where a party was denied an ASI. The case may also have not been well known in the West prior to publication, as it is not referenced in recent English-language academic literature on ASIs as transplants or otherwise,[50] nor does it appear in recent WTO proceedings.

## II. CHINESE ASI PRACTICE IN HISTORICAL CONTEXT

Chinese efforts to control foreign parallel patent litigation have a long history, stretching back to the late Qing dynasty in 1897, when two foreigners sued each other before the US Consulate in Shanghai over infringement of a US patent for the manufacture of cigarettes. During that era, US patents also had extraterritorial effect in China.[51] As the United States Consul General of Shanghai, acting in his judicial capacity, noted in *Mustard and Co. v. R. H. Wright et al.*, "The treaty between the United States and China provides that no American citizen residing in China can have his right adjudicated except in the consular courts of his country sitting in the Empire of China, such courts being United States Courts and governed by laws passed by the Congress of the United States." Furthermore, the Consul General noted, "the fact that the plaintiffs resided in China cannot except them" from "the legal principle announced as securing business certainty and safety" of a patent "granted and recorded" in the United States.[52] Disputes like these lend credence to Chinese arguments that patents were a tool of humiliation and extraterritorial oppression against China's own autonomy, "judicial sovereignty," and industrial growth. They also continue to be cited in the academic literature on legal transplants and ASIs.[53]

Another dispute at about the same time also revealed the difficulties foreigners faced in China's nontransparent and nascent patent regime. It involved the assignment of a Chinese patent to two Americans, who subsequently filed two patent applications in the United States based on this original application. Despite the intervention of the US State Department in support of the American assignees of the

[48] Lenovo v. Nokia, 2020 Yue 3 Min Chu 5105 (2020) 粤 03 民初 5105 号 (Published Mar. 9, 2022, Decided Jan. 27, 2021), https://ipc.court.gov.cn/zh-cn/news/view-1820.html.

[49] Nokia Press Release, Nokia and Lenovo Conclude Patent Cross-Licensing Agreement (Apr. 7, 2021), www.nokia.com/about-us/news/releases/2021/04/07/nokia-and-lenovo-conclude-patent-cross-licensing-agreement/.

[50] *Ping-Pong Olympics*, *supra* note 1; *Transplanting ASIs*, *supra* note 2.

[51] Mark Cohen, *An American Patent Dispute in the Qing Dynasty*, China IPR (July 2, 2012), https://chinaipr.com/2012/07/02/an-american-patent-dispute-in-the-qing-dynasty/.

[52] Mustard and Co. v. R.H. Wright et al., N. China Herald & Sup. Ct. & Consular Gazette, L. Reps. 38 (July 2, 1897), https://chinaipr2.files.wordpress.com/2012/07/1897-us-patent-case.pdf.

[53] *Transplanting ASIs*, *supra* note 2, at 1550.

patent, the Chinese government advised that there was no law affording foreign buyers of Chinese patents the authority to address infringement in China.[54] Although much has changed in China's IP system since that time, issues related to transparency, national treatment, extraterritoriality, fairness, and politics have continued to raise concerns.

China's concern with foreign IP assertions, including SEP litigation, began most prominently with debates over royalty payments for patents that read on optical media equipment. Peking University Professors Zhang Ping and Ma Xiao described the environment facing China two decades ago in their highly influential treatise "Standardization and Intellectual Property Strategies" (标准化与知识产权战略) (2005):

> In recent years, hot issues have multiplied on the topic of standardization and intellectual property. Beginning in early 2002 with the DVD patent royalties, continuing with Cisco suing Huawei, the appearance of the EVD standard, controversies over digital TV standards, the promulgation of TDS-CDMA standards, the WAPI standard running aground, up until INTEL's suit against DongJin in Shenzhen, we have seen too many cases of IP disputes arising from technical standards.[55]

In 2003, the former State Administration for Industry and Commerce (SAIC) commissioned the report "The Competition Restricting Behavior of Multinational Companies in China and Counter Measures." The report notes that multinationals "squeeze" the Chinese market by refusing to license IP.[56] The study reflected the view that had been widely adopted in China that "patent holdup" was an increasing problem for Chinese manufacturers. It also expressed an urgent need for an Antimonopoly Law (AML) to address this anticompetitive behavior.[57] The study provided an example of an unidentified company, presumably Cisco, that refused to license its IP to permit interconnectivity with its equipment. The report was released about the time that Cisco successfully won a preliminary injunction against Huawei in Texas in a trade secret dispute in June 2003.[58] Cisco and Huawei would ultimately settle their dispute in August 2004, which occurred shortly after the release of the final report.[59] Since that time, various other reports have surfaced

[54] Mark Cohen, *A New Winner: China's First Patentee in the U.S. and One of China's First Patentees in China*, China IPR (Sept. 11, 2005), https://chinaipr.com/2015/09/11/a-new-winner-chinas-first-patentee-in-the-us-and-one-of-chinas-first-patentees-in-china/.

[55] Ma Xiao & Zhang Ping, Standardization and Intellectual Property Strategies (标准化与知识产权战略) at Preface (2d ed. 2005).

[56] H. Stephen Harris, Jr. et al., Anti-Monopoly Law and Practice in China 230 (2d ed. 2011) [hereinafter *AML and Practice*].

[57] Dai Yan, *Monopoly Law Badly Needed, Report Says*, China Daily (May 25, 2004).

[58] This survey was undertaken after Cisco had sued Huawei for trade secret theft in Cisco Systems, Inc. v. Huawei Technologies, Co., 266 F. Supp. 2d 551 (E.D. Tex. 2003).

[59] *Cisco, Huawei Settle Lawsuit*, Wall St. J. (July 29, 2004).

regarding legal and extralegal threats placed by China on foreign companies to improve China's position in parallel litigation on SEPs involving a Chinese party.[60]

In another development in 2004, Wuxi Multimedia and Orient Power brought an unsuccessful lawsuit in the United States against the DVD patent pool, arguing that the pricing strategies of that pool violated the Sherman Act.[61] The claimants argued that the optical media licensing program was a vehicle for price-fixing and monopolization of the DVD-player market, and that it included so-called nonessential patents in the package license, which the plaintiffs claimed amounted to illegal tying. Prof. Huang Yong noted at the time that future legislation on monopolies needs to stipulate clear criteria for activities deemed anticompetitive, and the ongoing DVD suit could be an example for legislators to study.[62] This case was one of the few efforts by Chinese companies to bring their disputes overseas to companies seeking to license technology to China. The AML itself was finally enacted in 2007, after over a decade of discussion and legislative proposals. It has since been revised in 2022.[63] AML cases have since been used to establish FRAND rates in addition to civil litigation.

Another effort to address overseas litigation was made when China amended its Foreign Trade Law in 2004. The law authorized the Ministry of Commerce (MofCOM) to "take such measures as prohibiting the import of the relevant goods from being produced or sold by the infringer within a certain period."[64] The effort, as with ASIs, was often cloaked in the language of countering an alleged growing threat of foreign patent assertions.[65] As previously discussed, this was a transplant that was unnecessary, as China already had procedures in place to address imports that infringed Chinese patents.

In August 2005, still another effort was undertaken to develop legislation to compel licensing of SEPs by foreigners through Chinese standardization policy. Ms. Dai Hong of the High Technology Department of the Standardization Administration of China (SAC) noted at a conference hosted by the US government that "if a patentee refused to … permit exploitation of a patent, the Standardization Administration of China will suspend the implementation of the

60 *See, for example*, David L. Cohen, *A Short History of Vringo's Battle with ZTE*, Kidon IP (Aug. 2, 2018), https://kidonip.com/news/a-short-history-of-vringos-battle-with-zte/; Reuters Staff, *InterDigital Execs Fear Arrest, Won't Meet China Antitrust Agency*, Reuters (Dec. 16, 2013), www.reuters.com/article/us-interdigital-china-idUSBRE9BF0CW20131216.

61 Wuxi Multimedia, Ltd. v. Koninklijke Philips Elecs., N.V., No. 04cv1136DMS (BLM), U.S. Dist. LEXIS 9160 (S.D. Cal. Jan. 5, 2006).

62 *IPR Disputes Highlight Absence of Law*, China Daily (Feb. 2, 2005), www.chinadaily.com.cn/english/doc/2005-02/02/content_414284.htm.

63 *Anti-Monopoly Law*, NPC Observer, https://npcobserver.com/legislation/anti-monopoly-law.

64 *Foreign Trade Law of the PRC*, Ministry of Commerce Art. 29 (July 5, 2004), http://english.mofcom.gov.cn/article/policyrelease/Businessregulations/201303/20130300045871.shtml.

65 *Id.* at 34.

standard and will petition the Chinese patent office for a compulsory license."[66] Her comments were projected on a screen but were not otherwise recorded, leaving many in the audience to guess the intentions of SAC regarding compulsory licensing of SEPs to discourage SEP assertions, including the relationship with international standardization processes and China's emerging antitrust laws,[67] as well as the role of SAC in coordinating these procedures with other Chinese agencies.

Shortly after that time, China's State Intellectual Property Office (SIPO) also attempted to enter the fray. A draft of the proposed third amendment to China's patent law set forth an elaborate flowchart to compel licensing of patents in national standards based on an SAC recommendation for a compulsory license.[68] This process, however, was also ultimately never adopted by the National People's Congress in the amended patent law (2008), nor did it appear in more recent revisions of the patent law (2020).

In 2006, MofCOM also sought to convince WTO members that the incorporation of patents constituted a "technical barrier to trade" and should therefore be regulated by the WTO:

> China is of the view that, IPR issues in preparing and adopting international standards have become an obstacle for Members to adopt international standards and facilitate international trade. It is necessary for the WTO to consider negative impacts of this issue on multilateral trade and explore appropriate trade policies to resolve difficulties arising from this issue.[69]

At about this time, China's courts also began to explore their potential role in reducing royalty payments involved in standardization. The IP Division of the Supreme People's Court issued an "instruction letter" to the Liaoning High People's Court advising that "if … a patent has been included in a standard, the People's Court may deem that the patentee has licensed others to use its patents to implement standards. Such use does not constitute infringement under … the Patent Law." However, this practice was also not ultimately widely implemented.

During the period after the AML and before China's transplanting ASIs, Chinese courts occasionally undermined foreign cases through expedited decisions in their own jurisdiction. China's CPL mandates that first instance cases are required to be completed in six months, and second instance cases are required to be completed in

66 *AML and Practice*, *supra* note 56, at 236. Notes were taken by the author of the presentation *Overview of China's Perspectives on IP in Standards* (Aug. 23, 2005).

67 Marketa Trimble, *Patent Working Requirements: Historical and Comparative Perspectives*, 6 U.C. Irvine L. Rev. 483, 506 (2016).

68 *Id.*

69 Committee on Technical Barriers to Trade, Intellectual Property Rights Issues in Standardization, Communication from the People's Republic of China, WTO Doc. G/TBT/W/251 (May 25, 2005), Document 05-2126; see also addendum Background Paper for Chinese Submission to WTO on Intellectual Property Right Issues in Standardization (WTO Doc. G/TBT/W/251/Add.1 (Nov. 9, 2006), Document 06-5389).

three months (Arts. 152, 164). According to one database of 7,885 cases, first instance patent litigation in China was completed on average in 5.9 months.[70] By comparison, most US "rocket" dockets commit to trying a patent case within one year.[71] Appeals of patent cases at the Federal Circuit take considerably longer due to lengthy docketing and briefing periods and an estimated period of 180 days after oral argument before a final decision is rendered.[72] In many cases, a Chinese appellate court may render a final decision before discovery has been completed in US district courts or the USITC.

Pursuant to Article 277 of the CPL, litigation time limits are also suspended when a foreign party is involved. A Chinese court may leverage this flexibility to issue rulings at key junctures in a foreign court proceeding, effectively undermining foreign cases by rendering final judgments in advance of foreign decisions. Although adjustment of time frames may sometimes be necessary to accommodate foreign litigants, the unconstrained ability to adjust time frames raises concerns over national treatment under TRIPS. Such national treatment exceptions should be narrowly tailored to avoid undue discrimination.

A US district court enjoining a Chinese-backed defendant for infringement of an IP right or a USITC decision granting an exclusion order may be of little moment if the US plaintiff becomes a defendant in the parallel Chinese case, and especially if the US defendant in the Chinese case has sufficient market presence in China to be placed at risk of an adverse Chinese court or administrative decision.[73] A Chinese judgment could stop the US company's sales, manufacturing, and exports without incurring the attendant controversy of issuing an ASI.

Expedited civil procedures have long had the impact[74] of undercutting foreign parallel litigation, regardless of the availability of ASIs. A good example of the litigation race that China offers for SEP litigation is *Huawei v. Samsung*. The US and Chinese cases were filed at the same time by the same plaintiff (Huawei) on May 24–25, 2016. (Differences in dates were due to the international dateline.) On January 4, 2018, a judgment was issued by Shenzhen Intermediate People's

[70] CIELA, www.ciela.cn (last visited Apr. 11, 2021).

[71] Saurabh Vishnubhakat, *Reconceiving the Patent Rocket Docket: An Empirical Study of Infringement Litigation* 1985–2010, 11 J. MARSHALL REV. INTELL. PROP. L. 58 (2011).

[72] *The Life of an Appeal*, U.S. CT. APPEALS FED. CIR., https://cafc.uscourts.gov/wp-content/uploads/RulesProceduresAndForms/FilingResources/Life_of_an_Appeal_Narrative_and_flowchart.pdf (last visited June 20, 2022); *Case Filings*, U.S. CT. APPEALS FED. CIR., https://cafc.uscourts.gov/home/case-information/case-filings/ (last visited June 20, 2022).

[73] Certain Silicone Microphone Packages and Products Containing Same, Inv. No. 337-TA-888 (USITC) (for a discussion of this case, see Song Haining, *The Story of Battling Giants: Comments on Goertek Acoustics v. Knowles Electronics*, CCPIT PATENT & TRADEMARK LAW OFFICE (Oct. 2014), www.lexology.com/library/detail.aspx?g=82c14aac-7e05-40ed-985b-6dd884c7efd4); Ericsson Inc. v. Samsung Elecs. Co., No. 2:20-CV-00380-JRG, 2021 U.S. Dist. LEXIS 4392 (E.D. Tex. Jan. 11, 2021) (Chinese case: Samsung v. Ericsson).

[74] Mark Cohen, *China IP Time and the New York Minute*, CHINA IPR (Nov. 21, 2012), https://chinaipr.com/2012/11/21/china-ip-time-and-the-new-york-minute/.

Court granting an injunction against Samsung.[75] Meanwhile, in the United States, on April 13, 2018, an ASI was granted in favor of Samsung, prohibiting Huawei from enforcing the injunctions issued by the Shenzhen court. Judge Orrick acknowledged the slowness of his court:

> The Chinese actions have proceeded quicker than this one. In particular, the Shenzhen court has held trials on two of Huawei's SEPs and two of Samsung's SEPs. The trials addressed both FRAND issues and technical issues specific to each SEP. During these trials, the parties had full opportunities to present their evidence and argument [references omitted] ...
>
> We are scheduled to proceed to trial in December of this year.[76]

The ASI was granted by the US court after a decision had already been reached in the underlying dispute and at best would be limited to enforcement of the order. The case was settled while it was on appeal to the Guangdong High Court on or about April 14, 2019,[77] well before the start of the US trial, which was rescheduled to September 2019. With its limited duration, Judge Zhu noted Judge Orrick's ASI's "influence on comity could be ignored."[78]

Chinese judicial practices of expediting domestic litigation to undermine foreign parallel cases are also found outside of the SEP context.[79] This is also not surprising in light of the important role that the courts play in breaking through patent "monopolies" and "technological stiff necks" (bottlenecks).[80] These bottlenecks are often described in the Chinese media as patents or patent families, whether or not incorporated into standards, that are under the control of foreign entities. This type of language has also been more widely used to justify other actions that

[75] *Case of Huawei Sues Samsung et al. for Infringement of Invention Patents* (华为公司诉三星公司等侵害发明专利权纠纷案 (2016) 粤 03民初816, 840), China Ct. Trial Online (Jan. 11, 2018), http://tingshen.court.gov.cn/live/1759564.

[76] Huawei Techs. Co. v. Samsung Elecs. Co., No. 3:16-cv-02787-WHO, 2018 U.S. Dist. LEXIS 63052 (N.D. Cal. Apr. 13, 2018) (Order Granting Samsung's Motion for Antisuit Injunction, Re: Dkt. Nos. 234, 235, 240, 244, 277, 278).

[77] Guangdong High Court, *Huawei and Samsung Patent Infringement Dispute Cases Have Been Settled through Mediation* (广东高院：华为与三星专利侵权纠纷系列案调解结案), IPR Law (May 18, 2019), www.iprlaw.cn/index/news/show/id/7072.html; Florian Mueller, *Breaking News: Huawei and Samsung Settle*, FOSS Patents (Feb. 26, 2019), http://www.fosspatents.com/2019/02/breaking-news-huawei-and-samsung-settle.html.

[78] *Zhu Jianjun Article*, *supra* note 17.

[79] Mark A. Cohen, *Semiconductor Patent Litigation: Part 2 Nationalism, Transparency and Rule of Law*, China IPR (July 4, 2018), https://chinaipr.com/2018/07/04/semiconductor-patent-litigation-part-2-nationalism-transparency-and-rule-of-law/ (discussing timing of Veeco v. Amec parallel patent litigation).

[80] *Id.*

constrain China's techno-nationalist ambitions, such as high prices for patents[81] and trade secret theft.[82]

As Chinese judges are not part of an independent branch of the government, and Chinese courts are ultimately guided by Party policy, it is not surprising that Chinese judges also openly encourage utilization of Chinese judicial mechanisms to thwart Western technological assertions. Often this may occur through elevation of particular cases to a leading case for study, including awarding it a status as a "top 10," "innovative" case, or similar language. Sometimes, the court may directly exhort rightsholders to learn from the case as well. The Presiding Judge of the Guangdong High Court who heard the appeal in *Huawei v. InterDigital Corporation* (IDC) (2013), an AML case involving SEPs, advised Chinese companies that they should utilize the AML to "break through technical barriers in the development of space for their own gain."[83] A recent report from the Hubei Provincial High People's Court discussed these "bottle-necked, key core technologies, emerging industries" with specific reference to two SEP cases adjudicated in the provincial capital of Wuhan involving American interests. In both cases, the courts issued two ASIs to halt litigation in the United States and elsewhere overseas:

> Courts across the province heard key intellectual property cases, etc., involving bottleneck critical core technologies and newly emerging industries, in a fair and efficient manner in accordance with the law. Such as ... Samsung Company v. Ericsson Company's SEP royalty case ...
>
> In 2020, the Wuhan Intellectual Property Tribunal tried the ASI case filed by Xiaomi against the American IDC, ruling that IDC is prohibited from filing similar parallel lawsuits abroad, and thereby effectively safeguarded my country's high-tech enterprises' participation in intellectual property rights in transnational competition, and highlighted the wisdom and authority of China's judicial protection of intellectual property rights.[84]

These policies of thwarting foreign adjudications by protecting domestic entities were significantly elevated in January 2021, when Communist Party General Secretary Xi Jinping published an article in the leading Communist Party journal

[81] Hao Yuan, *Antitrust Aspects of "Unfairly High Patent Pricing" for Licensing Transactions in China*, China IPR (Mar. 29, 2020), https://chinaipr.com/2020/03/29/antitrust-aspects-of-unfairly-high-patent-pricing-for-licensing-transactions-in-china/.

[82] See United States v. Xiaorong You, No. 2:19-CR-14, 2022 U.S. Dist. LEXIS 80032, at *3 (E.D. Tenn. May 3, 2022) ("Defendant intended to 'break[] through both green and technical international trade barriers' to 'earn a share of the global market,' as well as 'break the international monopoly [on can coatings].'").

[83] Xu Qibin, *What Is the Meaning of Huawei's Victory*, S.E.U. (Oct. 30, 2013), www.seu.edu.cn/2013/1101/c124a52344/page.htm.

[84] Ke Xuewen and Lu Ming (eds.), *By the Establishment of Intellectual Property Courts and Quick Trial of Technical Cases Involving "Bottlenecks", the Hubei Courts Have Organized an Intellectual Property Network*, Hubei Daily (Oct. 27, 2021), https://hubeigy.chinacourt.gov.cn/article/detail/2021/10/id/6333102.shtml.

*Qiushi* ("Seeking Truth"). Secretary Xi called on China to "rigorously protect IP [to] safeguard indigenous Chinese R&D in core technologies in key fields."[85] Xi also renewed the call for China to form an efficient "early warning system for international intellectual property risks" and "increase assistance for overseas intellectual property rights protection of Chinese enterprises." Development of case law for lower courts to handle these types of cases was also especially critical, as Xi Jinping himself has propounded, "One case is better than a dozen documents."[86]

Whether in their policymaking or adjudication functions, Chinese courts have tended for some time to focus disproportionately on foreign-related IP cases due to their political sensitivity and their potential to disrupt domestic industrial plans, including those regarding technology, employment, and manufacturing.[87] These types of cases are identified by the court's own rules as requiring "special treatment" in their adjudication, including by formation of collegial panels, involvement of the court's leadership, or referral to Adjudication Committees for ultimate decision-making.[88] The Supreme People's Court (SPC) has also specifically elevated consideration of ASIs into an important research topic for the courts in order to protect China's "judicial sovereignty."[89] The endgame, to quote the Hubei Provincial High Court, is to protect China's role in "transnational competition."

Despite much high-level rhetoric, it is difficult at this time to ascertain whether China's changing policies toward overseas SEP assertions are durable long-term solutions to a perceived problem or short-term politicized responses and experiments. Efforts to date may be viewed as experimental in nature insofar as they are not fully codified into law. However, even codified transplants, such as China's short-term experiment with "Section 337" litigation in China's Foreign Trade Law, as previously discussed, may exist in name only. Economic changes may also drive changes in policy. An example of this shifting rhetoric is the change from China's official position that patents constitute a technical barrier to trade to one where China is seeking a larger share of the patent royalties, based on China's significant

85 Xi Jinping, *Comprehensively Strengthen the Protection of Intellectual Property Rights, Stimulate Innovation and Promote the Construction of a New Development Pattern*, QIUSHI (Jan. 31, 2021), http://en.qstheory.cn/2021-04/30/c_617533.htm.

86 Ding Yuejia (ed.), *General Security Xi Jinping Urges "One Case Is Greater than a Dozen Policy Documents"* (习近平总书记强调，"一个案例胜过一打文件"), LEGAL DAILY (June 25, 2021).

87 Mark A. Cohen, Presentation at Berkeley Law, When IP Systems Collide (Oct. 2015), www.law.berkeley.edu/wp-content/uploads/2015/07/Mark-Cohen-When-IP-Systems-Collide.pdf.

88 *Guiding Opinions on the Oversight and Management of "Four Types of Cases"* (关于进一步完善"四类案件"监督管理工作机制的指导意见), CHINA LAW TRANSLATE, Art. 3 (Nov. 11, 2021), www.chinalawtranslate.com/en/4-types-of-cases/.

89 Susan Finder, *Supreme People's Court's New Policy Document on Opening to the Outside World*, SUP. PEOPLE'S CT. MONITOR (Oct. 9, 2020), https://supremepeoplescourtmonitor.com/2020/10/09/supreme-peoples-courts-new-policy-document-on-opening-to-the-outside-world/.

holdings of SEPs.[90] China's ability to experiment with new laws and its continuous adaptation to changing economic and political circumstances in IP often makes it difficult to determine whether changes in Chinese legal practices are durable. IP especially has also been an area of experimentation for China's legal system in a wide range of areas, including in such areas as specialized IP courts, preliminary injunctions, publication of cases, and precedent.[91]

While ASIs are a nominal "transplant" from common law countries, it is only by also considering the differences between Chinese and other systems that one can begin to determine whether China's ASIs are also a type of "false friend" with foreign ASI practice. Some of the distinguishing features of China's ASI practice from common law jurisdictions include:

(a) Chinese ASIs are part of long-term efforts by the Chinese government to increase the value of Chinese technology and decrease the value of foreign technology "monopolies."
(b) Unlike common law countries, Chinese ASIs are exclusively extraterritorial in nature.
(c) Chinese ASIs are part of a national effort to increase the role of Chinese courts in establishing global judicial norms.
(d) Chinese ASIs have also precipitated other changes in the adjudication of SEPs to accommodate this more aggressive posture, including a greater willingness to set global FRAND rates, the extension of jurisdiction to foreigners based on situs of negotiations, recurring daily penalties for violations of ASIs, and the creation of a new civil cause of action for FRAND rate setting.
(e) China's ASI practices have been promoted and endorsed by the highest levels of China's political and judicial leadership.
(f) China's ASIs may be experimental in nature.

Despite these differences, the extent to which these policy differences impact how ASIs are administered in China is difficult to determine, since China does not publish all its final decisions and does not usually publish interim measures. China's emerging power in standards and its own domestic regulatory capacity, however, does raise concerns that it may yet become a significant "rules breaker" or even "rules faker," where it adjusts adherence to the international order to better advance

[90] National Security Commission on Artificial Intelligence, Final Report National Security Commission on Artificial Intelligence 471 (2021). Note that the author of this article contributed to this report.

[91] Mark Cohen, *Crossing the River by Feeling the IP Stones: How China's Civil Procedure System Benefits from Reforms Made in IP Civil Litigation*, China IPR (Nov. 8, 2012), https://chinaipr.com/2012/11/08/crossing-the-river-by-feeling-the-ip-stones-how-chinas-civil-procedure-system-benefits-from-reforms-made-in-ip-civil-litigation/.

its own short- or long-term interests.[92] China's rejection of complete transparency in ASIs may therefore be seen as a strategic tool to maintain maximum regulatory flexibility in this evolving area. This decision is not cost-free, as it may also minimize the predictability afforded to the affected public by transparent judicial decisions.

## III. TRANSPARENCY: HOW CHINA'S ASI REGIME FAILS

China's use of Article 100 of the CPL ("Behavior Preservation") has raised two major transparency concerns, one of which is long-standing: (1) the public availability of any final judicial decisions, as well as any behavior preservation orders that are final with respect to the issue at hand; and (2) the extent to which courts may act in an ex parte manner without disclosure to affected litigant(s) or other courts that may have an interest in the decision. Both aspects of transparency are governed by the international norms set by the TRIPS Agreement, which provides a useful, internationally recognized benchmark to judge regulatory compliance.

### A. *The Public Availability of Judicial Decisions*

Article 63 of the TRIPS Agreement is entitled "Transparency." It falls within Part V of the TRIPS Agreement regarding "Dispute Prevention and Settlement." It is clear from these descriptions that the drafters of Article 63 intended to promote transparency in large part to prevent disputes and encourage their settlement. Article 63 provides:

> 1. Laws and regulations, and final judicial decisions and administrative rulings of general application, made effective by a Member pertaining to the subject matter of this Agreement (the availability, scope, acquisition, enforcement and prevention of the abuse of intellectual property rights) shall be published ... in such a manner as to enable governments and right holders to become acquainted with them ....
>
> 3. Each Member shall be prepared to supply, in response to a written request from another Member, information of the sort referred to in paragraph 1. A Member, having reason to believe that a specific judicial decision or administrative ruling or bilateral agreement in the area of intellectual property rights affects its rights under this Agreement, may also request in writing to be given access to or be informed in sufficient detail of such specific judicial decisions or administrative rulings or bilateral agreements ....

During the summer of 2021, China received an Article 63 request from the European Union to disclose three SEP cases: *Xiaomi v. InterDigital* (Wuhan Int. Ct.), *OPPO v. Sharp* (Shenzhen Int. Ct.), and *Samsung v. Ericsson* (Wuhan Int.

[92] Sandra Lavenex, Omar Serrano, & Tim Büthe, *Power Transitions and the Rise of the Regulatory State: Global Markets in Flux*, 15 REGULS. & GOVERNANCE 445 (2021).

Ct.). The request asked for further information to enable rightsholders to "acquaint themselves with those decisions that are identified as typical, example cases."[93] The European Union filed the request independently of, and prior to, the initiation of formal consultations that are a prerequisite to the formal WTO dispute process.[94] China briefly responded to the Article 63 request that these cases "mentioned in the EU communication are cases for reference and have no *legal effect* of general application" (para. 4) [emphasis supplied]. This response did not differ significantly from an earlier Article 63 request that the United States filed in 2005, where China responded that it was not obligated to provide cases, as China "does not follow the common law system."[95] The United States also subsequently filed a dispute after it received this response. The EU request for consultations identified the same SEP cases, which China had failed to produce plus a fourth case, *ZTE v. Conversant* (Shenzhen Int. Ct). In addition, the European Union claimed substantive violations of WTO rules by reason of China's ASIs prohibiting access to non-Chinese courts, creating legitimate barriers to trade, and imposing excessively high fines for the owners. The EU request has since been joined by the United States, Canada, and Japan.

The lack of "legal effect" identified in the Chinese response in its response to the EU Article 63 request introduces surplus language not otherwise found anywhere in the TRIPS Agreement. The precise treaty language in Article 63, without any qualifiers, is "general application."

The issue of China's obligation to publish precedential or quasi-precedential cases also appeared prior to the request during the Trade Policy Review of China at the WTO between the European Union (October 20 and 22, 2021).

> EU Question No 80: Could China therefore clarify what is the status of these adjudication guidelines for deciding on an anti-suit injunction and daily penalties in light of the above reply?
>
> Reply: The "major cases," "typical cases," "typical technical cases" and the key points of decisions selected by the Chinese courts are reference cases with no universal application. These cases and the main points of decisions summarized on the basis of which reflect the judicial philosophy, trial ideas and decision methods of the Supreme People's Court in handling difficult, complex and new types of IPR cases. *Their role is to summarize trial experience, strengthen the*

[93] Council for Trade-Related Aspects of Intellectual Property Rights, Request for Information Pursuant to Article 63.3 of the TRIPS Agreement, Communication from the European Union to China, WTO Doc. IP/C/W/682 (June 7, 2021).

[94] European Union Permanent Mission to the WTO, *Request for Consultations by the European Union* (Feb. 18, 2022), https://trade.ec.europa.eu/doclib/docs/2022/february/tradoc_160051.pdf.

[95] Council for Trade-Related Aspects of Intellectual Property Rights, Communication from China, Response to a Request for Information Pursuant to Article 63.3 of the TRIPS Agreement, WTO Doc. IP/C/W/465 (Jan. 23, 2006) (available with other background materials from the author).

*promotion of the rule of law, and provide reference for judicial practice and legal teaching and research.*[96]

By its terms, Article 63 also does not explicitly refer to, nor require, that the cases are precedential. Nor does it limit its application solely to cases arising in a common law system. In this respect, the *travaux préparatoires* (negotiating history) of the TRIPS Agreement is of legal significance under the Vienna Convention on Law of Treaties (Art. 32). The negotiating history indicates that there was a specific rejection of a Swiss proposal on October 1, 1990, to substitute "precedential value," in favor of "general application." Since the conclusion of the TRIPS Agreement, the language around what constitutes cases of "general application" has undergone little further clarification.[97]

Application of Article 63 of TRIPS to Chinese SEP jurisprudence will ultimately entail careful consideration by a WTO panel of China's evolving practice in using cases to guide judges and rightsholders, including their specific application to China's recent ASI cases. This judicial practice is not unique to SEPs and has been widely used in IP. For example, a number of Chinese local courts recognized the significance of their "big" trade secret cases shortly after China amended its trade secret law.[98] Cases that have also sought to address the risks presented by parallel IP litigation have also had this recognition.[99] These cases generally fall within the category of "judicial normative documents." This category of documents includes "trial practice documents, guiding cases, and reference cases." According to Professor Susan Finder, "judicial normative documents are often cited by courts as a supplementary legal basis for a judgment and judges will recognize their validity and implement them in their judicial decision making."[100]

There is ample evidence that Chinese judicial institutions are utilizing precedents for various kinds of guidance, whether or not they are being cited in cases and despite a commitment in China to civil law norms.[101] Dr. Zhao Hong, a former member of the WTO appellate body and MofCOM official, has similarly noted that "though the legal theories or concepts of the two major legal families [civil and

96 Trade Policy Review Body, *Trade Policy Review China, Minutes of Meeting*, WTO Doc. WT/TPR/M/415/Add.1 (Dec. 22, 2021), at 428 (emphasis added).

97 Marketa Trimble, *Unjustly Vilified TRIPS-Plus?: Intellectual Property Law in Free Trade Agreements*, 71 Am. U.L. Rev. 1449 (2022).

98 Jerry Xia & Wang Yulu, *Analysis of Guiding Trade Secret Cases in China Published during the World IP Day in 2020*, *in* Mark Cohen, *An Update on Data Driven Reports on China's IP Enforcement Environment*, China IPR (July 13, 2020), https://chinaipr.com/2020/07/13/an-update-on-data-driven-reports-on-chinas-ip-enforcement-environment/.

99 Mark Cohen, *The SPC's "Top Two" Dueling IPR Cases*, China IPR (May 4, 2014), https://chinaipr.com/2014/05/04/the-spcs-top-two-dueling-ipr-cases/.

100 Susan Finder, *China's Translucent Judicial Transparency*, *in* Transparency Challenges Facing China 141, 164 (2018), https://papers.ssrn.com/sol3/papers.cfm?abstract_id=3344466.

101 Yuan Ye, *How "Case Law" Works in the Chinese Courts*, Sup. People's Ct. Monitor (May 29, 2022), https://supremepeoplescourtmonitor.com/2022/05/29/how-case-law-works-in-the-chinese-courts/.

common law] differ significantly, their practices actually have achieved similar effects in maintaining the consistency of decisions by judiciary bodies."[102] Widespread utilization of case law databases may also constitute a form of case "general application" by lawyers, litigants, policymakers, academics, and judges that use these services. As of May 22, 2022, the official Chinese judicial database alone had 87,284,333,307 hits since it first launched in 2013, with a library of over 132 million documents.[103] There are also private IP case law databases, which have additionally attracted significant usage for their search functions as well as the value-added services that they may provide.

Despite the great interest in cases and case law, SEP adjudication is not highly transparent. As a starting point, only about 46–54% of final patent decisions are published.[104] Only about 18% of SEP cases from 2010 to 2019 in the Chinese courts have been reported in a published decision of some kind.[105] Approximately 75% of these cases involved foreigners, and 96% of the cases were in the ICT sector. The data was sourced by contacting Chinese courts individually. It is especially difficult to estimate the percentage of interim behavior preservation measures that are published, as Chinese law does not require publication of nonfinal decisions. This general lack of transparency mandated that the opening paragraph to this article include the disclaimer that the discussion herein is based on "published data." Non-publication or unofficial publication of cases can occur for many reasons and limits the ability to draw authoritative conclusions based on published data.[106]

Since the early spate of ASI cases involving SEPs in China, there have also been several newer cases involving foreign SEP assertions, none of which have been officially published as of this writing (May 2022). The failures to publish cases may suggest a waning enthusiasm for ASIs, as they were previously granted. Many of these cases involve parallel litigation in other countries and may therefore be ripe for an ASI. These cases include *Coolpad v. Pantech*;[107] *Oppo v. IDC*;[108] *ZTE v. Tinno*

[102] Dr. Hong Zhao, Appellate Body Member, Farewell Speech (Nov. 30, 2020), www.wto.org/english/tratop_e/dispu_e/farwellspeechhzhao_e.htm.

[103] Supreme People's Court, China Judgments Online, http://wenshu.court.gov.cn (last visited June 20, 2022).

[104] Chris Bailey, Douglas Clark, Mark Cohen, & Aria Tian, *Chinese Patent Litigation Data: What It Tells Us and What It Doesn't*, Intell. Asset Mgmt. (Nov. 17, 2021), https://rouse.com/insights/news/2021/chinese-patent-litigation-data-what-it-tells-us-and-what-it-doesn-t.

[105] Lexfield, *Statistics of Chinese SEP Cases 2010–2019*, https://chinaipr2.files.wordpress.com/2020/07/statistics-of-chinese-sep-cases-in-2011-2019-lexfield9892.pdf (last visited June 20, 2022).

[106] Echo Xie, *Millions of Court Rulings Removed from Official Chinese Database*, S. China Morning Post (June 26, 2021), www.scmp.com/news/china/politics/article/3138830/millions-court-rulings-removed-official-chinese-database.

[107] Shenzhen Int. Ct., Mar. 2022. IPR Daily-Rene (ed.), *Patent War Resumes, Coolpad Sues South Korea NPE Pantech for Royalty Rates*, IPR Daily (Mar. 22, 2022), www.iprdaily.com/article/index/16197.html.

[108] Guangzhou IP Ct., Feb. 2022. Bing Zhao, *Oppo Suit in Guangzhou Sets Up Next Chance for Global FRAND Ruling in China*, Intell. Asset Mgmt. (Mar 9, 2022), www.iam-media.com/article/oppo-suit-in-guangzhou-sets-next-chance-global-frand-ruling-in-china.

*Mobile*;[109] and *Oppo v. Nokia*.[110] The delayed availability of the *Lenovo v. Nokia* decision is another indication of possible substantive changes in China's ASI practice as well as the continuing challenge of limited judicial transparency. A further indication of controversy around China's practice of ASIs is the lack of information on any ASI cases in Wuhan, which had been an initial center for ASI litigation. Perhaps the EU case will bring additional pressure on both transparency and ASI reform. One Chinese scholar who favors reform of China's ASI regime has pointed to the EU WTO case itself as an example of the evidence that the system and its implementation need to be "further perfected."[111]

While the lack of full transparency in judicial decisions in China is problematic with respect to these five SEP decisions, it is arguably even more problematic with respect to the SEP decisions to date or the several hundred thousand IP cases decided each year. It also remains impossible to address broader concerns, such as national treatment, without recourse to a complete judicial database where foreign and domestic litigants can be fully compared. Nonetheless, the relatively few ASI decisions and their legal significance underscore that the EU request has been limited in scope and should also be relatively easy to address for China.

### B. *Extended and Opaque Ex Parte Decision-Making*

Another transparency concern involves the disclosure of information to adversely affected parties in the issuance of an ASI. Concerns about a lack of transparency in ex parte ASIs have also been voiced by the Office of the United States Trade Representative in its annual Section 301 Report for 2021:

> Right holders have also expressed strong concerns about the emerging practice in Chinese courts of issuing anti-suit injunctions in standards essential patents (SEP) disputes, reportedly without notice or opportunity to participate in the injunction proceedings for all parties. Since the first issuance of such an anti-suit injunction in August 2020, Chinese courts have swiftly issued additional anti-suit injunctions in other SEP cases. Several of these anti-suit injunctions are not limited to enjoining enforcement of an order from a specific foreign proceeding but broadly prohibit right holders from asserting their patents anywhere else in the world.[112]

Article 50 of the TRIPS Agreement sets up a skeletal standard for notice to affected parties of a provisional measure:

[109] Florian Mueller, *ZTE Reportedly Goes on the Offense, Sues Unnamed Chinese Smartphone Maker over 4G Standard-Essential Patents: Possibly Tinno or Transsion?*, FOSS Patents (Sept. 24, 2021), www.fosspatents.com/2021/09/zte-reportedly-goes-on-offense-sues.html.

[110] Chongqing No. 1 Int. Ct., July 2021. Gregers Maller, *Oppo Files 5G Patent Infringement Suits against Nokia in China and Europe*, ScandAsia (Sept. 9, 2021), https://scandasia.com/oppo-files-5g-patent-infringement-suits-against-nokia-in-china-and-europe/.

[111] *Zhang Weiping Article*, *supra* note 3.

[112] United States Trade Representative, 2021 Special 301 Report 47–48 (2022).

> 50.2 The judicial authorities shall have the authority to adopt provision measures *inaudita altera parte* where appropriate, in particular where any delay is likely to cause irreparable harm to the rightsholder.
>
> 50.4 Where provisional measures have been adopted *inaudita altera parte*, the parties affected shall be given notice, without delay after the execution of the measures at the latest. A review, including a right to be heard, shall take place upon request of the defendant with a view to deciding, within a reasonable period after the notification of the measures, whether these measures shall be modified, revoked or confirmed.

Application of Articles 50.2 and 50.4 to *Samsung v. Ericsson*, as one example, suggests that China may not be affording adequate opportunities to be heard in ASI litigation. In that case, Samsung sued on December 7, 2020, in Wuhan (qisu/起诉). Ericsson filed its case in Texas on December 11, 2020. Samsung filed its request for an ASI on December 14, 2020. The ASI was issued on December 25, 2020. There is no record of service or notice having been delivered to Ericsson for the initiation of the case, or on the motion for an ASI, although a minimum of 11 days had passed since Samsung requested its ASI. Article 100 of the CPL would otherwise require the court to decide within 48 hours if there is an emergency. By these domestic and international standards, the 11-day delay undercuts the argument that the provisional measure was necessary due to the possibility that "any delay" in granting the ASI would cause irreparable harm. This period was also more than adequate for a court to deny a motion to grant an ASI *inaudita altera parte*. On December 17, 2020, Samsung notified Ericsson of the Chinese action but did not provide Ericsson with any of the filings from the Chinese action. The Chinese civil complaint was not provided to Ericsson until December 22, 2020, or three days before the ASI motion was granted.[113]

These are not, however, the only periods of opacity in a Chinese proceeding involving an ASI. A party initiates a case by "suing" (qisu/起诉). When a party "sues," it should file a complaint that meets the criteria set forth in Article 122 et seq. of the CPL, including setting forth the cause of action (Art. 124). The court has seven days to "accept and review" (shouli/受理) the complaint. After acceptance and review, the case will be "established" (li'an/立案). This process is generally not open to the public but could provide a starting point for notice to be delivered to an affected party. The Case Acceptance Division of the courts at one time also had primary responsibility for issuing preliminary injunctions in IP matters, the predecessor remedy to Article 100 (now 103) of the CPL.[114] It is quite possible that

[113] Ericsson Inc. v. Samsung Elecs. Co., No. 2:20-CV-00380-JRG, 2021 U.S. Dist. LEXIS 4392 (E.D. Tex. Jan. 11, 2021).

[114] Liu Nanping & Michelle Liu, *Justice without Judges: The Case Filing Division in the People's Republic of China*, 17 U.C. Davis J. Int'l L. & Pol'y 283 (2011); see also *Zhang Weiping Article*, *supra* note 3.

Samsung had given notice to the Wuhan Court of its intent to seek an ASI when it originally filed its complaint with the Wuhan Court, one week before case acceptance. If so, this may further militate against a finding that delay would cause "irreparable harm," given a delay that was 18 days from initial case filing, or 11 days from case acceptance.

Judge Gilstrap noted in the US counterpart to the Chinese case, *Ericsson v. Samsung*, that there is no Chinese "PACER"-type system making nonfinal judicial order publicly available (PACER, an abbreviation for Public Access to Court Electronic Records, is an electronic public access service for US federal court documents). Ericsson moved the court to require Samsung's counsel to "promptly send documents filed in the Chinese Action to Ericsson." Unfortunately, Judge Gilstrap denied this request to avoid the court "insert[ing] itself into matters of Chinese law or civil procedure" and because "it is not for this Court to require Samsung to operate in a foreign jurisdiction as though it were here."[115] As the TRIPS Agreement is not self-executing, Judge Gilstrap was not obligated to consider whether the lack of notice provided by China comported with China's TRIPS obligations in his comity analysis. Nonetheless, reference to TRIPS might have been helpful, as it could have helped the court avoid requiring imposition of a US-centric standard that Judge Gilstrap thought would otherwise be inappropriate. The perspective that China's procedures for ASIs should be understood in purely Chinese terms was also shared by former Supreme People's Court IP Tribunal Chief Judge Kong Xiangjun, who submitted an expert declaration in *Ericsson v. Samsung* regarding Samsung's lack of notice of its filing on December 14, 2020, that resulted in the court issuing its Christmas Day ASI: "Samsung's notice in Wuhan lawsuit is consistent with the common practice under civil proceedings in China, where Samsung may choose to notify or not, or may choose to notify the other party of the lawsuit at any point in time. It is in line with the common practice of Chinese litigation."[116]

Whether or not Chinese practice is in accordance with Chinese law, courts in the United States and in third countries have raised serious objections to China's lack of transparency in its ex parte decisions, including the failure to advise counsel of pending decisions. The Delhi High Court in *Interdigital Technology v. Xiaomi Corp & Ors.* (May 3, 2021), after reviewing six separate times when counsel for Xiaomi had appeared before the court without revealing that it was undertaking steps to take away the court's jurisdiction, stated that "the manner in which the defendants have acted borders on fraud, not only with the plaintiffs, but also towards this Court."[117] The Court also imposed a fine in the form of an indemnity against any penalty

[115] Ericsson, *supra* note 113.

[116] Declaration of Professor Kong Xiangjun, No. 2:20-CV-00380-JRG, ECF 26-12 (E.D. Tex. Jan. 1, 2021).

[117] Interdigital Tech. Corp. v. Xiaomi Corp., High Court of Delhi, I.A. 8772/2020 in CS (COMM) 295/2020 (May 3, 2021).

imposed by the Wuhan Court on IDC, the plaintiff in India.[118] Former Federal Circuit Chief Judge Paul Michel noted in his amicus filing in *Ericsson v. Samsung* that Samsung's behavior "raise[d] significant concerns about ... sufficient notice and due process."[119]

Foreign counsel may also bear some responsibility for this lack of transparency and unwillingness to inform foreign courts of pending ASIs. In another US case, Judge Sleet in Delaware, on hearing that he had been misled by ZTE into granting an ex parte ASI against Vringo's global patent campaign by not being informed of an ongoing SEP case in the Southern District of New York in violation of the Federal Rules of Civil Procedure, Rule 65, noted that Vringo could have been within its rights to "lay [the judge] low" for granting that motion based on these misrepresentations of counsel. Judge Sleet promptly retracted his prior ASI.[120]

"Submarine" ASIs of the type described previously raise difficult questions regarding how to accommodate two jurisdictions' differing procedures, cultures, and professional behavior.[121] Courts may not feel obligated to disclose key nonfinal decisions, service of process may not have been officially effected for initiation of the case, counsel may claim that there are violations of fundamental notions of due process or at least TRIPS obligations, and affected countries may complain of a lack of transparency. Judges may also raise concerns about the ethical responsibilities of counsel to inform bench and bar of developments affecting a court's jurisdiction. Hearings and deadlines may be timed to conflict with national holidays. The issues of civility and professional responsibility raised by such decisions have thus far been handled inconsistently by courts throughout the world and are worthy of further research.

## IV. CONCLUSION

China's ASI regime is distinct from those of common law systems in many aspects. The most important distinctions may be its exclusively extraterritorial orientation and its high degree of politicization and experimentation. China's lack of transparency also acts to shield China from an understanding by outsiders of its practices and maximizes China's regulatory flexibility. Whether or not the legal regime is a

[118] *Id.* ¶ 119.

[119] Brief of Amicus Curiae the Honorable Paul R. Michel (Ret) in Support of Plaintiff's Motion, Ericsson Inc. v. Samsung Elecs. Co., No. 2:20-CV-00380-JRG (E.D. Tex. Jan. 5, 2021), https://chinaipr2.files.wordpress.com/2021/01/2021.01.05-29-ntc-by-cj-paul-r-michel-ret-of-amicus-curiae-br-iso-ericsson-emergency-application-for-anti-main-document.pdf.

[120] Official Transcript of Teleconference Held on Feb. 10, 2015, ZTE Corp. v. Vringo Inc., No. 1:15-cv-00132, ECF 29 (D. Del. Feb. 11, 2015).

[121] *See, for example*, Richard Vary, *The Wuhan Submarine Surfaces at Christmas, to Be Met by a Texan TRO*, Bird & Bird Pat. Hub (Dec. 29, 2020), www.twobirds.com/en/patenthub/shared/insights/2020/global/the-wuhan-submarine-surfaces-at-christmas-to-be-met-by-a-texan-tro.

transplant,[122] it is not exempt from justified criticism, particularly if the transplant precipitates such undesirable consequences as not providing adequate respect for the jurisdictional priorities of other courts, limiting access by the public to key decisions, inducing attorneys to deceive their colleagues or a tribunal, or potentially depriving parties of adequate due process. These distinctions may suggest that other terminology, such as a "false friend," may be more appropriate in describing how China's ASIs function in practice.

Due to the continuing lack of transparency, it may be several years before we fully understand the impact of this particular transplant on China's legal system as well as the impact on the global ecosystem for licensing SEPs.

[122] *Transplanting ASIs*, *supra* note 2, at 1598.

11

# Patents and Competition

## *Commercializing Innovation in the Global Ecosystem for 5G and IoT*

*Hon. F. Scott Kieff and Thomas D. Grant*

### I. INTRODUCTION

For many decades, the products and services provided for and from our information and communication-related industries have required an ever-increasing number of technologies, many of which are patented, often from many firms, and often across several national borders. As the fifth generation (5G) of these communication tools gets deployed, users are enjoying vastly improved performance. 5G use cases include a broad range of applications, from enhanced mobile broadband for personal and autonomous communications, data processing, and entertainment devices, as well as massive levels of inter-machine communications needed for smart factories and cities, to ultrareliable and low-latency communications needed for potentially dangerous activities like self-driving cars and remote surgery. In turn, this is leading to newer uses and more complex interactions, such as those needed to support the Internet of Things (IoT) – as when home appliances and cars are directly communicating with each other, as distinct from the internet of devices that facilitate communications among humans, like personal computers, tablets, and phones. IoT use cases include smart homes, smart cities, telemedicine and telehealth, human and cyber security, building management, agriculture and aquaculture management, green energy management, enhanced and remote monitoring, and control of vehicles and other physical assets. To achieve these applications, a vast number of interactions and interconnections must take place, which in turn require immense transacting and private ordering, including a great deal of standard-setting.

Standards are the agreed-upon conventions that users of particular technologies follow to facilitate interoperability, like driving on the right or left side of the road to improve traffic flow and safety. Some standards are set informally through various coordination mechanisms, including passive ones, while many are set formally through active engagement among many participants in standard-setting organizations or standards development organizations (SSOs or SDOs). For 5G and IoT, standards are a significant part of the business ecosystem – such as standards for how cellular modems, Wi-Fi radios, or electronic memory operate – as well as a significant part of policy debates.

Early in the process of developing a given technology, an inventor might elect to seek patent protection instead of trade secret protection and might seek to advertise the invention or even encourage the development of standards to permit, require, or in some other way evolve to increase the value of the inventor's technology. Similarly, an implementer might, while totally unaware of a particular invention, inventor, or even standard, invest heavily in some capital expenditure, such as a multibillion dollar chip fabrication facility ("fab"), or set of commercial relationships. Concurrently, other parties may be contracting with the inventor or implementer to buy, sell, license, coinvest, codevelop, or co-deploy in any of the relevant asset markets, including the markets for technologies, employees, equipment, investment, and corporate control. Third parties are also making investment decisions along the way, often choosing to remain third parties by designing around and avoiding either the inventor or the implementer and their respective investments.

A significant amount of subsequent time after an invention is made is almost always then also involved, for two main reasons. One reason is that technological and business development are inherently time-consuming, expensive, risky endeavors with a mix of first-mover and second-mover advantages. It can take up to a decade or longer for a new technology to be brought to market in the form of, or as a component of, a particular product or service. Getting inventions put to use by consumers or even businesses often takes a large amount of subsequent development. Not many practical solutions – products or services – emerge fully developed and perfected like the mythological Athena from the head of Zeus, as if necessity alone were the instant mother of every invention and every end-use later developed for that invention.

The second reason is that nearly all modern patent systems are inherently premised on a significant government examination of the patent application to make an evaluation of the formal requirements of the patent application and a preliminary evaluation of the legal and technological substantive requirements of the patent application.[1] This process of patent prosecution by the applicant and patent examination by the patent office often takes about two years, or much longer.[2] The major contributor to this time delay is the effort needed to get even a rough assessment of the relevant technological field of art so that the patent law conditions of novelty, nonobviousness, and disclosure can be assessed against the benchmark of the state of this prior art.

[1] For a thorough discourse of US patent law, including the rules and procedures for obtaining, transacting over, and enforcing patents, see generally, JOHN M. GOLDEN ET AL., PRINCIPLES OF PATENT LAW (7th ed, 2018).

[2] USPTO, *Patent Pendency Data June 2022*, www.uspto.gov/dashboard/patents/pendency.html (last visited Aug. 3, 2022).

While all of that time is passing and all of these many actors are taking their own steps toward technology development, it is important to keep in mind that the patentees often will have filed their patent applications before moving too far into the marketplace or allowing too much time to lapse after inventing, in part because modern patent systems have rules that strongly encourage early filing, and in part because of what is often called the Arrow Information Paradox. Named after Nobel Prize-winning economist Kenneth Arrow,[3] the gist of the paradox is that it is hard to sell new information like an invention without giving potential buyers enough of a taste for them to formulate their level of appetite for it. But if they have been shown enough of it to really understand it, they wouldn't need to pay to make use of it. Having a patent application filed helps crack the paradox by turning someone using technology that may become patented into a potential infringer unless they strike a deal with the patentee for a sale or license of the patent.

In the middle of all of this wondrous complexity, a day – or decade – in the life of a commercial enterprise trying to implement a new technology can be viewed as quite hard. The technological and business challenges are exacerbated by the legal risks flowing from the reality that each issued patent gives its patentee a right – supported by the vibrant market for litigation financing and intermediaries like patent assertion entities (PAEs), which are sometimes called "patent trolls" – to threaten or actually bring various patent infringement lawsuits. These suits may include civil litigation in district court to allege patent infringement. Those district court actions typically seek damages, and these days, in some cases, also injunctions. Meanwhile, another type of patent infringement suit can be brought asking the United States International Trade Commission (ITC) to initiate an investigation that may lead to an order excluding the relevant articles from entry into the US market. The billions of dollars and vast human capital spent building that fab and those commercial relationships may be threatened, with each lawsuit or investigation alleging patent infringement typically costing up to ten or more million dollars in legal fees and associated expenses, often lasting five to seven years or longer in the case of district court litigation. And all of that is before appeals to the US Court of Appeals for the Federal Circuit, and possible appeal from there to the US Supreme Court. Even the mere threat or the mere initiation of these types of patent infringement proceedings can cause turbulent waves in the markets, whether they be the markets for the products the implementer wants to sell, the markets for commercial collaborators, or the markets for finance and corporate control. Critics of patent enforcement express great concern about the overall disruptive impact that patent enforcement and its threat can have on large technology-implementing companies like those of Silicon Valley fame.

[3] Kenneth J. Arrow, *Economic Welfare and the Allocation of Resources for Invention, in Rate and Direction of Inventive Activity, in* The Rate and Direction of Inventive Activity: Economic and Social Factors 609 (1962).

But there is another side to the coin, with implementers on one side and patentees on the other. Disruptive too is a day – or decade – in the life of a patentee trying to commercialize a patented invention. Implementers also have their own rights around patents owned by others, which are also supported by the vibrant market for litigation financing, including to bring district court litigation seeking a declaratory judgment that the patent is invalid – which involve similar time and expense to those of infringement litigation – as well as to bring or help others bring a plethora of post-grant review procedures in the patent office to cancel some or all of the claims of an issued patent. As with the turbulent waves that implementers face from the mere threat or initiation of infringement proceedings, patentees face similar market disruption from the mere threat or initiation of invalidation or cancellation proceedings. Especially for small early-stage ventures, this can cut off the vital access to financing that they need to even keep afloat as a going concern.

A great amount and variety of coordination mechanisms must be used well for all of this to occur relatively effectively and efficiently. In all of them, timing plays an important role, because there is a great deal of path dependency at stake, for almost everyone involved. One common coordination tool used in these settings is to have the SSOs deploy various approaches to facilitate the broad licensing of any patents that may be helpful or essential (so-called standard-essential patents, or SEPs) to practice a given standard. One such approach is to require parties to disclose pending patent applications, or to suggest which patents in a potentially large population of candidates are truly most likely to be adjudicated, infringed, and not invalid in a suit against those practicing the standard. Another such approach is to encourage or require that patentees participating in the SSO must make a commitment to license their patents on reasonable and nondiscriminatory (RAND) terms or fair, reasonable, and nondiscriminatory (FRAND) terms to those practicing the standard.

It may seem that many of these complexities, coordination challenges and opportunities, or risks or rewards is new, posing new questions calling for new policy responses. But that is not the case. They have each long been studied by scholars of the history of the interface between the patent and antitrust systems, in both the empirical economic literature and the legal literature.[4] The upshot from the

[4] For a sampling of this work, see, *for example*, Giles S. Rich, *The Relation between Patent Practices and the Anti-Monopoly Laws*, 24 J. Pat. Off. Soc'y 85, pts. 1–5, at 85, 159, 241, 328, 422 (1942) (broad legal exploration of the patent antitrust interface from the first half of the last century by the person who became one of the principal drafters of the 1952 Patent Act, which codified key approaches to that interface that remain only strengthened in the present iteration of the statute, and who sat as a federal appellate judge interpreting that statute until the end of the century); F. Scott Kieff & Troy A. Paredes, *The Basics Matter: At the Periphery of Intellectual Property*, 73 Geo. Wash. L. Rev. 174 (2004) (outlining how this approach to the patent antitrust interface facilitates commercialization and competition when applied to a range of more modern doctrinal and policy debates); F. Scott Kieff, Quanta v. LG Electronics: *Frustrating Patent Deals by Taking Contracting Options off the Table?*, 2007–2008 Cato S. Ct.

empirical economics literature is that a property rights approach to patents focused on commercializing innovation facilitates competition and access. This positive effect is evidenced by large decreases in quality-adjusted prices and steady ongoing entry even into markets for information and communications technologies with large numbers of patents and standards that are the focus of the debates about 5G and IoT.[5] The upshot from the legal literature, which is the focus of this chapter and explored in more detail later, is that such a property rights approach to the patent system gives clear guidance about which choices to make between particular versions of the detailed legal rules actually implemented across the patent system, from those governing patent validity to those governing patent transactions and patent enforcement. These different legal rules and mechanisms enable patents to

Rev. 315 (2008) (showing how modern case law can frustrate these goals); F. Scott Kieff & Henry E. Smith, *How Not to Invent a Patent Crisis*, in Reacting to the Spending Spree: Policy Changes We Can Afford 55 (Terry Anderson & Richard Sousa, eds., 2009); Richard A. Epstein et al., *The FTC, IP, and SSOs: Government Hold-Up Replacing Private Coordination*, 8 J. Competition L. & Econ. 1 (2012) (same for approaches of modern antitrust enforcement agencies); F. Scott Kieff, *Private Antitrust at the U.S. International Trade Commission*, 14 J. Competition L. & Econ. 46 (2018) (elucidating how ITC enforcement would operate better); Adam Mossoff, *The Rise and Fall of the First American Patent Thicket: The Sewing Machine War of the 1850s*, 53 Ariz. L. Rev. 165 (2011) (elucidating an infamous historical case); Jonathan Barnett, *The Great Patent Grab, in* The Battle over Patents: Historical Perspectives on Current Debates (Stephen Haber & Naomi Lamoreaux eds., 2021) (chapter focusing on current debates in recent edited volume exploring the field); Stephen Haber & Naomi R. Lamoreaux, *The Battle over Patents, Defining Ideas*, Hoover Inst. (Nov. 3, 2021), www.hoover.org/research/battle-over-patents (short essay reviewing the topic).

5 Alexander Galetovic, Stephen Haber, & Ross Levine, *An Empirical Examination of Patent Holdup*, 11 J. Competition L. & Econ. 549 (2015) (finding no evidence that SEP-reliant industries experience more stagnant quality-adjusted prices than non-SEP-reliant industries or that court decisions that reduce the excessive power of SEP holders accelerated innovation in SEP-reliant industries); Stephen Haber, *Patents and the Wealth of Nations*, 23 Geo. Mason L. Rev. 811 (2016) (exploring the aggregate social value of property rights in patents); Stephen Haber & Seth Werfel, *Patent Trolls as Financial Intermediaries? Experimental Evidence*, 149 Econ. Letters 64, 64 (2016) ("Our results indicate that PAEs served an intermediary function for two groups in our sample: subjects who identified as inventors rather than entrepreneurs, and subjects who were relatively more sensitive to financial losses"); Alexander Galetovic & Stephen Haber, *The Fallacies of Patent Holdup Theory*, 13 J. Competition L. & Econ 1 (2017) (showing serious flaws in the basic logic of patent holdup theory making it logically inconsistent and incomplete and inconsistent with economic fundamentals and evidence); Alexander Galetovic, Stephen Haber, & Lew Zaretzki, *An Estimate of the Average Cumulative Royalty Yield in the World Mobile Phone Industry: Theory, Measurement and Results*, 42 Telecomm. Policy 263 (2018) (empirical evidence that the royalty stack for patents in the mobile phone industry is about 3–6% rather than the 20–40% or higher estimated by critics of a property rights approach to patents); Alexander Galetovic, Stephen Haber, & Lew Zaretzki, *Is There an Anticommons Tragedy in the World Smartphone Industry?*, 32 Berkeley Tech. L.J. 1527 (2018) (same); Alexander Galetovic & Stephen Haber, *SEP Royalties: What Theory of Value and Distribution Should Courts Apply?*, 17 Ohio St. Tech. L.J. 189 (2021) (showing how well it works to price patent royalties using a common method that relies on information from the market about the value of comparable assets or their rental rates).

be so helpful in facilitating the vital coordination needed for the commercialization of new technologies.

## II. PROPERTY APPROACH TO PATENTS HAS SUPPORT ACROSS THE POLITICAL SPECTRUM

Debates about patents have long focused on the patent-antitrust interface.[6] Especially in this context, the views offered about patents by antitrust enforcers[7] have generally focused on the role that intellectual property (IP) in general and patents in particular can play, on the one hand, in providing beneficial incentives to create or invent, and, on the other hand, in enabling harmful concentrations of market power leading to increased prices and reduced output. Such discussions often then focus essentially on how much of the "good" is enough, how much of the "bad" is too much, and trade-offs between them.

In effect, those discussions highlight a direct tension between IP as a helpful incentive to create or invent and IP as the cause of deleterious anticompetitive monopoly effects. They then offer various approaches to legal regimes to address both sides of the tension. One set of approaches includes the use of other inducements or rewards for creation or invention in the place of, or in addition to, IP, such as regulatory exclusivity, tax credits, grants, prizes, and the like. A second set of approaches exempts particular fields of technology from eligibility for IP protection, such as those having to do with health care, software, or finance, usually with the expectation of significant, frequent, and ongoing updates to the boundaries of these exempted fields. A third set of approaches decreases the remedies available for IP infringement, including damages, injunctions, and exclusion orders. A fourth set of approaches directly addresses interactions between IP owners and IP users,

6 The discussion in this section is drawn from Letter from the Hon. F. Scott Kieff, Commn'r, U.S. Int'l. Tr. Comm'n, on the United Sates Federal Trade Commission's and the United States Department of Justice Antitrust Division's Joint Guidelines for the Licensing of Intellectual Property (Sept. 23, 2016), www.justice.gov/atr/file/897081/download; and F. Scott Kieff, *Pragmatism, Perspective, and Trade: AD/CVD, Patents, and Antitrust as Mostly Private Law*, 30 Harv. J.L. & Tech. 97 (2017).

7 Certain 3G Mobile Handsets and Components Thereof, Inv. No. 337-TA-613 (Remand), Reply Submission on the Public Interest of Federal Trade Commissioners Maureen K. Ohlhausen and Joshua D. Wright (USITC July 20, 2015); Certain 3G Mobile Handsets and Components Thereof, Inv. No. 337-TA-613 (Remand), Written Submission on the Public Interest of Federal Trade Commission Chairwoman Edith Ramirez (July 13, 2015); Correspondence from United States Federal Trade Commission Chairwoman Edith Ramirez to United States Trade Representative Michael Froman (July 15, 2013); Certain Wireless Communication Devices, Portable Music and Data Processing Devices, Computers and Components Thereof, Inv. No. 337-TA-745, Third Party United States Federal Trade Commission's Statement on the Public Interest (USITC June 6, 2012); Certain Gaming and Entertainment Consoles, Related Software, and Components Thereof, Inv. No. 337-TA-752, Third Party United States Federal Trade Commission's Statement on the Public Interest (USITC June 6, 2012).

including heightened antitrust scrutiny, compulsory licenses, and governmental takings of IP licenses or the entire IP rights themselves. Many other ideas are also offered.

A common theme across these approaches is to view IP more in the tradition of public law, or as regulatory entitlements, by focusing on the use of more extensive interactions between governmental bodies and private parties. The overarching goals across different perspectives in the literature are generally shared and laudatory: fostering access to creative or inventive technologies, competition, economic growth, and diverse and inclusive participation; improving both efficiency and fairness for all.

These shared goals also are championed by an intellectual approach to IP that is different than those briefly mentioned earlier. This different approach – a commercialization approach – has been embraced across the American political spectrum, including both the Carter administration and the Reagan administration,[8] as well as by celebrated jurists of the last century coming from diverse philosophical perspectives, including Circuit Judges Learned Hand, Jerome Frank, and Giles Rich,[9] who saw it as important to helping the economy and society.[10] The roots of a

[8] Judge Pauline Newman, *The Federal Circuit in Perspective*, 54 AM. U.L. REV. 821 (2005).

[9] Giles S. Rich, *The Relation between Patent Practices and the Anti-Monopoly Laws (II)*, 24 J. PAT. OFF. SOC'Y 159 (1942), *reprinted in* 14 FED. CIR. B.J. at pages 5, 21, 37, 67, and 87 (2004–2005) (five-part series of articles); Picard v. United Aircraft Corp., 128 F.2d 632, 643 (2d Cir. 1942) (Frank, J., concurring); Reiner v. I. Leon Co., 285 F.2d 501, 503 (2d Cir. 1960) (Hand, J.) (noting "[t]here can be no doubt that the Act of 1952 meant to change the slow but steady drift of judicial decision that had been hostile to patents"); Lyon v. Bausch & Lomb Optical Co., 224 F.2d 530, 536–37 (2d Cir. 1955) (Hand, J.) (noting "§ 103... restores the original gloss ... [A] legislature ... must be free to reinstate the courts' initial interpretation, even though it may have been obscured by a series of later comments whose upshot is at best hazy.").

[10] Some representative examples in the literature that are consistent with the commercialization approach include the following: Stephen Haber, *Patents and the Wealth of Nations*, 23 GEO. MASON L. REV. 811 (2016); Galetovic et al., *Empirical Examination of Patent Holdup, supra* note 5; Daniel F. Spulber, *How Patents Provide the Foundation of the Market for Inventions*, 11 J. COMPETITION L. & ECON. 271 (2015); Pierre Larouche et al., *Settling FRAND Disputes: Is Mandatory Arbitration a Reasonable and Nondiscriminatory Alternative?*, 10 J. COMPETITION L. & ECON. 581 (2014); F. Scott Kieff & Anne Layne-Farrar, *Incentive Effects from Different Approaches to Holdup Mitigation Surrounding Patent Remedies and Standard-Setting Organizations*, 9 J. COMPETITION L. & ECON. 1091 (2013); Epstein, *supra* note 4, at 1; Mark P. Gergen et al., *The Supreme Court's Accidental Revolution? The Test for Permanent Injunctions*, 112 COLUM. L. REV. 203 (2012); F. Scott Kieff, *An Inconvenient School of Thought*, 61 ALA. L. REV. 591 (2010); F. Scott Kieff, *On Coordinating Transactions in Intellectual Property: A Response to Smith's Delineating Entitlements in Information*, 117 YALE L.J. POCKET PART 101 (2007); Henry E. Smith, *Intellectual Property as Property: Delineating Entitlements in Information*, 116 YALE L.J. 1742, 1745, 1751–52 (2007); Kieff & Paredes, *supra* note 4; Naomi R. Lamoreaux & Kenneth L. Sokoloff, *Intermediaries in the U.S. Market for Technology, 1870–1920*, in FINANCE, INTERMEDIARIES, AND ECONOMIC DEVELOPMENT 209 (Stanley L. Engerman et al. eds., 2003); and B. Zorina Khan & Kenneth L. Sokoloff, *History Lessons: The Early Development of Intellectual Property Institutions in the United States*, 15 J. ECON. PERSP. 233 (2001).

commercialization approach to patents, in particular, reach back even further into American history, including Abraham Lincoln's view that the patent system "added the fuel of interest to the fire of genius, in the discovery *and production* of new and useful things."[11]

A commercialization approach to IP views IP more in the tradition of private law, as property rights, by focusing on the use of IP in interactions between private parties, including contracts. Centered on the relationships among private parties, this approach to IP emphasizes a different target and a different mechanism by which IP can operate. Rather than target individuals who are likely to respond to IP as incentives to create or invent in particular, this approach targets a broad, diverse set of market actors in general. This large group encompasses the creator or inventor as well as all those complementary users of a creation or an invention who can help bring it to market, such as investors (including venture capitalists), entrepreneurs, managers, marketers, developers, and owners of other key assets, tangible and intangible, including other creations or inventions. Another key difference in this approach to IP lies in the mechanism by which the IP assets and these private actors interact. This approach sees IP as a tool for facilitating coordination among these diverse private actors, in furtherance of their own private interests in commercializing the creation or invention.

This commercialization approach sees IP rights serving a role akin to "beacons in the dark," drawing to themselves potential complementary users of the IP-protected asset to interact with the IP owner and each other, exploring through the bargaining process the possibility of striking contracts with each other. Focusing on such a "beacon-and-bargain" effect can relieve the governmental side of the IP system of the need to amass the detailed information required to reasonably tailor a direct targeted incentive, such as each actor's relative interests and contributions, needs, skills, or the like. Not only is amassing all of that information hard for the government to do, but large, established market actors may be better able than smaller market entrants to wield the political influence needed to get the government to act, increasing risk of concerns about political economy, public choice, and fairness. Instead, each private party can bring its own expertise and other assets to the negotiating table while knowing – without necessarily having to reveal it to other parties or the government – enough about its own level of interest and capability when it decides whether to strike a deal or not.

Such successful coordination may help bring new business models, products, and services to market. It also can allow IP owners and their contracting parties to appropriate the returns to any of the rival inputs they invested toward developing and commercializing creations or inventions – labor, lab space, capital, and the like.

[11] Abraham Lincoln, Second Lecture on Discoveries and Inventions (Feb. 11, 1859), in 3 Collected Works of Abraham Lincoln 356, 363 (Roy P. Basler ed., 1953) (emphasis added and omitted).

At the same time, the government can avoid having to then go back to evaluate and trace the actual relative contributions that each participant brought to a creation's or an invention's successful commercialization – including, again, the cost of obtaining and using that information and the associated risks of political influence – by enforcing the terms of the contracts these parties strike with each other to allocate any value resulting from the creation's or invention's commercialization. In addition, significant economic theory and empirical evidence suggest this can all happen while the quality-adjusted prices paid by many end-users actually decline and public access is high. In keeping with this commercialization approach, patents can be important antimonopoly devices, helping a smaller "David" come to market and compete against a larger "Goliath."[12]

A commercialization approach thereby mitigates many of the challenges raised by the tension that is the focus of the other intellectual approaches to IP, as well as by their responses to that tension. Many of the alternatives to IP that are often suggested, such as rewards or tax credits, can face significant challenges in facilitating the private-sector coordination benefits envisioned by the commercialization approach. While such approaches often are motivated by concerns about rising prices paid by consumers and direct benefits paid to creators and inventors, they may not account for the important cases in which IP rights are associated with declines in quality-adjusted prices paid by consumers and other forms of commercial benefits accrued to the entire IP production team as well as to consumers and third parties, which are emphasized in a commercialization approach. In addition, a commercialization approach can embrace many of the practical checks on the market power of an IP right that are often suggested by other approaches to IP, such as antitrust review, government takings, and compulsory licensing, while at the same time showing the importance of maintaining self-limiting principles within each such check to maintain commercialization benefits and mitigate concerns about dynamic efficiency, public choice, fairness, and the like.[13]

To be sure, a focus on commercialization does not ignore creators or inventors or creations or inventions themselves. For example, a system successful in commercializing inventions can have the collateral benefit of providing positive incentives to those who do invent through the possibility of sharing in the many rewards associated with successful commercialization. Nor does a focus on commercialization guarantee that IP rights cause more help than harm in all circumstances. Significant

[12] *Picard*, 128 F.2d at 643 (Frank, J., concurring).

[13] While the details of the particular legal rules operating within these patent-checking legal systems of antitrust, government takings, and compulsory licensing are beyond the scope of this short overview chapter, the commercialization and property rights approach to patents that this chapter is exploring does leave ample room for those patent-checking systems to operate. For more on the commercialization approach to the details of those systems, *see, for example*, Kieff & Paredes, *supra* note 4; and Richard A. Epstein & F. Scott Kieff, *Questioning the Frequency and Wisdom of Compulsory Licensing for Pharmaceutical Patents*, 78 U. CHI. L. REV. 71 (2011).

theoretical and empirical questions remain open about how the system can be improved overall.

## III. PROPERTY APPROACH IS ROOTED IN THE EARLY US PATENT SYSTEM

Governments in many countries have used patent systems since the Renaissance. The British Empire used them like special monopolistic privileges given out by the Crown to its favorites. And even the British started to rein in that approach. Our Founders knew about this history and deliberately took a different approach. They thought it was so important to give Congress the power to create a patent system that they included it in the unamended text of the original Constitution. The early American patent system was designed carefully to work differently than the British one in that it was purposely restrained by objective facts and not open to political discretion. Economic historians credit those differences to the success of the early American patent system. By the mid-1800s, Charles Dickens was describing in his short story "A Poor Man's Tale of a Patent" how the unending bureaucracy of the British patent system not only failed to bring new inventions to market but also left inventors – as he wrote – "quite wore out, patience and pocket."

In modern debates about patent systems, there is really no need to speculate or invent new arguments. We've tried many approaches and seen many results. There's not much reason to expect the unexpected here. The more the patent system fills up with bureaucratic steps and administrative and policy discretion, the more they favor the large politically connected people and businesses, and the more both innovation and competition suffer. But the more the patent system turns on objective facts and clear and predictable rules, the more it increases the number of new technologies brought to market, the ability for diverse consumers to access those technologies, and the diversity in sizes among the businesses in the market. That system won't be against big business; but it won't so favor big business that it's against small and medium-sized businesses as well.[14]

So often in today's debates about patents, people ask us to imagine the old men in wigs with the technologies of the late 1700s and tell us that we have to update our patent system to deal with the new technologies of today and tomorrow. But that's where the genius of the American patent system comes into play. Rather than decide who gets a patent based on politics and fashion, we designed our patent system to

[14] For a longer sketch of the basic ideas about the benefits in fostering economic growth and inclusion from using objective rules for a property rights system, including for patents, see, generally, Stephen H. Haber et al., *On the Importance to Economic Success of Property Rights in Finance and Innovation*, 26 Wash U.J.L. & Pol'y 215 (2008). For more on the differences between the early US patent system and the British system, see, generally, B. Zorina Khan, The Democratization of Invention: Patents and Copyrights in American Economic Development, 1790–1920 (Claudia Goldin ed., 2005); Khan & Sokoloff, *supra* note 10.

turn on facts about the prior art and objective questions like novelty. So we don't need to update our patent system to deal with new technologies, because the only technologies that are patentable in our patent system are the ones that are new.

A similar attempt to question the core value of patents by looking to history is by invoking Thomas Jefferson's skeptical take on patents. Not only was Jefferson a leading figure in early American government in general, as a principal drafter of the Declaration of Independence, our first Secretary of State, our second Vice President, and our third President, he also was an inventor and ran our first patent office. Yet, when it came to broader views about patents, he was quite skeptical about property rights in ideas, writing:

> He who receives an idea from me, receives instruction himself without lessening mine; as he who lights his taper at mine, receives light without darkening me . . ..
>
> I know well the difficulty of drawing a line between the things which are worth to the public the embarrassment of an exclusive patent, and those which are not.[15]

While this certainly makes the skeptical case for having a patent system at all, once a patent system has been offered to inventors, the path dependency of their decision to abandon trade secrecy and instead seek patent protection on the expectation that patents will enjoy predictable enforcement as property rights leaves patentees especially vulnerable to holdup. Sticking with Jefferson's metaphor of a candle, it's important to bear in mind that blowing out someone else's light doesn't make yours brighter; it just darkens the scene for everyone.

## IV. COMMERCIALIZATION IS HELPED BY OBJECTIVE ADJUDICATION RATHER THAN POLITICAL ADMINISTRATION

Commercialization does not merely depend on the specific legal rules operating within the substantive fields of IP and antitrust themselves. It also is meaningfully helped by objective approaches to government decision-making and analysis more generally, such as those operating within the courts and agencies implementing the patent system. The ITC is a prime example of a tribunal that can provide objective adjudication for patents; and its success in this area is neither an accident nor hard to reproduce.

Much has been written about the vital need to have government agencies, including those in both the fields of IP and antitrust, conduct careful, scientific, fact-based, analysis, and decision-making, that accounts for diverse views and perspectives.[16] When the ITC celebrated its 100th

[15] Letter of Thomas Jefferson to Isaac McPherson (Aug. 13, 1813).

[16] Richard B. Stewart & Cass R. Sunstein, *Public Programs and Private Rights*, 95 Harv. L. Rev. 1193 (1982); David A. Hyman & William E. Kovacic, *Why Who Does What Matters: Governmental Design and Agency Performance*, 82 Geo. Wash. L. Rev. 1446 (2014).

anniversary,[17] it had occasion to remember the difficult task our Nation's first Treasury Secretary, Alexander Hamilton, had to manage when figuring out how to finance the operation of our new central government while at the same time hopefully helping or at least mitigating the harm to our then-fledgling domestic manufacturing industry.

For the first century of its existence, the federal government was financed essentially with tariffs on imports. There was no income tax back then. It took until 1913 for the Sixteenth Amendment to our Constitution to be ratified, giving the federal government the power to raise revenue from sources internal to the country such as via a tax on income.

Tariffs on imports can raise money for a national government. But that will only work to the extent that imported goods continue to flow into the country despite rising prices paid by purchasers. Tariffs also can protect domestic industries, including the then-fledgling manufacturing sector, from foreign competition in finished manufactured goods. But that will only work so long as the tariffs don't also cover imported inputs to domestic manufacturing processes. Tariffs also can trigger reciprocal tariffs that can hamper exports. It can be tricky to figure out the net impact of these several forces that point in opposite directions.

Although sometimes seen as an attempt at protectionism, Hamilton's effort brought a scientific approach to bear on these questions, which led him to compile a "Report on the Subject of Manufactures" as a study of this dynamic system and to offer more balanced recommended policy actions informed by such as study.[18] To be sure, Hamilton's report was just an initial effort; and the intense debates and problems surrounding the dynamic impact of tariffs continued for about a century until, together with slavery, they brought our country to war with itself in the Civil War.

By soon after the end of the Civil War, the confluence of two factors brought much-needed help. First was the evolution in the state of the art in economic science, including a much better understanding of how to gather data and analyze it. The second was the suggestion by Frank Taussig, Chairman of the Economics Department at Harvard, for a new approach to a government agency in this area.[19] That new agency model, attempted a few times after the Civil War, eventually became the ITC. It has a few key structural characteristics that being replete with checks and balances coerce behavior that is collaborative, independent, analytical,

[17] More about the ITC Centennial, including the entire freely available contents from a scholarly book on the topic, can be found online here: UNITED STATES INTERNATIONAL TRADE COMMISSION, A CENTENNIAL HISTORY OF THE USITC (Paul R. Baros ed., 2017), www.usitc.gov/documents/final_centennial_history_508_compliant_v2.pdf.

[18] Douglas Irwin, *The Aftermath of Hamilton's Report on Manufactures*, 64 J. ECON. HIST. 800 (2004).

[19] JOHN M. DOBSON, TWO CENTURIES OF TARIFFS, THE BACKGROUND AND EMERGENCE OF THE UNITED STATES INTERNATIONAL TRADE COMMISSION 86 (1976).

and professional, while punishing prerogative. While many of the Bi-Partisan-Commissions in the US government are led by an odd number of Presidentially-Nominated-and-Senate-Confirmed Commissioners (usually five), the ITC is designed for deadlock with an even number: six. While most of the other Commissions have a Chair who generally can serve until replaced by the President, the ITC Chair is required to switch person and party every two years, among the existing Commissioners. And, at the ITC, the Commissioner terms are longer than at many of the other commissions (nine years) and generally nonrenewable, thereby further reducing incentives for responsiveness to pressure from politics and intellectual fashion. This unleashes and empowers the vast talent of our several hundred staff of professional economists, industry experts, and lawyers to do the sometimes unthinkable within organizations: call the shots like they see them.

While the ITC is, like the federal courts, deliberately structured to be removed from the political influence of only one political party, the Department of Commerce's Patent Office, which operates the post-grant cancellation procedures for patents, as well as the Department of Justice's Antitrust Division (DoJ), are ordinary Executive Branch agencies directly responsive to the political leadership of the President. Similarly, the Federal Trade Commission, which also conducts antitrust enforcement like the DoJ, is only somewhat less directly responsive to political influence, because it is structured as a five-member agency with a Chair appointable and removable by the President, backed up by a majority in the President's party.

In addition to important differences in how these tribunals are structured internally, there also are important differences in how their basic substantive jurisdictional limits exacerbate the incentives for party advocates to engage in hyperbolic arguments. While the ITC and the federal courts have substantive power to simultaneously address issues relating to patent validity, patent infringement, remedy, and antitrust, the Patent Office, like the DoJ and FTC, do not. The Patent Office can only assess patent validity; and the DoJ and the FTC can only assess antitrust. When all four topics are in dispute within a single tribunal, each side of the case has powerful self-disciplining effect to make arguments more grounded in the record. The patentee has the selfish incentive when arguing about infringement and remedy to argue that the patent claims are broad (thereby sweeping in more infringements), but it also has the countervailing selfish incentive when arguing about validity and antitrust to assert that the patent claims are narrow (thereby avoiding the prior art and avoiding excessive market power). At the same time, the opposing party has the exact opposite set of mutually countervailing incentives. As a result, each side engages in self-restraint, providing the tribunal with a much more elaborate and thoughtfully presented (less hyperbolic) set of evidence and arguments.

The combined effects of more internal independence and less hyperbolic arguments from advocates help courts and the ITC reach more reasoned determinations that are more transparently grounded in the record. Simply put, they enjoy the greater opportunity to be more informed by more diverse opinions and perspectives, and they face more discipline to ground their opinions in the public factual record.

One example of this politically diverse and independent approach of the ITC acting at the IP-antitrust interface is the several views that emerged engaging the specific factual record of the actual negotiating and litigation behavior of actual parties to an IP dispute in the *Amkor v. Carsem* "encapsulated integrated circuits" case involving the standard-setting organization called "JEDEC." In that case, four of the six Commissioners provided additional views exploring various procedural safeguards akin to waiver and estoppel to maximize fairness and the ways that specific conduct of both the IP owner and the IP user can give rise to symmetrical concerns about holdup and reverse holdup.[20] Similar symmetrical concern for such procedural and substantive nuances is elaborated in the European Court of Justice's (ECJ) *Huawei v. ZTE* decision, which may suggest the emergence of an international norm, at least for those parts of a government designed to operate more removed from the direct influence of only one political party.[21]

## V. INTELLECTUAL PROPERTY IS NOT THAT DIFFERENT FROM TANGIBLE PROPERTY

With tangible property like land, a car, or a cell phone, the property right includes a right for the owner to use the thing covered by the property right. If you own land, a car, or a cell phone, you can basically use it without needing permission from other people. The government likely will regulate your use in many ways; but if the government so restricts your use that you can't use it at all, then you probably have a claim against the government for just compensation due from their taking.

With intangible property like IP, the main and basically only right that the owner gets is the right to exclude other private people or businesses from infringing. That means that if the IP owner can't actually enforce that right to exclude, there's not much incentive for infringers to avoid infringement or to negotiate for a license or purchase of the IP. That's why the right to exclude is so important for IP.

With tangible property like land, a car, or a cell phone, everyone can easily tell if someone is using it, because you can see them on the land, in the car, or holding the phone. With intangible property like a mortgage, a share of stock, a bond, or a patent, we have to read the detailed written words to know what the thing is, what its

20 Certain Encapsulated Integrated Circuit Devices and Products Containing Same, USITC Inv. No. 337-TA-501, Commission Opinion (USITC Apr. 28, 2014), www.essentialpatentblog.com/wp-content/uploads/sites/64/2014/05/2014.04.28-Encapsulated-Integrated-Circuits-. . .-ITC-337-TA-501-sm.pdf. (with additional views of Aranoff, Broadbent, Kieff, and Pinkert).

21 Huawei Technologies Co. Limited v. ZTE Corp., Case C-170/13, Judgment of the Court (Fifth Chamber, July 16, 2015), http://eur-lex.europa.eu/legal-content/EN/TXT/?uri=CELEX%3A62013CA0170.

boundaries are, and who is using it. The right to exclude is the only way IP owners can easily keep track of who is using their IP and at the same time easily interact with those users so that everyone can make informed choices about whether to infringe, negotiate for a license or sale of the IP, or design around the IP to avoid infringement.

For many forms of both tangible and intangible property, we can look to a government registry to tell who owns it. Those registries are not perfect, and it can take real time to comb through them to find what you might be looking to target or avoid. But much of that work gets done by owners of those assets when they knock on your door or write you a letter and tell you why you should consider taking a license under their patent or buying their patent. You surely won't take their word for it that you should, but you also know now what your lawyers should read and consider before you decide to invest billions of dollars in a new product line that might infringe some of those patents.

Users of patented technologies complain they are too often surprised to learn they are infringers because patents can be hard to interpret. While some legal instruments are harder to interpret than others, the legal rules for each kind of instrument set the standard. In many of the patent cases that have made headlines over recent years – like *eBay v. MercExchange*, *TiVO v. Echostar*, and *i4i v. Microsoft*,[22] the patents were adjudicated to have fully met each of patent law's disclosure requirements – including enablement, written description, and definiteness. While we should always consider the pluses and minuses of making objective disclosure rules like these somehow more demanding on the patentee, there will always be a zone of uncertainty between what a patent does and does not cover.

But however uncertain things may be in some settings, we shouldn't forget that in *eBay*, *TiVO*, and *i4i*, the patent infringements were adjudicated to have been willful. That means the infringer knew or should have known its conduct was wrong. It stretches the definition of surprise to reach a case where a lawyer's legal advice would have told – or did tell – the user of the patented technology that its use would be adjudicated to be infringement.

The intangible nature of patent rights is not a reason to allow parties willfully to ignore those rights. Quite the opposite: It is the very reason society has predictable rules and reliable and transparent procedures for the enforcement of those rights. Imagine you own an electronics shop and come in one morning to find a broken window and some items strewn across the floor, but it's hard to tell exactly what has been stolen. At least with tangible property like phones or chips or any physical goods in a shop you can count your inventory and see what was stolen. But, even with a theft like this of physical goods, your insurance company will still require a report by the police investigators verifying what's been taken before they cut a check

[22] Microsoft Corp. v. i4i Ltd. 564 U.S. 91 (2011); eBay, Inc. v. MercExchange, LLC, 547 U.S. 388 (2006); NTP, Inc. v. Research in Motion, Ltd., 418 F.3d 1282 (Fed. Cir. 2005).

to cover the loss. With intangible property like a patent, you also need a process to sort out the facts, and perhaps even more so, because you might only imperfectly know exactly what a thief has taken. Patent infringement suits help everyone in the market figure how many of each specific type of inventory has been taken, whether the USPTO somehow made an error, and what remedies are appropriate. That's why patentees often go both to the court and the ITC at the same time. The court has the full panoply of remedies in its tool belt, but takes much longer, usually several years, to reach a judgment. The ITC has many fewer remedy tools in its tool belt, but goes faster, and carries out at least as full and fair an adjudication that helps everyone in the market get a full and crisp picture of what actually has happened. It also happens to be the only venue in the US patent system today where a patentee who wins an adjudication of patent infringement has a reasonable likelihood of securing an injunction-like remedy, which at the ITC would be either an exclusion order to keep particular goods from being imported into the country or a cease and desist order preventing particular parties from taking particular actions.

Several long-standing doctrines of tangible property offer important lessons about how property rights in patents can accommodate the real apprehensions that implementers of patented technologies may be truly surprised to learn they are infringers. Property law doctrines that govern cases of a mistaken improver of another's personal property or a mistaken building encroachment on another's real property operate to protect both the interest of the mistaken infringer and the interests we all share in protecting property – including the interests of non-mistaken owners, second parties who may have invested in transacting with the owners, and non-mistaken third parties who may have invested in avoiding the property, such as by designing around the patented technology. These property doctrines go far in vindicating the accidental infringer's holdup or hassle costs due mostly to path dependency (why tear down the big building built an inch over the property line?). They also go far in vindicating the autonomy interests (and emotional interests in expressing exasperation) of the property owner, second parties who elected to transact with that owner, and third parties who paid to get better surveys and design around that owner's lot. The way property law meets those dual goals is not by making the property interest invalid or unenforceable, merely because of "innocent" infringement or evidence of "spite" in the suit. After all, aren't we all allowed to reveal emotional pique when our autonomy and financial interests are aroused due to the unilateral acts of an infringer, however "innocent" or "accidental" the infringement? The way we meet those dual goals when it comes to building encroachments and mistaken improvements is that we tailor the remedy.[23] And, of course, the tailoring of the remedy to account for

[23] Wetherbee v. Green, 22 Mich. 311 (1871) (on showing that good faith innocent mistake was made in taking raw wood belonging to another and working it into parts for barrels, the mistaken improver was allowed to keep the property and pay damages to the true owner).

"spite" would include all of the usual tools our legal system uses to police bad faith litigation, including Rule 11 of the Federal Rules of Civil Procedure, that even the pro-patent Federal Circuit showed it would use against bad faith patentees who brought baseless litigation suits as in cases like *Judin v. US*, which was decided at a time when critics of a property rights approach to patents saw the court as being too property rights oriented.[24]

Other long-standing legal doctrines set a very different backdrop than suggested by patent critics concerned about implementers surprised to learn they are infringers. Implementers may very well be surprised by a patent if they did not copy the patent or derive from it. But whether an implementer has copied or derived from a patent is not relevant to the basic question of patent infringement, because the infringement doctrine of the patent system is deliberately different than the infringement doctrine in the copyright system. Infringement determinations in the patent system focus on the scope of the written claim in the patent, while infringement determinations in the copyright system focus on copying of or deriving from an earlier work of creative expression. Implementers similarly may have excellent legal title to the physical goods they use when infringing a patent, whether by bona fide gift or good faith purchase. But legal title to physical assets is no bar to infringement under any of the intellectual property regimes, just as it is no bar to enforcement of the many regulatory regimes. Implementers may also be surprised about infringement if they thought some third party had a better claim to the patent's title than the patentee. But our property law systems have since Roman law rejected the doctrine known as the *jus-tertii* defense that otherwise would stave off a property enforcement action brought by an owner due to the possibility of some third party with a potentially better claim to ownership.

At the same time, shifting the balance back toward patent enforcement from antitrust enforcement doesn't eliminate antitrust enforcement. There remains plenty of room for ordinary antitrust enforcement where there is actual other evidence – other than the mere presence of a patent – of actual market power. There also remains plenty of room for antitrust action where the patent was procured with knowing fraud as in *Walker Process*, or when the patent enforcement is "a mere sham to cover what is actually nothing more than an attempt to interfere directly with the business relationships of a competitor" as in

[24] Judin v. United States, 110 F.3d 780, 781 (Fed. Cir. 1997). For more on how fee-shifting in both directions, for bad faith litigation, can work better than post-grant review procedures in furthering the combined goals of both sides of the patentee-implementer debate, see, generally, F. Scott Kieff, *The Case for Preferring Patent Validity Litigation over Second Window Review and Gold Plated Patents: When One Size Doesn't Fit All, How Could Two Do the Trick?*, 157 U. Pa. L. Rev. 1937 (2009); Kieff & Layne-Farrar, *supra* note 10; F. Scott Kieff & James E. Daily, *Benefits of Patent Jury Trials for Commercializing Innovation*, 21 Geo. Mason L. Rev. 865 (2014).

*Handguards* and is both objectively baseless and subjectively motivated to cause harm to the market as in *PRE*.[25]

## VI. RECENT US PATENT SYSTEM INNOVATIONS HARM MORE THAN HELP INNOVATION

The major changes to the patent system over the past two decades, some through Congress and some through the courts, have harmed innovation, competition, and national security. They have all operated to do basically two things. One is to drastically shift many of the specific legal rules about patent validity, patent infringement, and patent transactions from generally turning on objective facts applicable the same way to everyone to generally turning on subjective discretion finely tailored to each different user. The second is to drastically add to the number and strength of administrative and bureaucratic procedures available to keep a patent from being enforced in court or at the ITC.[26] It's as if we read Dickens's story "A Poor Man's Tale of a Patent" and decided that the tragic caricature of a broken patent system that he was telling everyone to avoid was something we should actually seek out and put into place.

Whenever a commercial law system is so finely adapted to policy preferences of politically motivated government actors and subjectively tailored by them to each different use and user, the only kinds of businesses that can engage that system are the huge politically powerful ones. Property rights are at their worst when they are created and changed and erased at the discretion of the government, and when private actors have to include the government in every decision about whether to bundle, divide, or license or sell the property rights. That kind of system forces market actors to constantly deal with the government, and that always favors big players with more political power. That just concentrates wealth and power.

When a property rights system is working well, the rules of the game are predictable, applicable to everyone, and private actors are generally given broad flexibility to bundle or divide and license and sell the property rights among themselves. That kind of system forces market actors to constantly deal with each other. That drives competition, innovation, economic growth, and jobs.

So, what are some principal reasons that large companies might like a patent system chocked full of weak patents? The big picture is that large firms have other

[25] Pro. Real Estate Inv'rs, Inc. v. Columbia Pictures Indus., Inc., 508 U.S. 49, 60 (1993); Walker Process Equip. v. Food Mach. & Chem., 382 U.S. 172, 177 (1965); Handgards v. Ethicon, 601 F.2d 986, 996 (9th Cir. 1979).

[26] For more on changes to the patent system through case law, *see, for example*, F. Scott Kieff, *Removing Property from Intellectual Property: (Intended?) Pernicious Impacts on Innovation and Competition*, 19 Sup. Ct. Econ. Rev. 25 (2011). The major statutory change in this direction was the 2001 America Invents Act, which created the extensive post-grant review procedures inside the patent office.

ways to earn rents than relying on strong patents and many ways to earn rents that rely on weak patents. The BigCos can use weak patents to extract a range of significant economic benefits from various market regulators. For example, regulators focused on antitrust, consumer safety, the environment, or food and drug administration may see a BigCo's portfolio of weak patents as sufficient evidence of innovation to justify more regulatory relief or leeway over pricing. Similarly, regulators focused on tax treatment of inter-business and international transfer pricing may see a BigCo's portfolio of weak patents as appropriate offsets against sources of income that otherwise would be taxed at higher rates. Importantly, these BigCo benefits from portfolios of weak patents often can be extracted even if the patents are very weak, because unlike any of these regulatory or tax authorities, alleged infringers have strong incentives, expert access to technological facts, and expert ability to evaluate the host of issues that ordinarily arise in the context of a possible or actual patent infringement litigation in district court or at the ITC. But little firms have a vital need for strong patents and little use for weak patents, and the little firms face significant costs from the BigCos raising all available arguments against even the strongest of patents. Hence BigCos love a patent system full of lots of weak patents. The Goliaths can then be sure no Davids will show up with a fatal slingshot.

It is so tempting to think that IP is just about money and that money damages, or maybe even tax credits or other direct targeted incentives are all that is needed. Why gum up the works of the market with so many injunctions?

The better question is to ask why it's best to have the government figure out everyone's relative contribution, or merit, and trace it all the way through a complex commercialization process, and then pay each person her due, which presumably is just enough to entice them away from their other options to do each specific step, and no more and nothing else.

This ardent search for scientific evidence of the true value of an infringed patent is what unfortunately led even the distinguished Judge Posner to strike the economics experts of both sides of a patent case from offering damages testimony, because he viewed them as insufficiently grounded in scientific evidence or historical fact.[27] But where an infringer has decided to infringe rather than buy title or license to the patent, we also know the search for historical or scientific evidence of what price would have met the needs of both a willing buyer and a willing seller is entirely fictional. It will always be a frustrating search for scientific and historical fact when the specific topic has already been demonstrated to be a figment of imagination.

Commercializing innovative, creative, and distinctive goods and services requires a ton of coordination among a ton of private actors spread out across the marketplace. We are talking about much more than inventing or creating. We are talking about bringing it all the way to market. That takes a complex dance among inventors or creators, entrepreneurs, venture capitalists, managers, manufacturers, marketers,

[27] Apple v. Motorola, 869 F. Supp. 2d 901 (N.D. Ill. 2012).

distributors, and owners of other key assets, tangible and intangible, including other creations or inventions.

When IP is governed by a predictably enforced set of rules and backed up by a right to exclude rather than a mere right to some payment, it stands like a beacon in the dark, drawing to itself all of those many different actors in the commercialization process. They can decide on their own to strike whatever deals with each other that they like, or not, and to practice the IP subject matter or design around. The government then needs to merely enforce whatever deals they strike. No player needs to reveal to any other player or to the government what outside options it is considering, or what internal economics it faces. That keeps the government far away from the need to do any fine-grained analysis of the specific merit and incentives that may have been best tailored to each step in the long and complex process of commercialization. The government doesn't need to trace contributions or allocate values. All the government has to do is enforce any valid IP rights to exclude, and any contractual rights to payment negotiated by the parties on their own terms.

Justice Thomas is correct that a public right like a public franchise to build a toll bridge is something that requires intense scrutiny.[28] He's also correct that the patent system that has evolved over the past 20 years has gone way too far in that same direction. But, of course, that's just one more reason to steer course back to the patent system we had in the 1980s and 1990s. That was a patent system that brought us a massive increase in the number of new pharmaceuticals and new medical devices brought to market, while at the same time supporting both large pharmaceutical companies as well as a large pool of small and medium-sized biotechnology companies.

Notice also that the US patent system of the 1980s and 1990s was the product of both political parties in the United States. It also was unique to the United States. While many of the people and inventions and companies were located in Europe or Japan, only the patent system in the United States was operating so strongly at that time and that strong US patent system supported commercialization and competition for the world. It even supported the lesser-developed countries of the world. As those countries started to enforce these biopharma patents the same way as in the United States, distribution into areas of high poverty actually increased immensely while prices in those poverty-stricken areas did not increase beyond the small amounts associated with local regulation and distribution.[29]

[28] Oil States Energy Servs. v. Greene's Energy Grp., 584 U.S. 1365, 1375 (2018) (a case focusing on the administrative tribunals inside the USPTO).

[29] US International Trade Commission, *Economics Impact of Trade Agreements Implemented under Trade Authorities Procedures, 2016 Report* at 80, (Publication Number: 4614, Investigation Number: 332-555, June 2016), www.usitc.gov/publications/332/pub4614.pdf (citing Mark Duggan et al., *The Market Impacts of Pharmaceutical Product Patents in Developing Countries: Evidence from India*, 106 AM. ECON. REV. 99 (2016)) ("2005 implementation of patent protection for pharmaceutical products in India increased average prices only slightly, and also had little impact on quantities").

## VII. CONCLUSION

Today's technology and business professionals working to bring to market all the great promise of 5G and IoT are making an amazing contribution to our society today and tomorrow. Today's legal and policy professionals wrestling on all sides of the debates about the patent-antitrust interface raise great questions, in good faith, with the shared goal of fostering a better and more diverse and more inclusive society for us all, today and tomorrow, fostered by innovation and competition. While the debates are of the moment, they are also echoes of those long waged at least across the past century and a half. A prudent policymaker of today can save a great deal of time, effort, and unintended consequences for all, by bearing in mind the ideas explored here, that are extracted from those historical debates and that have enjoyed great support from leaders across our domestic political spectrum.

# Index

Printed in the USA
CPSIA information can be obtained
at www.ICGtesting.com
LVHW011257070124
768339LV00008B/727

9 781009 274272